POWER ELECTRONICS AND APPLICATIONS SERIES

Muhammad H. Rashid, Series Editor
University of West Florida

PUBLISHED TITLES

Complex Behavior of Switching Power Converters
Chi Kong Tse

DSP-Based Electromechanical Motion Control
Hamid A. Toliyat and Steven Campbell

Advanced DC/DC Converters
Fang Lin Luo and Hong Ye

Renewable Energy Systems: Design and Analysis with Induction Generators
M. Godoy Simões and Felix A. Farret

FORTHCOMING TITLES

Uninterruptible Power Supplies and Active Filter
Ali Emadi, Abdolhosein Nasiri, and Stoyan B. Bekiarov

RENEWABLE ENERGY SYSTEMS

Design and Analysis
with Induction Generators

RENEWABLE ENERGY SYSTEMS

Design and Analysis
with Induction Generators

M. Godoy Simões
Felix A. Farret

CRC PRESS

Boca Raton   London   New York   Washington, D.C.

Library of Congress Cataloging-in-Publication Data

Simões, Marcelo Godoy.
    Renewable energy systems: design and analysis with induction generators / M. Godoy
Simões and Felix A. Farret.
        p.   cm. — (Power electronics and applications series)
    Includes bibliographical references and index.
    ISBN 0-8493-2031-3 (alk. paper)
    1. Renewable energy sources. I. Farret, Felix A. II. Title. III. Series.

TJ808.S56 2004
621.31'36—dc22   2004041370

Visit the CRC Press Web site at www.crcpress.com

© 2004 by CRC Press LLC

No claim to original U.S. Government works
International Standard Book Number 0-8493-2031-3
Library of Congress Card Number 2004041370
Printed in the United States of America 1 2 3 4 5 6 7 8 9 0
Printed on acid-free paper

For my mother, Cecília, and my sister, Maysa,

who always believed in my idealism.

For my wife, Luzia, and my children, Ahriel and Lira,

who support me now (M.G.S.).

For my family who, with their love and affection.

helped me write every line I put in this work,

Gerry, Matheus, and Patrick (F.A.F.).

Foreword

It is indeed an honor and a privilege to write the foreword for this important and timely book. We are now at a critical crossroad of our global energy scenario. Energy has been the lifeblood of the continual progress of human civilization. Since the industrial revolution of two centuries ago, global energy consumption has increased by leaps and bounds to improve our living standards, particularly in the industrialized nations of the world.

Currently, about 87% of total energy is generated from fossil fuels (coal, oil, and natural gas), 6% is generated in nuclear plants, and the remaining 7% comes from renewable sources (mainly hydro and wind power). Unfortunately, the world has limited amounts of fossil fuel and nuclear power resources. According to current estimates, natural uranium for nuclear power will last only about 50 years; oil will last no more than 100 years; gas, 150 years; and coal, 200 years. Will the wheels of our civilization come to a screeching halt after the 23rd century when fossil and nuclear fuels become totally exhausted?

Not only that, our overdependence on fossil and nuclear fuels is causing environmental pollution and safety problems, which are now becoming dominant issues in our society. There is, in addition, the urban pollution problem due mainly to internal combustion engine vehicles. The impact of environmental pollution on global warming and the resulting climatic changes can have disastrous consequences in the long run.

At this juncture, we are turning more and more to environmentally clean and safe renewable energy sources like wind, photovoltaic, and fuel cells. The world has enormous resources of wind energy. It has been estimated that tapping barely 10% of the wind energy available could supply all the electricity needs of the world. Recent technological advances in variable-speed wind turbines, power electronics, drives, and controls have made wind energy competitive with coal and natural gas power.

One third of the world's population lives outside the power grid system. For them, wind and photovoltaic energy are very important. The U.S. wind energy potential is so large that it can meet more than twice current electricity consumption. Currently, 1% of U.S. power demand is supplied by wind, which will be expanded to 5% by the year 2020. In Denmark, 13% of electricity is provided by wind, which will be expanded to 50% by 2030. Germany is the world leader in wind energy installed capacity (4500 MW), and the United States is the second (2554 MW). Developing countries such as India and China have huge expansion programs in wind generation.

One of the problems of wind energy is that its availability is sporadic, and, therefore, it needs to be backed by other power sources. The concepts of a future hydrogen economy and hydrogen fuel cell cars are based on conversion of wind power and storing energy in the form of hydrogen. This book emphasizes wind generation systems using induction generators (induction machines acting in generating mode). It is the right topic at right time.

Photovoltaic systems have the additional advantage of being static and barely requiring repair and maintenance. However, photovoltaic power is typically five times more expensive than wind power. Currently, there is a tremendous research and development effort to develop low-cost photovoltaic panels for widespread terrestrial applications in the future.

Solar power conversion efficiency is typically 16%, and its availability is also sporadic. The primary fuel for fuel cell energy is hydrogen or a fossil fuel (gasoline or methanol) with a reformer. Only in the former case can it be defined as renewable energy. Fuel cells are static and have high conversion efficiency (60%). Currently, they are bulky and expensive and have poor transient response. Fuel cells show tremendous promise for the future, particularly for electric cars, although a tremendous amount of research and development is needed to arrive at this goal.

This book highlights induction generators, particularly the cage type, for application in renewable energy systems. Although both induction and synchronous machines have been used widely, induction machines appear to be the right choice for variable-speed renewable energy systems because they are robust and economical. They have easily supplied excitation from the stator, good transient behavior, and mild short-circuit fault. Generally, a pulsewidth modulated inverter in the front end supplies the variable-frequency variable-voltage power to the induction generator along with its excitation need.

The drawbacks of induction generators, however, compared to synchronous machine, are slightly lower efficiency and inability to be operated at unity or leading power factor. Doubly fed induction generators with slip power control using cascaded converters or cycloconverter have been used in Scherbius drives. An example of a high-power Scherbius drive is the Kansai Power Company's 400-MW variable-speed pumped-storage hydropower station in Kansai Power (Japan). Toshiba has recently built flywheel energy storage units in the utility system using this principle. Therefore, this timely book enables the reader to comprehend the technology required in the implementation of induction-generator–based renewable energy systems.

Dr. Bimal K. Bose
Condra Chair of Excellence in Power Electronics
The University of Tennessee, USA

Preface

Induction generators have been used since the beginning of the 20th century. However, seemingly abundant fossil fuel production led to their almost complete disappearance by the 1960s. In the last few years, attention to environmental issues and international policy has made people conscious of the importance of distributed generation and the need for development of renewable and alternative energy solutions.

This book is written for the reader whose goal is to understand the technology of induction generators. Topics such as the process of self-excitation, numerical analysis of stand-alone and multiple-induction generators, requirements for optimized laboratory experimentation, application of modern vector control, optimization of power transference, use of doubly fed induction generators, computer-based simulations, and the social and economic impact of induction generators are presented in order to take the subject from the academic realm to the desks of practicing engineers and undergraduate, and graduate students.

Part of our intent is to suggest ideas and point further directions; in several cases the reader is referred to other sources for more detailed development. Much of the material presented here has not appeared in book form before. As teachers and researchers, we realize the importance of feedback, and we appreciate comments and suggestions for improvements and hearing about what might have been valuable in the material we have presented.

This book is organized in 13 chapters.

Chapter 1 presents some definitions and the characteristics of primary sources and industrial, commercial, residential, and remote sites and of rural loads, with highlights for the selection of the suitable electric generator.

Chapter 2 presents the steady-state model of the induction generator with classical steady-state representation, parameter measurements, and peculiarities of the interconnection to the distribution grid.

Chapter 3 expounds on transient modeling of induction generators with a novel numerical representation of state space modeling that permits generalization of the aggregation of generators in parallel, an important subject for wind farms.

Chapter 4 introduces in detail the performance of self-excited induction generators, voltage regulation, and the mathematical description of self-excitation.

Chapter 5 presents some general characteristics of the induction generator with regard to torque vs. speed, power vs. output current, and the relationship of air-gap voltage to magnetizing current.

Chapter 6 discusses the construction features of induction machines, particularly generator sizing, design, and manufacturing aspects.

Chapter 7 presents power electronic devices; requirements for injection of power into the grid; interfacing with renewable energy systems; and AC-DC, DC-DC, DC-AC, and AC-AC conversion topologies as they apply to the control of induction machines used for motoring and generation purposes.

Chapter 8 describes the fundamental principles of scalar control of induction motors/generators and how control of the magnitude of voltage and frequency achieves suitable torque and speed with impressed slip.

Chapter 9 presents vector control techniques in order to calculate stator current components for decoupling of torque and flux for fast-transient closed-loop response.

Chapter 10 presents contemporary optimization techniques for peak-power tracking control of induction generators. In particular, hill-climbing control and fuzzy control are emphasized.

Chapter 11 covers wound-rotor induction generators as applied to high-power renewable energy systems with important pumped-hydro and grid-tied applications.

Chapter 12 describes simulation approaches to transient response of self-excited induction generators in several environments, steady-state analysis, and vector-control–based induction motoring/generating systems.

Chapter 13 discusses economic, social, and impact issues related to alternative sources of energy with appraisal and investment considerations for the decision maker.

The Authors

M. Godoy Simões, Ph.D., earned the B.Sc. and his M.Sc. degrees from the University of São Paulo, Brazil, and his Ph.D. degree from the University of Tennessee, in 1985, 1990 and 1995, respectively. In 1998, he also received a D.Sc. degree (Livre-Docência) from the University of São Paulo.

Dr. Simões joined the faculty of the University of São Paulo from 1989 to 2000 and Colorado School of Mines in April 2000. He has been working to establish research and education activities in the development of intelligent control for high-power electronics applications in renewable and distributed energy systems.

Dr. Simões is currently serving as IEEE Power Electronics Society Inter-society chairman. He is associate editor of *Energy Conversion* as well as editor of *Intelligent Systems of IEEE Transactions on Aerospace and Electronic Systems*. He is also associate editor of *Power Electronics in Drives* of *IEEE Transactions on Power Electronics*. He has been actively involved in the Steering and Organization Committee of the IEEE 2005 International Future Energy Challenge. Dr. Simões is IEEE Senior Member, EPE and IEE Member and was the recipient of a National Science Foundation (NSF) Faculty Early Career Development (CAREER) in 2002. It is the NSF's most prestigious award for new faculty members, recognizing activities of teacher-scholars who are considered most likely to become the academic leaders of the 21st century.

Felix A. Farret, Ph.D., earned his B.Sc. in 1972 and M.Sc. in 1979 in electrical engineering from the Federal University of Santa Maria (UFSM), Brazil. He also received a M.Sc. in electrical engineering from the University of Manchester, England, in 1981, and a Ph.D. degree from Imperial College, University of London, England, in 1984. Dr. Farret's Ph.D. research was concerned with the injection of supplementary control signals to minimize uncharacteristic harmonics in high voltage direct current (HVDC) transmission systems.

After working at the Companhia Estadual de Energia Eléctrica (CEEE) of Rio Grande do Sul, Brazil, during the two years of implantation of the Thermopower Station of Candiota II, in 1974, he moved to UFSM. He did research work in high power and high frequency in the Osaka Prefectural Industrial Research Institute, Japan, in 1975. Since then, he has been teaching in the Department of Electronics and Computing and doing research in the field of micropower plants and industrial electronics, mostly related to power converters for injection of energy into the public network.

Dr. Farret was a visiting professor in 2002/2003 at the Colorado School of Mines, in the area of integration of alternative sources of energy with special emphasis on the use of fuel cells, induction generators and power electronics. Recently, he became coordinator of the Center of Studies in Energy and Environment (CEEMA-UFSM) working in the utilization of fuel cells for load management in distribution networks and stand-alone power generation.

Acknowledgments

This book took us more than a year to write. Many presentations, discussions, meetings, debates, and hundreds of e-mails helped move us toward our goal of finishing this book. Many colleagues assisted us in the process. We would like to thank people who patiently read and commented on the drafts, gave practical suggestions, supported us in developing simulations and experiments, people who inspired us and made us find veiled relationships in the intricacy of the topics.

In particular we would like to thank Luiz Pedro Barbosa, Ivan Chabu, Sudipta Chakraborty, Lineu Belico dos Reis, Jeferson Marian Correa, Colin Grantham, Cigdem Gurgur, Hua Jin, Enes Gonçalves Marra, Bhaskara Palle, José Antenor Pomilio, Dawit Seyoum, and Robert Wood.

We would like also to thank Prof. Muhammad H. Rashid for first approaching us with the suggestion of preparing this book and the external reviewers who read our proposal, Frede Blaabjerg, Leon Freris, David A. Torrey, and Edson H. Watanabe. They made relevant suggestions at the outset of this project and believed in our competence, motivating us to carry out the venture. We appreciate the photographs contributed by WEG Motores and Equacional Elétrica e Mecânica LTDA. We want to thank Luzia J. Hirai Ornelas for the cover design. We also want to show our gratitude to three wonderful people from CRC Press who diligently worked with us on all the important details, Nora Konopka, Samar Haddad, and Jessica Vakili.

Marcelo Godoy Simões
Golden, Colorado, USA

Felix Alberto Farret
Santa Maria, Rio Grande do Sul, Brazil

Table of Contents

1   Principles of Renewable Sources of Energy and Electric Generation 1
1.1   Scope of This Chapter 1
1.2   Legal Definitions 2
1.3   Principles of Electrical Conversion 3
1.4   Basic Definitions of Electrical Power 5
1.5   Characteristics of Primary Sources 7
1.6   Characteristics of Remote Industrial, Commercial, Residential Sites and Rural Energy 9
1.7   Selection of the Electric Generator 11
1.8   Interfacing Primary Source, Generator, and Load 12
1.9   An Example of a Simple Integrated Generating and Energy-Storing System 14
1.10 Problems 16
References 17

2   The Steady State Model of the Induction Generator 19
2.1   Scope of This Chapter 19
    2.1.1   Interconnection and Disconnection of the Electric Distribution Network 19
    2.1.2   Robustness of the Induction Generator 21
2.2   Classical Steady-State Representation of the Asynchronous Machine 21
2.3   Generated Power 25
2.4   Induced Torque 27
2.5   Representation of Induction Generator Losses 30
2.6   Measurement of Induction Generator Parameters 32
    2.6.1   Blocked Rotor Test ($s = 1$) 33
    2.6.2   The No-Load Test ($s = 0$) 35
2.7   Peculiarities of the Induction Machine Working as a Generator Interconnected to the Distribution Network 36
2.8   The High Efficiency Induction Generator 39
2.9   The Doubly Fed Induction Generator 40
2.10 Problems 40
References 41

3   The Transient Model of the Induction Generator 43
3.1   Scope of This Chapter 43
3.2   An Induction Machine in a Transient State 43

3.3   State Space Based Induction Generator Modeling 44
      3.3.1   A No-load Induction Generator 45
      3.3.2   State Equations of SEIG with Resistive Load, R....................46
      3.3.3   State Equations of SEIG with an RLC Load...............47
3.4   Partition of the SEIG State Matrix with an RLC Load.................49
3.5   Generalization of the Association of Self-Excited Generators 55
3.6   The Relationship between Torque and Shaft Oscillation....................58
      3.6.1   The Oscillation Equation..58
3.7   Transient Simulation of Induction Generators 61
      3.7.1   An Example of the Transient Model of the Induction
              Generator 63
      3.7.2   Effect of RLC Load Connection......................................63
      3.7.3   Loss of Excitation 63
      3.7.4   Parallel Connection of Induction Generators......................64
3.8   Using This Chapter to Solve Typical Problems....................65
3.9   Problems..66
References..67

4   The Self-Excited Induction Generator 69
4.1   Scope of This Chapter...69
4.2   Performance of the Self-Excited Induction Generator 69
4.3   Voltage Regulation...72
4.4   Magnetizing Curves and Self-Excitation 74
4.5   Mathematical Description of the Self-Excitation Process 77
4.6   Series Capacitors and Composed Excitation of the Induction
      Generator..83
4.7   Problems..84
References..86

5   General Characteristics of the Induction Generator 89
5.1   Scope of This Chapter...89
5.2   Torque–Speed Characteristics of the Induction Generator....................89
5.3   Power Versus Current Characteristics 92
5.4   The Rotor Power Factor in Rotation 94
5.5   The Nonlinear Relationship Between Air-Gap Voltage V_g and
      Magnetizing Current I_m 95
      5.5.1   Minimization of Laboratory Tests 98
5.6   An Example of Determination of the Magnetizing Curve and the
      Magnetizing Reactance 102
5.7   Voltage Regulation..104
5.8   Characteristics of Rotation 105
5.9   Problems...109
References.. 111

6   Construction Features of the Induction Generator 113
6.1   Scope of This Chapter 113
6.2   Electromechanical Considerations 113
6.3   Optimization of the Manufacturing Process 119
6.4   Types of Design 120
6.5   Sizing the Machine 123
6.6   Efficiency Issues 125
6.7   Problems 128
References 130

7   Power Electronics for Interfacing Induction Generators 131
7.1   Scope of This Chapter 131
7.2   Power Semiconductor Devices 131
7.3   Power Electronics and Converter Circuits 135
      7.3.1   Regulators 135
      7.3.2   Inverters 136
      7.3.3   Protection and Monitoring Units 136
7.4   DC to DC Conversion 139
7.5   AC to DC conversion 143
      7.5.1   Single-Phase Full-Wave Rectifiers, Uncontrolled and
              Controlled Types 143
              7.5.1.1   Center-Tapped Single-Phase Rectifier 143
              7.5.1.2   Diode-Bridge Single-Phase Rectifier 144
              7.5.1.3   Full-Controlled Bridge 145
              7.5.1.4   Half-Wave Three-Phase Bridge 149
              7.5.1.5   Three-Phase Full-Wave Bridge Rectifier
                      (Graetz Bridge) 152
              7.5.1.6   Three-Phase Controlled Full-Wave Bridge
                      Rectifier 152
              7.5.1.7   The Effect of Series Inductance at the Input AC
                      Side 153
              7.5.1.8   The Three-Phase Half-Converter 156
              7.5.1.9   High-Pulse Bridge Converters 156
7.6   DC to AC Conversion 156
      7.6.1   Single-Phase H-Bridge Inverter 157
      7.6.2   Three-Phase Inverter 160
      7.6.3   Multi-Step Inverter 160
      7.6.4   The Multi-Level Inverter 161
7.7   AC to AC Conversion 163
      7.7.1   Diode-Bridge Arrangement 165
      7.7.2   Common-Emitter Antiparallel IGBT Diode Pair 165
      7.7.3   Common-Collector Antiparallel IGBT Diode Pair 166
7.8   Problems 166
References 167

8 Scalar Control for Induction Generators **169**
8.1 Scope of This Chapter 169
8.2 Scalar Control Background 169
8.3 Scalar Control Schemes 173
8.4 Problems 177
References 178

9 Vector Control for Induction Generators **179**
9.1 Scope of This Chapter 179
9.2 Vector Control for Induction Generators 179
9.3 Axis Transformation 180
9.4 Space Vector Notation 184
9.5 Field-Oriented Control 186
    9.5.1 Indirect Vector Control 187
    9.5.2 Direct Vector Control 190
9.6 Problems 194
References 195

10 Optimized Control for Induction Generators **197**
10.1 Scope of This Chapter 197
10.2 Why Optimize Induction-Generator-Based Renewable Energy
     Systems? 197
10.3 Optimization Principles: Optimize Benefit or Minimize Effort 198
10.4 Application of Hill Climbing Control (HCC) for Induction
     Generators 199
10.5 HCC-Based Maximum Power Search 202
    10.5.1 Fixed Step 202
    10.5.2 Divided Step 203
    10.5.3 Adaptive Step 204
    10.5.4 Exponential Step 204
10.6 Fuzzy Logic Control (FLC)-Based Maximum Power Search 209
    10.6.1 Description of Fuzzy Controllers 216
       10.6.1.1 Speed Control with the Fuzzy Logic Controller
                FLC-1 216
       10.6.1.2 Flux Intensity Control with the Fuzzy Logic
                Controller FLC-2 219
       10.6.1.3 Robust Control of Speed Loop with the Fuzzy
                Logic Controller FLC-3 224
    10.6.2 Experimental Evaluation of Fuzzy Optimization Control 225
10.7 Chapter Summary 233
10.8 Problems 234
References 234

11 Wound-Rotor Induction-Generator Systems **237**
11.1 Scope of This Chapter 237
11.2 Features of WRIG 237

11.3   Sub- and Supersynchronous Modes 239
11.4   Operation of WRIG 242
11.5   Problems 254
References 254

12   Simulation Tools for the Induction Generator 257
12.1   Scope of This Chapter 257
12.2   Design Fundamentals of Small Power Plants 258
12.3   Simplified Design of Small Wind Power Plants 259
12.4   Simulation of the Self-Excited Induction Generator in PSpice 262
12.5   Simulation of the Self-Excited Induction Generator in Pascal 271
12.6   Simulation of the Steady-State Operation of an Induction
       Generator Using Microsoft Excel 272
12.7   Simulation of Vector-Controlled Schemes Using
       MatLab/Simulink 277
       12.7.1 Inputs 279
       12.7.2 Outputs 279
       12.7.3 Indirect Vector Control 280
       12.7.4 Direct Vector Control with Rotor Flux 283
       12.7.5 Direct Vector Control with Stator Flux 285
       12.7.6 Evaluation of the MATLAB/Simulink Program 288
12.8   Simulation of a Self-Excited Induction Generator in PSim 290
12.9   Simulation of a Self-Excited Induction Generator in MATLAB 292
12.10 Simulation of a Self-Excited Induction Generator in C 294
12.11 Problems 296
References 297

**13   The Economics of Induction Generator-Based Renewable
     Systems 299**
13.1   Scope of This Chapter 299
13.2   Optimal and Market Price of Energy in a Regulatory
       Environment 300
13.3   World Climate Change Related to Power Generation 302
13.4   Appraisal of Investment 306
       13.4.1 Benefit-Cost Ratio 310
       13.4.2 Net Present Value (or Discounted Cash Flow) 310
       13.4.3 Internal Rate of Return 310
       13.4.4 Payback Period 311
       13.4.5 Least Cost Analysis 311
       13.4.6 Sensitivity Analysis 311
13.5   Concept Selection and Optimization of Investment 312
13.6   Future Directions 314
13.7   Problems 314
References 315

APPENDIX A   Introduction to Fuzzy Logic.................................... 317
A.1 Natural Vagueness in Experimental Observations 318
A.2 Fuzzy Logic Operations...321
A.3 Rules of Inference 323
A.4 Fuzzy Systems...326
A.5 An Example of Fuzzy Logic Control for Air Conditioning
     Systems..327
A.6 Conclusion 330
References..330

APPENDIX B   C Statements for Simulation of a Self-Excited
     Induction Generator 333

APPENDIX C   Pascal Statements for Simulation of a
     Self-Excited Induction Generator 339

Index 347

1

Principles of Renewable Sources of Energy and Electric Generation

1.1   Scope of This Chapter

Induction generators were used from the beginning of the 20th century until they were abandoned and almost disappeared in the 1960s. With the dramatic increase in petroleum prices in the 1970s, the induction generator returned to the scene. With such high energy costs, rational use and conservation implemented by many processes of heat recovery and other similar forms became important goals. By the end of the 1980s, wider distribution of population over the planet, as improved transportation and communication enabled people to move away from large urban concentrations, and growing concerns with the environment led to demand by many isolated communities for their own power plants. In the 1990s, ideas such as distributed generation began to be discussed in the media and in research centers. The general consciousness of finite and limited sources of energy on earth and international disputes over the environment, global safety, and the quality of life have created an opportunity for new, more efficient, less polluting power plants with advanced technologies of control, robustness, and modularity.

In this new millennium, the induction generator, with its lower maintenance demands and simplified controls, appears to be a good solution for such applications. For its simplicity, robustness, and small size per generated kW, the induction generator is favored for small hydro and wind power plants. More recently, with the widespread use of power electronics, computers, and electronic microcontrollers, it has become easier to administer the use of these generators and to guarantee their use for the vast majority of units where they are more efficient up to around 500 kVA.

The induction generator is always associated with alternative sources of energy. Particularly for small power plants, it has a great economic appeal. Standing alone, its maximum power does not go much beyond 15 kW. On the other hand, if an induction generator is connected to the grid or to other sources or storage, it can easily approach 100 kW. Very specialized and custom-made wound-rotor schemes enable even higher power. More recently, power electronics and microcontroller technologies have given a decisive boost to induction generators because they enable very advanced and inexpensive types of control, new techniques of reactive power supplements, and asynchronous injection of power into the grid, among other features. This chapter will begin the discussion of these aspects of induction generator technology. The details will be dealt with in later chapters.

1.2 Legal Definitions

In most countries a small power plant is one that generates up to 10,000 kW of power from some source of energy found in nature. For a long time, this classification was applied mostly to hydropower plants, but more recently the term small power plant has come to include three different ranges of power generation: micro, mini, and small, as shown in Table 1.1.[1-3] In this book, it is applied to plants that use all other forms of energy. Micropower plants (up to 100 kW) are distinguished from the others because in many countries they are much simpler, legally as well as technically, than mini power plants (between 100 kW and 1,000 kW), small power plants (between 1,000 kW and 10,000 kW), and large power plants (larger than 10,000 kW). Table 1.1 also lists the heads of waterfalls necessary for the various power ranges as recommended by Eletrobrás, the Brazilian energy agency, OLADE, the Latin-American organization for development and education, and the U.S. Department of Energy (DOE).[5]

Mini, small, and large power plants are usually commercial and usually supply several consumers. Although power plants of these sizes can be asynchronous or synchronous, isolated or directly or indirectly connected to the network, there is a growing concern with the quality and frequency of the voltage they generate. As mini, small, and large power plants supply

TABLE 1.1

Classification of Small Hydroelectric Power Plants

Size of the Power Plant	Source	Power (kW)	Water Head (meters)		
			Low	Medium	High
Micro	OLADE	up to 50	15	15 to 50	50
	Eletrobrás	up to 100	as above	as above	as above
	DOE	100	10	Probl. 1.7	Probl. 1.7
Mini	OLADE	50 to 500	20	20 to 100	100
	Eletrobrás	100 to 1,000	as above	as above	as above
	DOE	100 to 1,000	Probl. 1.7	Probl. 1.7	Probl. 1.7
Small	OLADE	500 to 5,000	25	25 to 130	130
	Eletrobrás	1,000 to 10,000	as above	as above	as above
	DOE	>1,000	Probl. 1.7	Probl. 1.7	Probl. 1.7

energy to a considerable number of people, they tend to be governed by complex legal regulations and technical standards. Although their technology is well established, large units can be responsible for a large portion of the total energy being generated in their area, hence there is an increased possibility of blackouts.

Micropower plants, on the other hand, are practically free of legal regulations. They have a strong dependence on the load connected across their terminals, which may affect their voltage and frequency control. Because of their small size, they lend themselves to scheduled investments. In general, they have reduced losses and reduced transmission and distribution costs. Furthermore, aggregation of generators allows for programmed interruptions of loads for maintenance. They can supply only so much to stand-alone or interconnected loads. When connected to power systems, they can be considered more as loads than as generating points.

With respect to the amount of generated power, three different situations can be considered: (1) sufficient energy for the consumer's own consumption and for sale to or interchange with the distributor; (2) lower generation than the consumer's own consumption to avoid the sale of excess energy (from the point of view of the public network, the generated power is seen more as an apparent reduction of the effective load); and (3) surplus generation for accumulation of energy (water, battery, fuel, coal) to be used for demand reduction in the peak hours of the public network.

As can be inferred from the discussion above, the design of a micropower plant should not follow the same steps as that of a larger plant only on a smaller scale. Mini, small, and large power plants are fundamentally different because they are strictly for commercial or community applications, which require rigid voltage and frequency controls. The necessary investments, technical implications, and operational costs are higher and very difficult for an individual consumer to assume. The bureaucratic formalities are so complicated as to be almost insurmountable everywhere in the world.[4]

Most micropower plants, on the other hand, are of the stand-alone type. They are often the only possible solution for a remote site. When they operate connected to the public network, their purpose is to reduce energy bills. There should be a negotiation with the local energy distributor with regard to cost of a kilowatt-hour generated, safety conditions, voltage tolerances, and frequency, among other matters, but to begin operations, most micropower plants need only a formal communication to the competent authorities.

1.3 Principles of Electrical Conversion

Electrical energy conversion is either direct or indirect. Primary energy is said to be converted directly when it is transformed into electricity without going through any intermediate forms of energy. Examples are photovoltaic arrays and thermoelectric. Direct conversions are usually static. Surprisingly, they are not necessarily the most efficient. Indirect conversions like hydro and wind power use one or more intermediary processes.

Direct energy conversions are the most desirable because they are nonpolluting. They are silent, although they are also costly and not very efficient. An example is thermal energy converted directly into electrical power. Chemical energy is also directly converted into electricity through the use of batteries and fuel cells. Electromagnetic energy is converted into electricity in photovoltaic panels. Mechanical energy is converted into electricity in conventional AC or DC generators or in fluid dynamic converters such as electro-gas-dynamic (EGD) or magneto-hydro-dynamic (MHD) systems. Thermal conversion of energy is limited by the Carnot cycle, which is expressed as a relationship between sources of high and low temperatures $1 - T_L/T_H$. The other direct conversions are not limited by this externally reversible heat-engine cycle. The attention of many researchers throughout the world is still focused on them.

The thermoelectric generator produces a very clean type of energy, and it has a number of advantages over the conventional thermodynamic heat-engine system, based on the Seebeck effect, the Peltier effect and the Thompson effect, which describe the generation of voltage by the application of temperature to pairs of different metals or thermocouples. Thermoelectric generators are rugged, reliable, compact, and without moving parts. These generators are confined to very low-power applications because their efficiency is extremely low, something around 5 to 10%.

Batteries and fuel cells convert chemical energy into electrical energy. They are similar in that both normally contain two electrodes separated by an electrolyte solution or matrix. In a fuel cell, the fuel reactant, usually hydrogen or carbon monoxide, is fed into porous electrodes and oxygen or air is fed into other porous electrodes. Some batteries can be reloaded and are used to store energy. Energy can also be stored in fuel cells by storing the fuel used to activate the reactions in the cell. Actually, batteries store energy, and fuel cells generate it based on chemical reactions.

Solar energy is directly converted into electrical energy in a photovoltaic cell. These cells are based on the p–n junction of semiconductors, usually silicon, in which there is a transient charging process that establishes an electrical field close to the junction. The concentration of doping impurities in the material of the n-type is made so high that when it is combined with the p-type material some of the electrons from the n-type material flood the nearby holes of the p-type material, making the n material positively charged and the p material negatively charged in the area close to the junction. This charging process continues until the electric field or junction potential inhibits further charges to be moved away, making the electron and hole flow the same in both directions of the semiconductor material. An external voltage generated by photons can alter this balance of electrical charges. Photons react with the valence electrons near the p–n junction to produce a forward biasing voltage and, thus, electrical current. The problems of photovoltaic energy are low efficiency (less than 20% in good quality commercial systems) and high cost. Furthermore, any terrestrial solar conversion system needs to be integrated with either an energy-storage system or with another type of conversion system to supply energy when solar energy is not available. This feature limits even more the use of photovoltaic cells in large power schemes.

The principle of a rotating electrical generator is based on the relative movement between a coil and a magnetic core. If a coil is rotated between a pair of poles (a field) of an electromagnet or a permanent magnet, the output from the rotor will be either alternating or direct current, depending on whether a slip ring (AC) or segmented commutator (DC) is employed. The field windings of a conventional DC generator require a DC power supply, and they can be excited in several ways. Some of these machines must be excited by an external DC power generator. Others are self-excited (shunt, series or compound), needing only a small amount of residual magnetic flux present in the core of the machine to start the excitation process. AC-current generators or alternators also require a DC power supply for the field winding, usually coming from a small DC generator (exciter) connected to the same alternator shaft. These generators can be synchronous or asynchronous; they are synchronous if their frequency is an integer multiple of the frequency of the grid. When connecting a generator to an AC power grid, care must be taken that the alternator is in phase with the alternating current of the grid at the instant of connection. If the alternator is out of phase, the grid will cause immediate acceleration or deceleration of the rotor and high peaks of voltage and current. Severe damage to the machine may result. Asynchronous generators do not suffer so drastically from this limitation, but they may cause stabilization problems to the network, since their power frequency is not a fixed value.

1.4 Basic Definitions of Electrical Power

Since production of electrical power is the goal, it is important to understand some fundamentals of electric power. Let's consider the variables: V_{dc}, I_{dc}, V_{rms} and I_{rms} where: V_{dc} and I_{dc} are the average values of voltage and current and V_{rms} and I_{rms} are the effective values of voltage and current in the load including the harmonic components. The DC load power (lossless rectifier) and AC load power in the source, respectively, are defined as:

$$P_{dc} = V_{dc}I_{dc} \qquad P_{ac} = V_{rms}I_{rms}$$

$$\eta = \frac{P_{dc}}{P_{ac}} \qquad V_{ac} = \sqrt{V_{rms}^2 - V_{dc}^2} = \sqrt{\sum_{i=1}^{n} V_i^2}$$

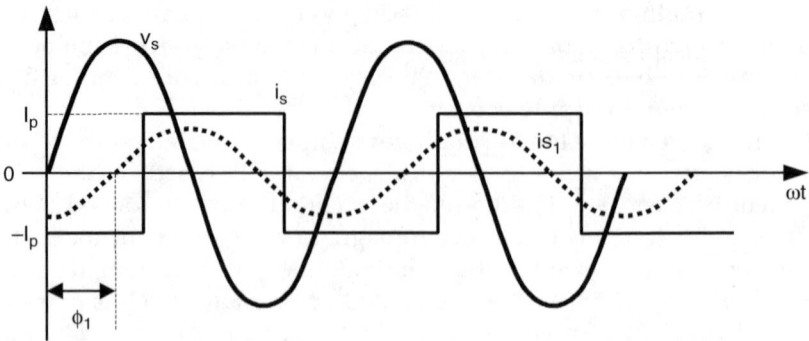

FIGURE 1.1
Input voltage and current waveforms.

where η is the efficiency and V_i is the effective value of the harmonic component of order i.

$$\text{Form factor:} \quad FF = \frac{V_{rms}}{V_{dc}}$$

$$\text{Ripple factor:} \quad RF = \frac{V_{ac}}{V_{dc}} = \sqrt{\left(\frac{V_{rms}}{V_{dc}}\right)^2 - 1} = \sqrt{FF^2 - 1}$$

Circuits using transformers need some additional definitions:

$$\text{Transformer utilization factor:} \quad TUF = \frac{P_{dc}}{V_s I_s}$$

where V_s and I_s are the effective values, respectively, of the voltage and current waveforms of the transformer secondary.

Displacement power factor, $\cos\phi_1$ is the cosine of the angle ϕ_1 between the fundamentals of voltage V_{s1} and current I_{s1} (Figure 1.1).

The harmonic factors or the total harmonic distortion (*THD*) for voltage and current, respectively, are:

$$THD_{(v)} = \left(\frac{V_{rms}^2 - V_{rms1}^2}{V_{rms1}^2}\right)^{1/2} = \left[\left(\frac{V_{rms}}{V_{rms1}}\right)^2 - 1\right]^{1/2}$$

$$THD_{(i)} = \left(\frac{I_{rms}^2 - I_{rms1}^2}{I_{rms1}^2}\right)^{1/2} = \left[\left(\frac{I_{rms}}{I_{rms1}}\right)^2 - 1\right]^{1/2}$$

TABLE 1.2

Ideal Parameters of a Rectifier

$\eta = 100\%$	$V_{ac} = 0$	$RF = 0$	$SCR \to \infty$
$TUF = 1$	$THD_{(v)} = THD_{(i)} = 0$	$PF = DPF = 1$	

where V_{rms1} and I_{rms1} are the rms values of the fundamental components of the input voltage V_s and current I_s, respectively.

$$\text{Power factor: } PF = \frac{V_{rms1}I_{rms1}}{V_{rms}I_{rms}}\cos\phi_1 \approx \frac{I_{rms1}}{I_{rms}}\cos\phi_1$$

The power factor is strongly influenced by the harmonic content of the voltage and current because if the harmonic distortion factors V_{rms1}/V_{rms} and I_{rms1}/I_{rms1} are substituted by the inverse of the total harmonic distortion factors and simplified as a whole we get:

$$PF = \frac{\cos\phi_1}{\sqrt{\left(THD_{(v)}^2 + 1\right)\left(THD_{(i)}^2 + 1\right)}} \cong \frac{\cos\phi_1}{\sqrt{\left(THD_{(v)}^2 + THD_{(i)}^2 + 1\right)}}$$

The above approximation applies to normally acceptable cases of total harmonic distortion of voltage and current that rarely go beyond 10%.

$$\text{Crest factor: } CF = I_{s(peak)}\big/I_s$$

$$\text{Short circuit ratio: } SCR = SCC\big/P_{dc}$$

where SCC is the short circuit capacity of the AC source, that is, the AC open circuit voltage (ideal source of voltage) multiplied by the short circuit current of this source.

So, the resulting short circuit current is only limited by the source impedance. Any conversion system is considered usually weak when $SCR < 3$ and strong when $SCR > 10$. The ideal parameters for a rectifier are shown in Table 1.2. Other basic electric definitions are in Table 1.3.

1.5   Characteristics of Primary Sources

Electrical energy systems can be static, rotating, or energy storing. Systems differ with respect to their primary source, whether they are stand-alone or connected to the grid, their degree of interruptibility, and the quality and

TABLE 1.3

Basic Electric Definitions for Sinusoidal Waveforms

Peak voltage	$V_p = \dfrac{V_{max} - V_{min}}{2}$	Active power	$P = S.\cos\phi$
rms voltage	$V_{rms} = \dfrac{V_p}{\sqrt{2}}$	Reactive power	$Q = S.\sin\phi$
Peak current	$I_p = \dfrac{I_{max} - I_{min}}{2}$	Impedance	$Z = \dfrac{V_{rms}}{I_{rms}}$
rms current	$I_{rms} = \dfrac{I_p}{\sqrt{2}}$	Resistance	$R = Z.\cos\phi$
Frequency	$f = \dfrac{1}{T}$	Reactance	$X = Z.\sin\phi$
Apparent power	$S = V_{rms}.I_{rms}$		

cost of their output. In general, the electrical power can offer power flux reversion, phase changing, direct current transformation, line-frequency and line-voltage transformation, isolation, and stabilization.

The interruptibility of power sources and dependence on each source are strongly related to the costs of the overall power supply; all aspects of these factors should be taken into consideration at the design stage. Such factors can be listed as: allowable outage time, allowable transfer time between normal and standby supplying alternatives, initial costs, operation costs, and maintenance costs.

The vast majority of sources of energy in the world depend on rotating generators: hydro, wind, diesel, general internal combustion engines (ICE), steam turbines, and micro turbines. Solar and fuel cells are static sources and depend on purely electronic systems to convert their electricity.

Energy storage systems may also be classified into static and rotating. Some of these sources are: battery, water dam, heat, flywheel, fuel tank, charcoal field, pneumatic container (air or gas), super capacitor, and injection to grid. Superconducting magnetic energy storage (SMES) takes advantage of a new generation of high temperature superconducting materials. SMES can potentially be used in large and small stationary products. SMES's are currently commercialized for power quality conditioners. It seems that static sources of energy will play a very important role alongside the rotating types since they tend to be cleaner, noiseless, lighter, and less dependent on non-renewable sources of energy like coal, petroleum and natural gas.

An inertial flywheel is a heavy wheel coupled to the shaft of a turbine-generator group; it works as a regulator of the variations of load in the generating unit, within determined regulation conditions imposed previously, the GD^2 (moment of inertia) of the generator and of the turbine.[5] Manufactured of melted steel or of melted iron, flywheels are considered the best way of storing pure or mechanical energy in terms of simplicity, speed in the delivery and absorption of energy, obtaining of high power, limitless number of cycles (charge and discharge), high reliability, and low maintenance cost. Flywheel technology has been advanced recently by the use of light materials, vacuum operation, and sealed joints capable of withstanding high pressure differences.

Nontechnical factors like politics and economics are also very important and have to be considered. Many nonrenewable sources of energy cause high degrees of pollution and are located in remote places of extreme political and economical instability. So it is very important to find and develop new alternative sources of energy, preferably renewable ones. To do that, it is important to consider the following factors: development time, cost, environmental effects, expectations to meet future demands, robustness, and access routes.

A very simple example of the principles uniting hydro storage of energy and rotating systems of energy generation in the form of an induction generator and wind energy is presented in Section 1.9.

1.6 Characteristics of Remote Industrial, Commercial, Residential Sites and Rural Energy

The availability of energy is closely related to the following:

- Supply of electricity, telephone, gas and water
- Increase and improvements in the networks of transportation and communications
- Construction of schools, hospitals and clinics
- Access to media (radio, television, newspapers, etc.)
- Increased job opportunities
- Access to agricultural and health services
- Rural exodus

Because of the overall importance of energy and the scarcity of investment capital, it is vital to make maximum use of available energy resources. Maximum use is not difficult to achieve, but it takes planning and management. Energy planning is a capital investment decision. An important part of energy planning is the sizing of the loads associated with electric power plants.[6,7]

The initial studies leading to sizing plant loads should encompass a sufficiently long time frame. Hydroelectric power plants should not be sized only for the present load or for the immediate future. A power plant capacity has to absorb possible regional economic growth.

Oversizing a project implies idle investments; undersizing leads to repressed demand with a consequent reduction in the quality of the supply and premature aging of the equipment. However, undersizing may be necessary if the capital for a larger plant is not available.

The determination of the power necessary to serve the users of the electric load is made based on the relationship of the hourly electric loads of all their components. In that list of loads should be lighting, appliances, and other electric equipment that consumes energy, characterized by different power and consumption periods. The largest load found defines the necessary power generation for the power plant.

Distributing each component power demand on the installation throughout the operating hours of the day and adding the power demand of those components in each operating period yields the load peak. So, an electric load system, working within a schedule and consumption period, is characterized by a coefficient called Demand Factor, defined as:

$$F_d = \frac{D_{av}}{D_{max}}$$

where:

D_{av} is the average demand (average load consumed)

D_{max} is maximum demand (maximum load consumed)

The closer the demand factor F_d is to 1.0, the better the energy consumption distribution of the system. In places where there is intense consumption variation during certain periods, it is necessary to estimate the load distribution in those periods and compare it with the distribution in a normal period of consumption. The highest peak found in the two periods will define the rated power of the power plant to be installed. If there is available potential, the rated power can be increased when necessary as a safety measure, anticipating a change in the load programming or even to provide for load growth in the future.

Furthermore, to determine the power to be installed, the necessary electric load should be compared with the electric potential of the primary source of energy. One of the following three hypotheses should be considered.

If the available power is larger than the power demanded, one can take advantage of the entire potential, obtaining a surplus of energy. Or, in the case of a hydroelectric power plant, the net fall or the water flow can be reduced, for the sake of economy, adjusting the potential to the power demanded. The decrease of water flow reduces the cost of the receiving systems and pipelines, while a decrease in the raw fall reduces the height of the dam.

In the cases of wind or solar energy, such economies have no meaning. Electronic controls that can take maximum advantage of those sources are to be preferred. For biomass micropower plants and other fuels, the production of energy should also be strictly limited to what is required for consumption, due to their high costs and to the ease of adding new units when there is an increase in the demand.[7-9]

Where there is a state or private monopoly of electric power distribution, with no possibility of sale of excess of production, it is convenient to determine the year-round minimum consumption of energy. Otherwise, the public network can acquire the demand above that limit, drastically reducing project costs. but when the public and private demand peaks coincide, it can overload the public network.

If the available potential equals the desired power, the whole energy potential should be taken advantage of. If the available potential is smaller than the desired power, consumption must be rationalized by selection of loads according to service priorities or by making up the deficit with energy originating from other sources. The interconnection with the public network can be a good way out for such situations. Brownouts can be also implemented. That is, the available rated voltage for the consumer can be slightly reduced. It is known that consumers do not even notice a 5 or 10% reduction in supply voltage, which can obtain a 10 to 20% reduction in consumption. The reason for this economy is that the supplied energy is reduced as the square of the voltage reduction.

1.7 Selection of the Electric Generator

This book focuses on the induction generator rather than other types of energy converters because of its importance for rotating energy conversion from hydro and wind systems. Having selected the installation site of the power plant, the next step is to select the turbine rating, the generator, and the distribution system. In general, the distribution transformer is sized to the peak capacity of the generator and the capacity of the available distribution network. As a rule, the output characteristics of the turbine power do not follow exactly those of the generator power; they have to be matched in the most reasonable way possible. Based on the maximum speed expected for the turbine and taking into account the cubic relationship between the fluid speed and the generated power, the designer must select the generator and the gearbox so as to match these limits. The most sensitive point here is the correct selection of the rated speed for the generator. If it is too low, the high speed of the primary source fluid, say wind, will be wasted; if it is too high, the power factor will be harmed. The characteristics of the commercially available turbines and generators must be matched to the requirements of the project with regard to cost, efficiency, and maximum generated power in an iterative design process.

Several types of generators can be coupled to the rotating power turbines: DC and AC types, parallel and compound DC generators, with permanent magnetic or electrical fields, synchronous or asynchronous, and, especially,

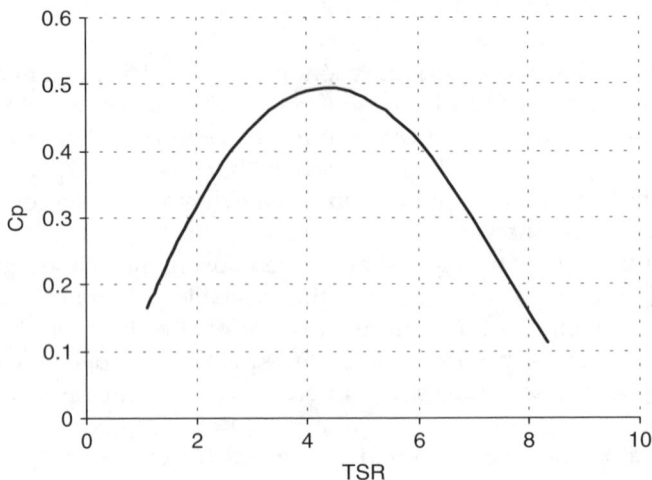

FIGURE 1.2
Maximized value of C_p for wind turbines.

induction generators. The right choice of the generator depends on a wide range of factors related to the primary source, the type of load, and the speed of the turbine, among others.

In the case of wind energy, there is a control variable known as power coefficient, C_p, which is of utmost relevance. The maximum value of C_p occurs approximately at the same speed as the maximum power in the power distribution curve, and it is defined as the relationship between the tip speed ratio (TSR) of the rotor blade and the wind speed. So, the turbine speed must be kept optimally constant at its maximum possible value to capture the maximum wind power. It suggests that to optimize the annual energy capture at a given site, it is necessary for the turbine speed (tip speed ratio) to vary to keep C_p maximized as illustrated in Figure 1.2 for a hypothetical turbine. Obviously, the design stress must be kept within the limits of the turbine manufacturer's data since the torque is related to the instant power by $P = T\omega$.

Because of the way it works as a motor or generator, the possibility of variable speed operation, and its low cost compared to other generators, the induction machine offers advantages for rotating power plants in both standalone and interconnected applications.

1.8   Interfacing Primary Source, Generator, and Load

As previously discussed, to think of generation of small amounts of electricity is, as a practical matter, to think of producing energy for one's own

FIGURE 1.3
Basic elements for electric power generation.

individual consumption. Due to the types of loads used today by society in general, only in rare situations can such uses be thought of in commercial or collective terms. A small generator can be a solution for rural areas or for very poor and distant centers of electric power distribution where there is no other option and the loads do not go beyond a few kilowatts. Electric showers, often used in developing countries, air conditioning, electrical stoves, and other modern appliances with high and concentrated consumption of electric power should not be connected to a small generator. So, we need to think in terms of a spectrum of power supplies from small (few Watts) to large (close to 100 kW). Small could mean:

- Small data transmission equipment or telephony for distant places
- Equipment for scientific experiments or unassisted collection of data in remote areas
- Power supply for electric equipment in places subject to dangerous conditions: fire, explosions, poisonous, irradiation, conflict areas, cliffs, proximity to wild life, and so on

Large could mean:

- Irrigation
- Facilities in farms with high conditions of comfort
- Reduction in the electric bills of companies or other large consumers
- Storage of energy during off-peak hours of consumption of the public network for reduction of the peak hours of larger demand[4,5]

In all its manifestations, primary energy comes in nature in raw form and has to be captured and adapted to be useful. The process of conversion of energy as nature provides it to us into electric power can be described in three stages: the primary energy, the conversion system, and the electric load (Figure 1.3).[8,9] Among the alternative primary sources of energy, the most common, in order of current importance, are:

- Hydraulic energy represented by the movement of waters
- Wind energy present in the movements of air masses
- Thermal energy from the burning of fuels
- Solar energy (light and heat) found in the sun's rays

- The energy of gases or of biomass generated from the decomposition of plants and animals or from the burning of organic matter like trees and organic garbage
- Electrochemical energy represented by fuel cells and batteries

Energy conversion takes advantage of primary energy in its raw natural state and transforms it into electricity in a useful and efficient form suited to its purpose. The rotating power converter is usually called the primary moving machine because it is the one involved directly in the process of transforming energy. Recently a variety of types of electronic or static power converters have emerged. Power converters are discussed in Chapter 7. The loads of all converters should be adapted to their uses. They can be of direct or alternating current, static or rotating, and electrochemical or mechanical, depending on the electric principle of operation as discussed in Section 1.7.

1.9   An Example of a Simple Integrated Generating and Energy-Storing System

A wind power shaft turbine of P_{tw} = 2 kW supplies energy for pumping water to a height h = 10 m through a hydraulic pump of efficiency η_{bh} = 60% and a mechanical transmission system of efficiency η_{ts} = 80%, as depicted in Figure 1.4. What is the storage time of energy under these conditions and what is the average output power being supplied?

Of all the wind energy in nature, only a small average portion is available during one day at a power plant, taking into account periods of high and low wind intensities. This average is represented, in effective terms, by the characteristic factor of the place, F_c, whose typical value is around 0.30. Besides these losses, we must discount losses in the wind turbine that are reflected in its efficiency, usually around η_{tw} = 0.35. So, the available wind power at a given place is:

$$P_{wind} = \rho \pi R^2 V^3 / 2 \qquad (1.1)$$

where:
ρ is the air density in kg/m³ (1.2929 kg/m³),
V is the wind speed in m/s,

R is the length of each blade of the turbine from the center of the rotor shaft to the blade tip measured in meters.

The net power available on the wind power turbine shaft for the hydraulic pump, for simplicity, is considered, in this example, as being:

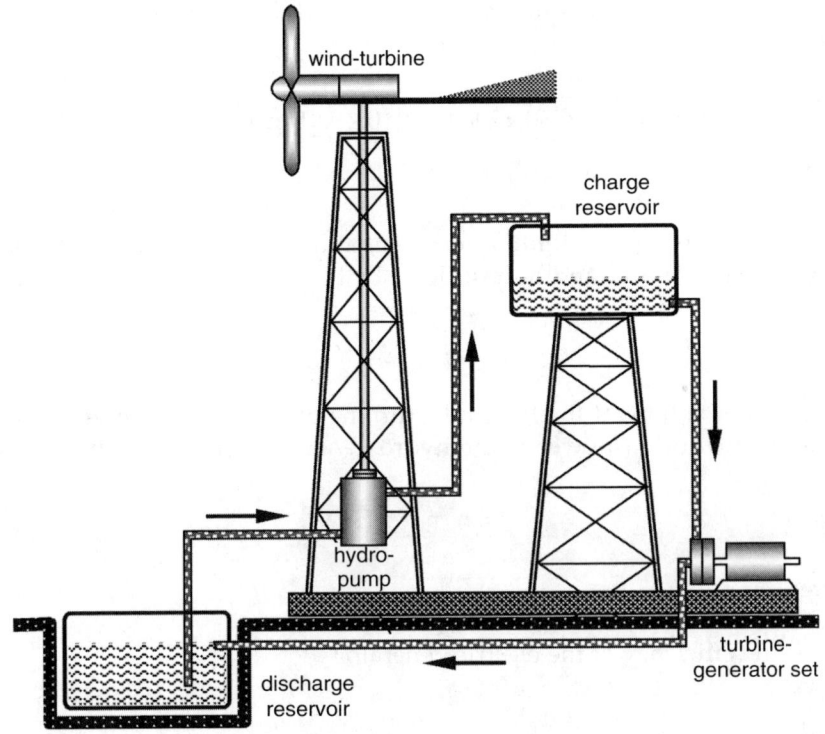

FIGURE 1.4
Regularization of the wind power energy by hydraulic storage.

$$P_{te} = P_{wind} = 2.0 \text{ kW}$$

The power P_{bh} is supplied from the useful power in the rotor of the wind power turbine, P_{te}, and expressed as:

$$P_{bh} = \eta_{ts}P_{tw}$$

and, then, the average flow of the hydraulic pump will be given by:

$$Q_{bh} = F_c\eta_{bh}\eta_{ts}P_{tw}/Hg\gamma \text{ (in m/s)} \tag{1.2}$$

where:
P_{bh} is the power of the hydraulic pump
η_{bh} is the efficiency of the hydraulic pump
H is the level difference between the lower and the upper reservoir
γ is the specific mass of the water (= 1000 kg/m³)

Using the data of this example and Equation 1.2 the flow of the hydraulic pump is:

$$Q_{bh} = 0.30 \times 0.60 \times 0.80 \times 2000 \ \text{W}/\left(10 \times 9.81 \ \text{m}/\text{s}^2 \times 1000 \ \text{kg}/\text{m}^3\right)$$

$$= 2.936\ell/\text{s} = 10,570\ell/\text{h}$$

To avoid an average accumulation of water in the reservoirs, it is necessary that the average input and output flows be the same, therefore:

$$Q_{bh} = Q_{th}$$

where Q_{th} is the flow of the hydraulic turbine of the generating group.
The electric output power of the hydro generator set is given by:

$$P = \eta_{th}\eta_{ge}Q_{th}Hg\gamma \tag{1.3}$$

where:
  η_{th} is the efficiency of the hydraulic turbine
  η_{ge} is the efficiency of the electric generator

or:

$$P = 0.90 \times 0.95 \times 2.936 \times 10^{-3} \ \text{m}^3/\text{s} \times 10 \ \text{m} \times 9.81 \ \text{m}/\text{s}^2 \times 1000 \ \text{kg}/\text{m}^3 = 247 \ \text{W}$$

For those values, if an ordinary water reservoir of 1000 liters is used with a calculated flow of 2.936 liters/s (that is, steadily generating 247 Watts), there will be an autonomy of 1000/2.936 = 5.67 minutes (without wind). That means, it can permanently light four ordinary bulbs of 60 Watts, if the wind is not absent for more than 5 minutes.

1.10   Problems

1.1 Classify small power plants according to their power ranges and propose briefly two basic examples of applications for each.

1.2 Given a potential site for wind power of 10 kW and a terrain elevation difference of 15 m and using the factors given in the example in section 1.9, calculate the volume of reservoir water required to allow 30 minutes of steady energy without wind.

1.3 Why are micropower plants so different from larger ones? In what cases are they recommended?

1.4 Discuss which of the conventional alternative sources of energy is more suitable for use with induction generators to provide a steady supply of energy with minimum equipment.

1.5 Calculate the power factor of an electronic system subject to a reasonable amount of harmonics given as $THD_{(v)} = 1.0\%$ and $THD_{(i)} = 4.4\%$, the displacement power factor is 0.92. What would be the calculated error if harmonic components were not taken into consideration? If the effective output voltage was harmonic-free, but if a harmonic content of 10% was allowed in the effective output current, how much power would be measured with no power filters?

1.6 A transformer winding was designed to be fed by an AC voltage of 110 V and to produce an effective current of 10.5 A. By mistake, the transformer was fed from this same voltage but rectified in a full wave converter. Consider a lossless rectifier and no voltage drops across the connections. Estimate what would be the apparent power applied to the transformer in both cases if its primary resistance was R = 1.8 Ω.

1.7 Collect information in the Internet to complete Table 1.1.

References

1. *Small Hydro Power Plants Handbook (Manual de pequenas centrais hidroelétricas)*, Eletrobrás/DNAEE, Brasília, Brazil, 1985.
2. Legislation on small power plants, *Energy Power Magazine (Revista Força Energética)*, 2, No. 5, p. 7–8, Dec. 1993, MEPB Publishers, Porto Alegre, Brazil.
3. *Diário Oficial (Official Publication), Section 1. extra-edition*. Brasília, Brazil, No. 129-A, ano CXXXIII, July 1995.
4. Krämer, K.G., Schneider, P.W. and Styczynski, Z.A., Use of energy in the power network and options for the Polish power system. In: *Seminar of the Institute of Power Transmission and High Voltage Technique, University of Stuttgart*, Feb. 1997. Sponsored by the European Community Directorate-General XVII, Energy, Ver. 29, Jan. 1997.
5. Stroev, V.A. and Gremiakov, A.A., Energy storages in electrical power systems. Electrical Power Systems Department, Moscow Power Engineering Institute. In: *Seminar of the Institute of Power Transmission and High Voltage Technique, University of Stuttgart*, Feb. 1997. Sponsored by the European Community Directorate-General XVII, Energy, Ver. 29, Jan. 1997.
6. Inversin, A.R., *Micro-hydropower Plant Handbook*, National Rural Electrification Cooperative Association (NRECA), 1986.
7. REA Staff, *Modern Energy Technology (Vols. I and II)*, Research and Education Association (REA), New York, 1975.
8. Culp Jr., A.W., *Principles of Energy Conversion*, McGraw-Hill Book Company, New York, 1979.
9. Decher, R., *Energy Conversion: Systems, Flow physics and Engineering*, Oxford University Press, New York, 1994.

2

The Steady State Model
of the Induction Generator

2.1 Scope of This Chapter

The induction machine offers advantages for hydro and wind power plants because of its easy operation as either a motor or a generator, its robust construction, its natural protection against short circuits, and its low cost compared to other generators. Only the case of connection of the induction generator to the infinite bus — the point of the distribution network with no voltage drops across the generator terminals for any load situation (ideal voltage source) — will be considered in this chapter.

Many commercial power plants in developed countries were designed to operate in parallel with large power systems, usually supplying the maximum amount of available primary energy for conversion in their surroundings (wind, solar, or hydro). This solution is very convenient because the public network controls voltage and frequency while static and reactive compensating capacitors can be used for correction of the power factor and harmonic reduction. Aspects related to voltage regulation, stand-alone generation and output power, are further discussed below.

2.1.1 Interconnection and Disconnection of the Electric Distribution Network

When interconnected to the distribution network, the induction machine should have its speed increased until equal to the synchronous speed. The absorbed power of the distribution network in these conditions is necessary to overcome the iron losses. The energy absorbed by the shaft to maintain itself in synchronous rotation is necessary to overcome mechanical friction and air resistance. If the speed is increased, a regenerative action happens, but without supplying energy to the distribution network. This happens when the demagnetizing effect on the current of the rotor is balanced by a stator component capable of supplying core losses. In this situation, the generator is supplying its own iron losses. From this point on, the generator begins to supply power to the load.

The electrical frequency is the number of times per second a rotor pole passes in front of a certain stator pole. If p is the number of poles, the synchronous speed is determined by:

$$n_s = 120 f_s / p \tag{2.1}$$

where:

n_s is the synchronous mechanical speed in rpm
f_s is the synchronous frequency in Hz

When the induction machine is operating as a generator, the maximum current is reached when the power transferred to the shaft with load varies within a slip factor range given by:

$$s = \left(n_s - n_r\right)/n_s \tag{2.2}$$

where:

s is the slip factor
n_r is the rotor speed in rpm.

When considering interconnection to or disconnection from the electric distribution network, some matters affecting efficiency and electric safety should be observed:

- Efficiency. There is a rotation interval above the synchronous speed at which efficiency is very low. If the choice of the moment of interconnection is between synchronous rotation and satisfactory efficiency, great precision is not necessary in the sensor that authorizes the connection.

- Safety. It is necessary to determine the maximum rotation at which disconnection from the distribution network should happen, so that the control system acts to brake the turbine under speed controlled operation. The disconnection should be planned for the electric safety of the generator in case of a flaw in the control brake. Besides, if the plant is interconnected to the distribution network, it is important to switch it off during maintenance, as much for the sake of the local team as for the electric power company team. Although IEEE Standards have been recently released to the public, interconnection guidelines may vary among local utilities and should be checked during the design of the installation.[1-5]

2.1.2 Robustness of the Induction Generator

From the discussion above it is clear that the induction generator accepts constant and variable loads, starts either loaded or without load, is capable of continuous or intermittent operation, and has natural protection against short circuit and overcurrents through its terminals. That is, when the load current goes above certain limits, the residual magnetism falls to zero and the machine is de-excited. When this happens, there are three methods available for its remagnetization: (1) maintain a spare capacitor always charged and, when necessary, discharge it across one of the generator phases; (2) use a charged battery; or (3) use a rectifier fed by the distribution network.

An induction generator with a squirrel cage rotor can be specifically designed to work with wind or hydro turbines, in other words, with a larger slip factor, with more convenient deformation in the torque curve, winding sized to support higher saturation current, increased number of poles, and so on. Such steady-state conditions must be foreseen at the design stage of the power supply.

2.2 Classical Steady-State Representation of the Asynchronous Machine

An equivalent circuit of the induction machine, also known as the per-phase equivalent model[21] is represented in Figure 2.1. In this figure R_1 and X_1 are the resistance and leakage reactance respectively of the stator; R_m and X_m are the loss resistance and the magnetizing reactance; and R_r and X_r are the resistance and the reactance of the rotor. This model is limited to the case of sinusoidal and balanced excitation.

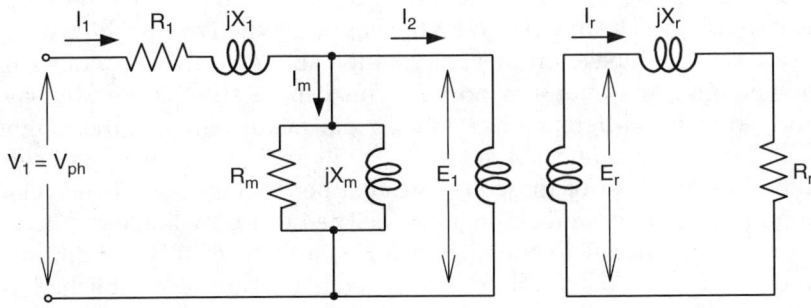

FIGURE 2.1
Transformer model of the induction generator.

The induction generator does not differ much in construction from the induction motor dealt with in Chapter 6. In its operation, it resembles the transformer except that the transformer secondary is a stationary (nonrotating) part. For this reason, it is common to use the transformer model to represent the induction generator. However, it should be observed that in spite of the fact that the magnetizing curves of both machines are similar in form, in the characteristic BxH (or ΦxNI or VxI) of these machines, the slope and the saturation area of the mmf curve of the induction generator, is much less accentuated than that of a good quality transformer. This is due to the air gap in the induction generator that reduces the coupling between the primary and secondary windings. This means that the high reluctance caused by the air gap increases the magnetizing current required to obtain the same level of magnetic flux in the core, and X_m will be much smaller than it would be in a transformer.

The internal primary voltage of the stator, E_1, and the secondary voltage of the rotor, E_r, are coupled by an ideal transformer with an effective transformation ratio, a_{rms}. This a_{rms} value is easy to determine for a motor with a cylindrical rotor; it is the relationship between the number of stator phase turns to the number of rotor phase turns modified by the differences introduced by the pitch and distribution factors. On the other hand, for the squirrel cage rotor, as in most induction generators, determining the transformation rate becomes extremely difficult because there are no distinct windings. In any situation, the voltage induced in the rotor, E_r, generates a current that circulates through the short-circuited rotor.

The equivalent circuits of the induction generator and of the transformer differ fundamentally in that in the induction generator, the rotor voltage is subject to a variable frequency making E_r, R_r, and X_r also variable. Everything depends on the slip factor, that is, on the difference between the rotor speed and the speed of the rotating magnetic field of the stator. As previously mentioned, with a rotation below the synchronous speed the machine works as motor; above the synchronous speed ω_s, it works as generator.

To better understand the induction phenomenon in the asynchronous generator, it should first be understood what happens when the machine rotation goes from zero (or blocked rotor) to a rotation above the synchronous speed, ω_s, the voltage across stator windings. This can be summarized by saying that the larger the difference of speed or slip factor between the magnetic fields of the rotor, ω_r, and the stator, ω_s, the larger the induced voltage on the rotor. The smallest of these voltages happens when there is no relative rotation between rotor and stator, $\omega_r = \omega_s$. The induced voltage on the rotor is directly proportional to the slip factor, s, and from this definition, the induced voltage on the rotor for any speed, E_r, can be established with respect to the blocked rotor, E_{r0}, that is:[14–16,21]

$$E_r = sE_{r0} \qquad (2.3)$$

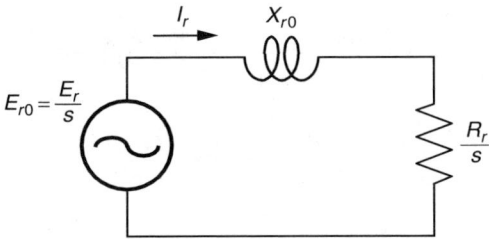

FIGURE 2.2
Equivalent circuit of the rotor.

where s is the slip factor given by:

$$s = \frac{f_s - \frac{p}{2}\frac{n_r}{60}}{f_s} = \frac{n_s - n_r}{n_s} \tag{2.4}$$

where:

n_r is the rotor speed (rpm)
$f_s = pn_s/120$

The rotor frequency, f_r, is also related to the electric frequency, $f_e = f_s$, through the expression:

$$f_r = \frac{p}{120}(n_s - n_r)\frac{n_s}{n_s} = sf_s \tag{2.5}$$

Notice that $(n_s - n_r)$ is the relative speed between the stator and rotor magnetic fields. The circulating current in the rotor will depend on its impedance, the resistance and inductance of which alter slightly due to the skin effect for the usual values. However, only the inductance is affected in a more complex way by the slip factor, depending on the self-inductance of the rotor and on the frequency of the induced voltage and current. The reactance is usually presented as the following function of the slip factor:

$$X_r = 2\pi f_r L_r = 2\pi s f_s L_r = sX_{r0} \tag{2.6}$$

where X_{r0} is the blocked rotor reactance.

Figure 2.2 displays the equivalent circuit of the rotor impedance $Z_{r0} = R_{r0} + j\omega_s L_{r0} = R_{r0} + jX_{r0}$ as a function of the slip factor given by:

$$I_r = \frac{\overrightarrow{E_r}}{\overrightarrow{Z_r}} = \frac{sE_{r0}}{R_r + jsX_{r0}} = \frac{E_{r0}}{\frac{R_r}{s} + jX_{r0}} \qquad \text{or}$$

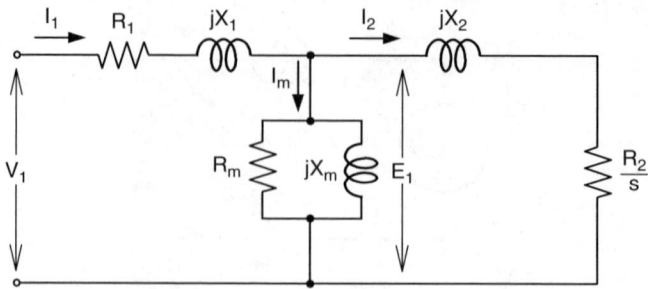

FIGURE 2.3
Equivalent circuit per phase of the induction generator.

$$Z_r = \frac{\overrightarrow{E_{r0}}}{\overrightarrow{I_r}} = \frac{R_r}{s} + jX_{r0} \tag{2.7}$$

For small slip factors ($s \to 0$) the rotor impedance becomes predominantly resistive and the rotor current can be considered to vary linearly with s. With very large slip factors, the rotor impedance plus skin effect will impose a non-linear relationship.

Similarly to what is done in the representation of transformers, the model presented in Figure 2.1 can be converted to the model shown in Figure 2.3, converting the secondary parameters per phase to primary parameters per phase by using the transformation ratio a_{rms}:

$$E_1 = a_{rms} E_{r0} \tag{2.8}$$

$$I_2 = \frac{I_r}{a_{rms}} \tag{2.9}$$

$$Z_2 = a_{rms}^2 \left(\frac{R_r}{s} + jX_{r0} \right) \tag{2.10}$$

or, yet:

$$Z_2 = \frac{R_2}{s} + jX_2 \tag{2.11}$$

being Z_2, R_2 and X_2 the rotor values referred to the stator.

2.3   Generated Power

The power balance in an induction machine or any machine can be expressed as:

$$P_{out} = P_{in} - P_{losses} \qquad (2.12)$$

The output power, P_{out}, can be given in the form of three balanced voltages and three balanced currents, each one shifted by 120° and expressed in line values as:

$$P_{out} = \sqrt{3} V_\ell I_\ell \cos \phi \qquad (2.13)$$

Like any other electric machine, an induction generator has inherent losses that can be combined in the following expression:

$$P_{losses} = P_{statorCu} + P_{iron} + P_{rotorCu} + P_{frict+air} + P_{stray} \qquad (2.14)$$

The copper losses in the stator are obtained by $P_{statorCu} = I_1^2 R_1$. The iron losses are due to the hysteresis current (magnetizing) and Foucault current (current induced in the iron). It should be said at this point that the stator and rotor iron losses usually appear mixed, although in the latter they are much smaller and more difficult to distinguish. So the loss resistance, R_m, in the equivalent circuit of the induction generator, practically represents these losses with the other mechanical losses. Subtracting these losses from input power, the air gap losses remain dissipated in the power transfer from the stator to the rotor. The rotor copper losses, $P_{rotorCu} = I_2^2 R_2$, remaining due to the transformation of mechanical into electrical power should be subtracted from the power received by the rotor. In these losses are included the friction losses and the ones resulting from the rotor movement against the air and spurious losses. Spurious losses can be combined with other losses like high frequency losses from stator and rotor dents and parasite currents produced by fast flux pulsations when dents and grooves move from their relative positions.[14,21] Finally, the output power is the net power in the shaft, also called shaft useful power.

As rotation increases, friction losses increase in ventilation (against the air) and spurious losses. In compensation, losses in the core decrease up to synchronous speed; these are called rotating losses. Such losses are considered approximately constant because some of their components increase with rotation and others decrease.

When voltage appears across the terminals of the induction generator for a given constant rotation, represented by the equivalent circuit of Figure 2.3, there is an average equivalent impedance per phase given by:

$$Z = R_1 + jX_1 + \cfrac{1}{\cfrac{1}{jX_m} + \cfrac{1}{R_m} + \cfrac{1}{\cfrac{R_2}{s} + jX_2}}$$

(2.15)

Quantitatively, a current I_1 circulates through this impedance per phase that produces losses in the stator winding of the three phases whose total is given by:

$$P_{statorCu} = 3I_1^2 R_1$$

(2.16)

The iron core losses are given by:

$$P_{iron} = \frac{3E_1^2}{R_m}$$

(2.17)

The three-phase power transferred from the rotor to the stator through the air gap can be obtained, from Equations 2.13, 2.16 and 2.17:

$$P_{airgap} = \sqrt{3}V_\ell I_\ell \cos\phi + 3I_1^2 R_1 + \frac{3E_1^2}{R_m}$$

(2.18)

On the other hand, from Figure 2.3, the only possible dissipation of the total power for the three phases corresponding to the secondary part of the circuit is through the rotor resistance as:

$$P_{airgap} = 3I_2^2 \frac{R_2}{s}$$

(2.19)

However, from the rotor equivalent circuit shown in Figure 2.2, the resistive losses are given by:

$$P_{rotorCu} = 3I_r^2 R_r$$

(2.20)

In an ideal transformer there are no losses; the rotor power remains unaltered when referred to the stator and, therefore, from Equation 2.19:

$$P_{rotorCu} = 3I_2^2 R_2 = sP_{airgap}$$

(2.21)

The mechanical power converted into electricity, or the power developed in the shaft for a negative s, is the difference between the power that goes through the air gap and what is truly dissipated in the rotor, or:

$$P_{converted} = 3I_2^2 \frac{R_2}{s} - 3I_2^2 R_2 \tag{2.22}$$

that simplified becomes:

$$P_{converted} = P_{mec} = 3I_2^2 R_2 \frac{1-s}{s} = (1-s)P_{airgap} \tag{2.23}$$

2.4   Induced Torque

Converted torque is instantaneous torque defined in terms of power in the shaft. It results from a force, F, that is concentrated on the shaft surface of radius R exerting a torque on it with respect to the center of its cross section. In Figure 2.4, we see the work, dW, that this force exerts to move any point on the shaft to a distance $d\ell$:

$$dW = Fd\ell$$

being $d\ell = Rd\theta$ and, as a consequence, $dW = FRd\theta$.

By definition, the torque is $T = FR$. Therefore, $dW = Td\theta$. Deriving both sides of this equality with respect to time, we get:

$$P = T\omega \tag{2.24}$$

This induced torque in the machine is defined as the torque generated by the conversion of the mechanical power into internal electric power, which

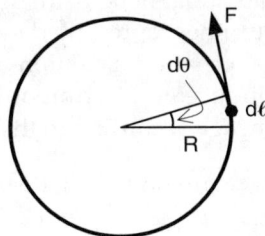

FIGURE 2.4
Torque on the shaft of a rotating machine.

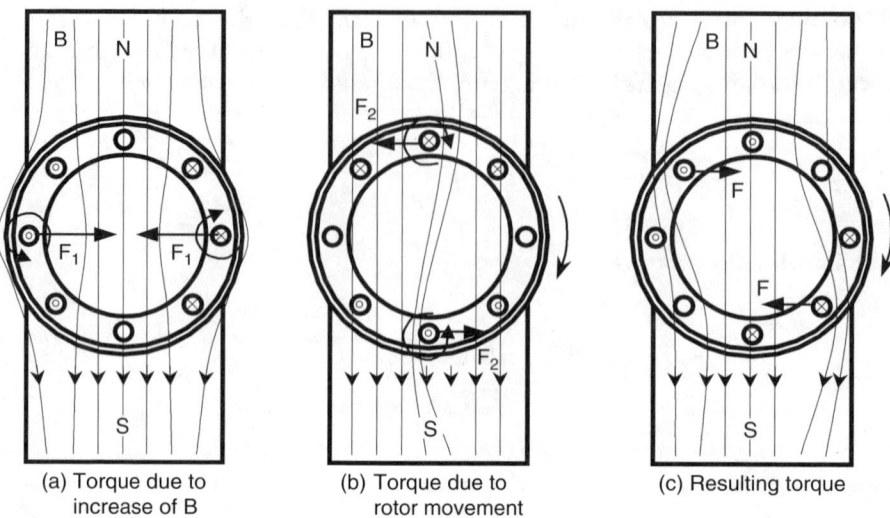

(a) Torque due to increase of B

(b) Torque due to rotor movement

(c) Resulting torque

FIGURE 2.5
Magnetic torque of the induction generator.

differs from the available torque T in the shaft because of the negative torques of friction and movement against the air. In such case, this torque is given by:

$$T_{converted} = \frac{P_{converted}}{\omega_r} \tag{2.25}$$

Recalling from Equation 2.4 that: $\omega_r = (1-s)\omega_s$, from Equations 2.23 and 2.25 we can deduce a relationship between power and torque that does not change except for ω_s and see that it is therefore more directly useful to estimate the power transferred through the air gap as:

$$P_{airgap} = \omega_s T_{converted} \tag{2.26}$$

Two components of electrical torque are induced in the rotor: one due to the variable magnetic field of the stator and other due to the rotor movement. Let us suppose that a sinusoidal magnetic field, Φ, is applied on the squirrel cage rotor, as shown by the flux lines represented in Figure 2.5. This sinusoidal field is characterized by an alternating elevation and reduction of the magnetic flux whose variations induce an electromotive force (emf) on the conductive segments of the rotor cage according to the Faraday's Induction Law:

$$e = -\frac{d\phi}{dt}$$

The induced emf on the rotor provokes a current obeying the right hand rule and, in turn, a magnetic intensity around each conductor as represented in Figure 2.5(a). The interaction between the magnetic intensities of stator and rotor generates on each conductor segment, a force F whose direction goes toward the center of the shaft and the net torque is null:

$$F = I\ell B$$

On the other hand, for a constant magnetic density, B (in *Webers*/m²), in the stator, when the rotor is driven by an external force at a tangent linear speed v (in m/s), according to the Faraday's law of electromagnetic induction, it will induce an electromotive force on each conductor segment of length ℓ of the rotor (in meters) given by:

$$e = B\ell v$$

This induced electromotive force, in turn, produces a current through the rotor segments trying to oppose the movement that has generated it, according to Lenz's law and shown in Figure 2.5(b). Again, as one can see, the net torque opposes the torque that has generated it.

Dynamically, that is, when the field is variable and the rotation is different from zero at the same time, both torques that previously had the same magnitude and opposed directions now are unbalanced with the highest torque in the direction of the initial rotor movement, according to Figure 2.5(c). The magnetic field resulting from the stator is distorted although not represented in Figure 2.5.

Let us suppose now that the machine is rotating without load. Let us designate by B the resulting flux that will only exist due to the magnetizing current I_m, that, in turn, is proportional to the stator effective voltage, $E_1 \cong V_{ph}$. These values will be constant for a constant V_{ph}. Actually, under load, there is a small voltage drop across the stator winding due to load variations (current I_1) through the stator impedance Z_1, which is very small, as a rule. Therefore, without load, the magnetic flux can be represented as in Figure 2.6(a) where δ is also the angle between I_r and I_m. The power factor angle, θ_r, of the rotor circuit is given by:

$$\theta_r = \tan^{-1}\left(\frac{2\pi s f_s L_2}{R_2}\right) \tag{2.27}$$

The torque induced by the rotor movement alone is given by:

$$T = k B_r B \sin\delta \tag{2.28}$$

where:
  k is a proportionality factor.
  $\delta = \theta_r + 90°$

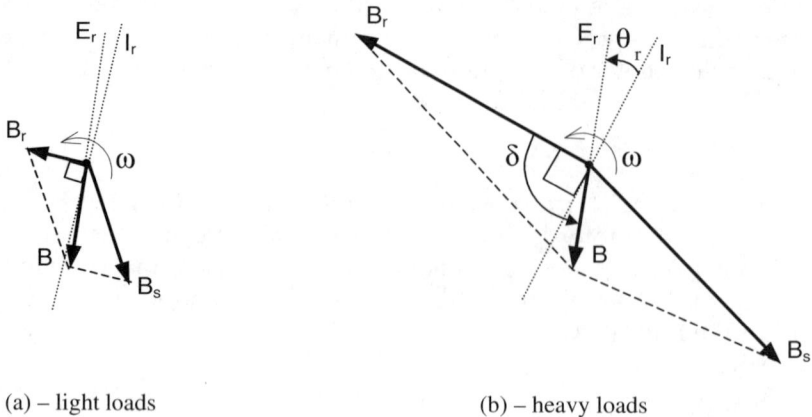

(a) – light loads   (b) – heavy loads

FIGURE 2.6
Magnetic fields of the induction generator.

The torque may be determined from the rotor current, the total magnetic field, and the power factor; it is represented in Figure 2.6.

As the magnetic field is quite reduced in the rotor, the induced torque will also consume power only for the losses. However, if there is current through the stator winding, there will be an increase in the slip factor and a larger relative movement between rotor and stator. The larger the slip factor, the larger will be the induced power on the rotor through E_r and, therefore, the larger will be I_r and B_r. The angle between the rotor current and its magnetic field increases. The frequency of the rotor increases according to $f_r = sf_s$ increasing, therefore, the rotor reactance, causing a larger delay of I_r and of the consequent magnetic field, regarding V_r, according to the diagram of Figure 2.6(b). The magnetic field B_r increases as much as δ (>90°) almost having a torque compensation effect according to Equation 2.27 with predominance of the first on the second.

With the increase of the exerted torque on the rotor, there will be a point, known as the disrupting point or point of maximum torque, where the term $\sin\delta$ in Equation 2.27 begins to decrease in larger proportions than B_r can increase, tending to a reduction in the power supplied until total collapse occurs.

2.5   Representation of the Induction Generator Losses

The stator power transferred from the rotor through the air gap is consumed in driving the rotor and in copper losses whose values can be represented independently in the machine equivalent model. This value can be deduced

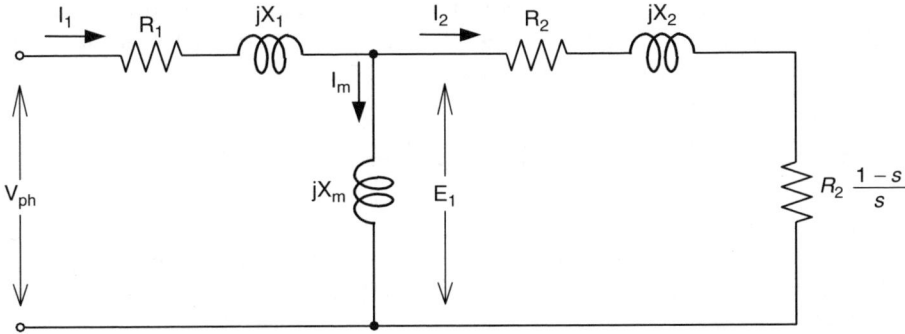

FIGURE 2.7
Separate representation per phase of the copper losses and the power converted in the rotor.

directly from Equation 2.23, being defined as an equivalent resistor per phase to represent the input power:

$$R_{2eq} = R_2 \frac{(1-s)}{s} \tag{2.29}$$

So, the equivalent circuit of Figure 2.3 can be transformed as shown in Figure 2.7 with the copper losses separated from the converted power and neglecting losses, R_m.

To estimate the generator efficiency we use:

$$\eta = \frac{P_{out}}{P_{in}} 100\% \tag{2.30}$$

or, using Equations 2.13 and 2.14:

$$\eta = \frac{P_{in} - P_{losses}}{P_{in}} 100\% = \frac{-3I_2^2 R_2 \frac{1-s}{s} - P_{losses}}{-3I_2^2 R_2 \frac{1-s}{s}} \tag{2.31}$$

The efficiency of the induction generator is not lower than that of the synchronous generator, although some manufacturers claim otherwise.[2] Even so, as we will see later on, induction generators have been used in practice only for low power, more commonly in hydro and wind electric generation.

As has already been said, from the active power per phase supplied to the machine, P, we should discount the stator copper losses, $I_1^2 R_1$, the hysteresis losses, parasites currents, friction and air opposition, represented by E_1^2/R_m. The amount of these losses varies with the frequency, temperature, and operating voltage. However, this model is sufficiently approximate for the rated values for most applications, which, as a rule, do not stray far from the rated values. Thus we may have:

$$P_h = I_1^2 R_1 + \frac{E_1^2}{R_m} \tag{2.32}$$

Recall that s is negative for a generating operation. The net output power, P_{out}, of the machine can be obtained from Equations 2.13, 2.21 and 2.31, as:

$$P_{out} = P_{mec} - P_{losses} = -3I_2^2 R_2 \frac{1-s}{s} - 3I_1^2 R_1 - \frac{3E_1^2}{R_m} - 3I_2^2 R_2 - P_{frict+air} - P_{stray}$$

or, simplified:

$$P_{out} = P_{airgap} - P_{st-losses} = -3I_2^2 R_2 \frac{1}{s} - 3I_1^2 R_1 - \frac{3E_1^2}{R_m} - P_{frict+air} - P_{stray} \tag{2.33}$$

2.6   Measurement of Induction Generator Parameters

Evidently, measurement of induction generator parameters follows the same methods as for induction motors found in the classic texts and standard test procedures about electrical machines (for example, IEEE Std. 112). They can be useful in models and control algorithms under variable load. Similarly to the methods used in transformers, these parameters should be precisely established in laboratory tests of short circuit and open circuit conditions. The losses are combined in Table 2.1.

TABLE 2.1

Standard Loss Distribution in an Induction Machine (IEEE112)

Type of Loss	Description
Friction and windage	Mechanical loss due to bearing (and brush) friction and windage
Core	Loss in iron at no load
Stator ($I_1^2 R_1$)	$I_1^2 R_1$ loss in stator windings
Rotor ($I_2^2 R_2$)	$I_2^2 R_2$ loss in rotor windings (and brush-contact loss of wound rotor machines)
Stray load	Stray loss in iron and eddy-current losses in conductors

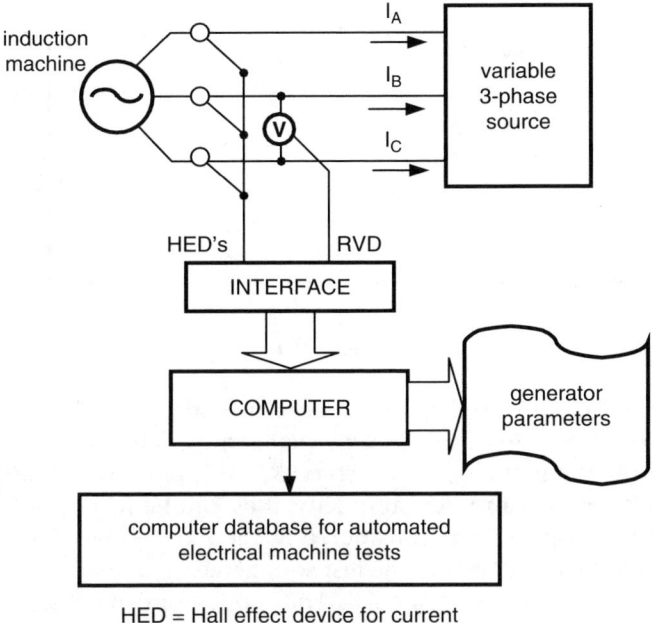

FIGURE 2.8
Automatic test to obtain induction machine parameters.

For the conventional short circuit and no-load tests discussed in any basic text on alternate machines and defined by IEEE Std. 112, we need two watt meters, three ampere meters, and a voltmeter, under the usual hypothesis that the three-phase source is well designed for amplitude and phase unbalances. If it is available, a computerized bank of tests can be used with sensors for the instantaneous values of voltage and current, as sketched in Figure 2.8. With these values, direct measure of the average value of power can be skipped; it can be obtained by processing only two variables (instantaneous voltage and current). Other alternatives also exist.

To test the machine without load (open circuit) we measure the rotating losses and the magnetizing current. The blocked rotor test measures the circuit parameters. In other words, if $s = 1$ in Figure 2.3 (short circuit), R_2/s becomes equal to R_2 or almost zero, as much as X_2.

2.6.1 Blocked Rotor Test ($s = 1$)

The blocked rotor test is the equivalent of the above test for a short-circuited secondary transformer. The voltage, usually very low, is slowly increased across the terminals of the motor with the rotor locked until the current reaches the rated value being measured, then, voltage, current, and power.

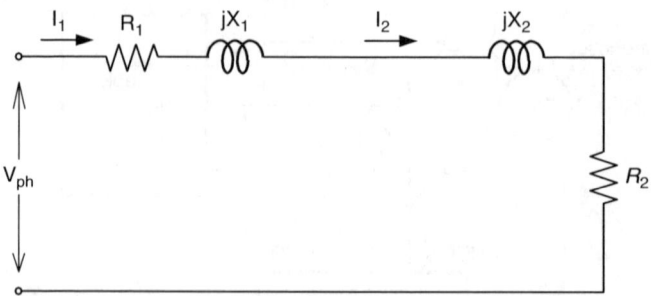

FIGURE 2.9
Approximate circuit of the induction generator.

The equivalent circuit used for $s = 1$ is shown in Figure 2.9 where R_m and X_m have been subtracted from the complete equivalent circuit of Figure 2.3. Notice that the magnetizing parameters, R_m and X_m, are a lot larger, respectively, than R_2 and X_2 and, for simplicity, they can be neglected.

An important aspect to be considered in the blocked rotor test is that the rotor assumes the frequency of the test source when, in reality, this is around 1 to 3 Hz for slip factors of 2 and 4%. The slip factor can be larger in motors designed for low noise, as in air conditioning fan motors. To compensate for this phenomenon the use of 15 Hz is recommended (25% of rated frequency and at rated current according to IEEE Std. 112).[1,21]

Once V_ℓ, I_ℓ and P_{in} have been measured in the blocked rotor test, Equations 2.13 enables us to obtain the power factor of the motor. The module of the total impedance is:

$$\left|Z_{blocked}\right| = \frac{V_{ph}}{I_\ell} = \frac{V_\ell}{\sqrt{3}I_\ell} \qquad (2.34)$$

From the power factor and Equation 2.34 we obtain the values of summations of resistances and reactances (at the test frequency) of the stator and rotor defined as:

$$R_1 + R_2 = \left|Z_{blocked}\right| \cos\phi \qquad (2.35)$$

$$X_1 + X_2 = \left|Z_{blocked}\right| \sin\phi \qquad (2.36)$$

To obtain R_1, usually, an approximate measure is used in DC current from ordinary multitesters across the stator terminals. As conventional multitesters use DC current for measurements of resistances, of course, they will not include in the measure any reactance or reaction of the rotor and only R_1 will oppose the passage of current. Here the effects of heating, frequency and,

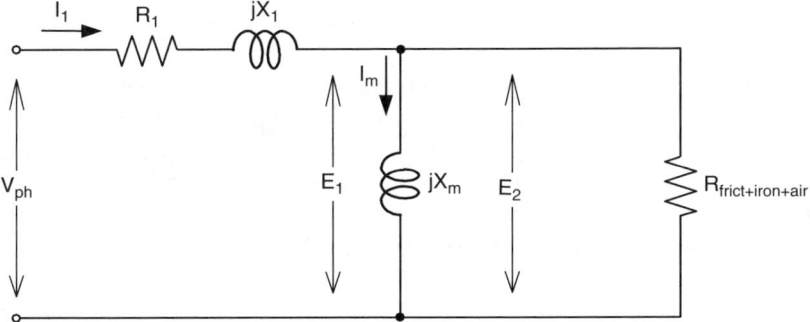

FIGURE 2.10
Equivalent circuit of the induction generator load.

most importantly, skin effects on the motor are not included. In the three-phase windings, the measure across the motor terminals always will measure the resistance of two windings in series. As a consequence:

$$R_1 \approx \frac{V_{cc}}{2I_{cc}}$$
(2.37)

With this value of the stator winding resistance, R_2 can be determined from Equation 2.35.

As the sum of reactances in Equation 2.36 is proportional to the measured frequency, if the rated frequency is different from the test frequency then the results should be multiplied by $F = f_{rated}/f_{test}$.

In practice, commercial motors make individual contributions of the rotor and stator to the total reactance of the machine. The reactance X_1 contributes something between 30 and 50% and X_2, between 50 and 70%.[1,21]

2.6.2 The No-Load Test (s = 0)

The no-load test begins by making the machine work as an induction motor, freely rotating under the rated voltage without any load on the shaft. In this way, the power dissipation just overcomes friction and air opposition; these are mechanical losses. The slip factor will be minimal (of the order of 0.001 or less). With this slip factor, and from Figure 2.7, we can see that the resistance corresponding to the converted power is much larger than the rotor losses in the copper or in the rotor reactance. The losses are concentrated in friction, core iron, and air windage opposition, $R_{frict+iron+air}$. The equivalent circuit will be transformed, then, as displayed in Figure 2.10.

Notice that $R_{frict+iron+air}$ is parallel to X_m for $s \approx 0$, but much larger, making $I_2 \approx 0$; therefore, the copper losses in the rotor are negligible. As a result, it can be established that:

$$P_{in} = 3I_1^2 R_1 + P_{iron} + P_{frict+air} + P_{stray} \tag{2.38}$$

Since the current necessary to establish the magnetic field in induction motors is quite large due to the high reluctance of the air gap, the resistances due to iron losses, R_{iron}, and that of the converted power, $R_2(1 - s)/s$, both in parallel with X_m, the magnetizing reactance becomes much smaller than that total parallel resistance. The power factor will be very small. This circuit will be predominantly inductive and expressed as:

$$\frac{V_{ph}}{I_{10}} \cong (X_1 + X_m) \tag{2.39}$$

where I_{10} is the stator current with the motor without load.

2.7 Peculiarities of the Induction Machine Working as a Generator Interconnected to the Distribution Network

As stated above (Equation 2.22), when $n_r > n_s$ the developed mechanical power becomes negative. In other words, when $n_r > n_s$ the asynchronous machine begins to consume mechanical power working as an electric power generator. In electromagnetic terms, for rotations above the synchronous, the relative speed between the rotating flux and the rotor itself changes sign for speeds below the synchronous. Therefore, the rotor voltage and current change direction (change sign) and, therefore, V_1 and E_1 do the same. Equations transformed by the sign change and describing this phenomenon from Figure 2.3 can be combined as below:[2,6]

$$E_1 - V_1 + I_1 R_1 + jI_1 X_1 \tag{2.40}$$

$$E_2 = I_2 \frac{R_2}{s} + jI_2 X_2 \tag{2.41}$$

$$E_1 = E_2 = -I_m Z_m = (I_1 - I_2) Z_m \tag{2.42}$$

In Equation 2.41 the slip factor s is negative and the representation of these equations is shown in Figure 2.11.[1,6] As s is negative in the generator, the secondary emf E_2 is leading by 90° with respect to the magnetic field B (and opposed to the direction of the secondary voltage if it is a motor). For the

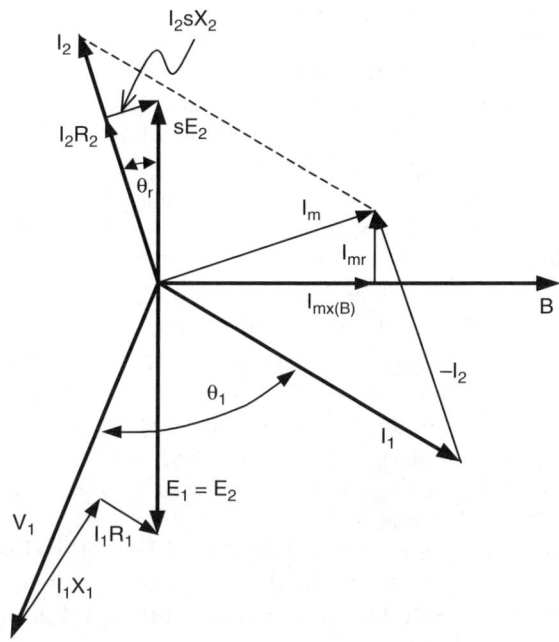

FIGURE 2.11
Vector diagram of the induction generator.

same reason, the voltage drop jI_2sX_2 is lagging regarding the current I_2 in 90° and I_2 is leading with respect to E_2. The stator voltage E_1 is leading V_{ph}.

In the induction generator, the magnetizing current produces the magnetizing field B. For this, synchronous generators or other electrical sources are used to feed external circuits in association with induction generators. Some interesting conclusions can be drawn from the observations above. As the current I_m reaches values around 20 to 40% of I_r and supplies the generator at the circuit voltage, the excitation power (in kVA) will also reaches 20 to 40% of the generator's rated values. Therefore, if two to four induction generators of same output are installed in a distribution network, their excitation needs will pull all the capacity of a synchronous generator of similar capacity, recalling that the excitation power of synchronous generators is smaller than 1%. Such differences in the need for excitation power are the main argument against the induction generators. Besides that, the current I_m is lagging behind the voltage by approximately 90° and, consequently, the parallel operation of induction generators with synchronous generators reduces the power factor even when the external load is purely active. We see that the synchronous generator should supply the lagging current as much for the usual case of inductive loads as for the induction generator. For this reason, the induction generator has a double disadvantage with respect to the synchronous: it is not capable of supplying lagging current for the load and, on top of that, it will draw lagging current from the distribution network.

Connecting induction generators to the distribution network is a quite simple process, as long as interconnection and protection guidelines are followed with the local utilities. The rotor is made to turn in the same direction in which the magnetic field is rotating, as closely as possible to synchronous speed to avoid unnecessary speed clashes. A phenomenon similar to the connection of motors or transformers in the distribution network will happen.

The active power supplied by an induction generator to the circuit, similarly to what happens with the synchronous generators, can be controlled by speed variation, in other words, controlled by the mechanical primary power.

In the case of the stand-alone operation of induction generators, the magnetizing current can be obtained from the self-excitation process. The generator can supply a capacitive current, because, as seen in Figure 2.11, the current I_1 is leading V_{ph}. This is due to the fact that the mechanical energy of rotation can only influence the active component of the current; it does not affect the reactive component. Therefore, the rotor does not supply reactive current. As a component of the magnetizing current in the main field direction, $I_{mx(B)}$, is needed to maintain this flux, another reactive current is also needed to maintain the dispersion fluxes. Such currents should be supplied by a leading current with respect to the stator current.

As the induction generator depends on the parallel synchronous machine to get its excitation, the short circuit current that it can supply depends on the voltage drop produced across the terminals of the synchronous generator. A very intense short circuit transient current arises, however, of extremely short duration. If the voltage across the terminals goes to zero, the steady-state short circuit current is zero. A small current is supplied in the case of a partial short circuit since the maximum power the induction generator can supply with fixed slip factor and frequency is proportional to the square of the voltage across its terminals (see Equation 5.1). The incapacity of sustaining short circuits greatly reduces possible damage caused by electrical and mechanical stresses. As a consequence, it allows the use of short circuit power of smaller capacity and cost when compared to the case when only synchronous generators provide all the capacity of the facilities.

If the induction generator is self-excited on a stand-alone basis with respect to the distribution network, a very high load current or a short circuit across its terminals alters the value of the effective exciting capacitance or it may remove it altogether. In that case, there is no excitation possibility; voltage and current collapse immediately to zero.

The induction generator is friendly to oscillations, as long it does not have to work at synchronous speed. All the load variation is accompanied by a speed variation and a small phase displacement, much the same as with the synchronous generator. The mechanical speed variations of the primary machine driving the generator are so small that they only produce minor variations of load.

In short, the advantages of the induction generator are: (a) a robust and solid rotor; (b) the machine cannot feed large short circuit currents and is free of oscillations; (c) the rotor construction is suitable for high speeds; (d) it does not need special care with regard to synchronization; and (e) voltage and frequency are regulated automatically by the voltage and frequency of the distribution network in parallel operation.

Its disadvantages are: (a) the power factor is determined by the slip factor and has very little to do with the power factor of the load when working in parallel with synchronous machines; (b) synchronous machines need to supply with lagging reactive power as much of the load as the induction generator and, consequently, they should work with a power factor higher than the load.

2.8   The High Efficiency Induction Generator

A high efficiency induction generator is a commercially available high efficiency induction motor, except for some peculiarities. Therefore, the same care must be taken in design, materials selection, and manufacturing processes for building a high efficiency generator.

The main advantages of the high efficiency induction generator compared to the conventional induction generator are better voltage regulation, less loss of efficiency with smaller loads, less oversizing when generators of lower power cannot be used, reduced internal losses, and, therefore, lower temperatures, less internal electric and mechanical stress, and, thus, increased useful life. The constraints are the need for larger capacitors for self-excitation. High efficiency induction generators should not be used for self-excited applications.

The efficiency of the high efficiency generator compared to the conventional ones differs by more than about 10% for small powers (up to 50 kW) and about 2% for higher powers (above 100 kW). It is therefore highly recommended for micropower plants. Rated efficiencies are normalized, and they should have guaranteed minimum values declared by the manufacturer on the plate of the machine for each combination of power times synchronous speed.

The investment return time (payback period) for these high efficiency generators will be smaller in an inversely proportional way to the local tariff values in the cases of injection into the public distribution network. This investment will be increased by the number of operating hours, increased by the increased difference in the efficiency with respect to the conventional ones, and smaller for small load power with respect to the rated power of the machine. The investment return time can go from 0.9 to 3 or 4 years for micropower plants.

High efficiency generators are better suited to stand the harmful effects of the harmonic generated by nonlinear loads (power converters) because they have higher thermal margin and smaller losses.

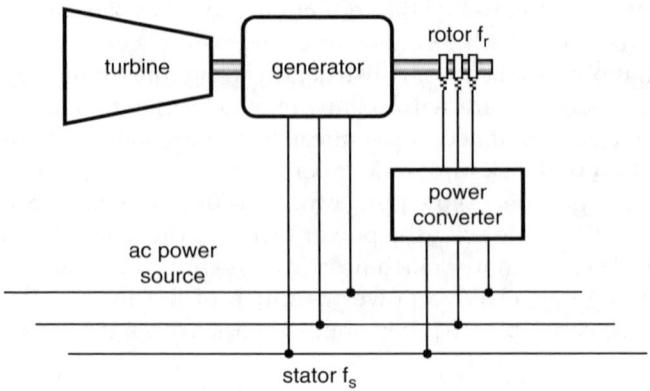

FIGURE 2.12
Doubly fed induction generator.

2.9 The Doubly Fed Induction Generator

The doubly fed induction generator, also known as the Scherbius variable speed driver, is a wound rotor machine with slip rings to allow control of the rotor winding current. The rotor circuit is connected to an external variable-frequency source via slip rings and the stator is connected to the grid network as illustrated in Figure 2.12. The speed of the doubly fed induction generator, which is usually limited to a 2:1 range, is controlled by adjusting the frequency of the external rotor source of current. These machines have not been very popular due to the easy wearing out of the slip rings. More recently, with the development of new materials and power electronics, the doubly fed induction generator has made a comeback for up to several hundreds of kW ratings. Power converters whose cost for large units becomes irrelevant usually make up the need for a variable frequency source for the rotor. More details are given in Chapter 11.

2.10 Problems

2.1 Determine the necessary primary mechanical torque of a turbine for driving an induction generator of 1.0 kW, 2.3 kV, 60 Hz, 12 poles and $s = -0.03$.

2.2 What is the slip factor of a 100 hp, 6 pole, 60 Hz 3-phase induction generator to achieve its rated power when the rotor current is 10 A and its rotor resistance is 0.407 Ω? What would be the turbine torque under these conditions?

2.3 A 10 hp, Y-connected, 60 Hz, 380 V, 3-phase induction generator is used to drive a small micropower plant. The generator's load is 500 W for a 4 A line current at 1740 rpm. The rotor resistance is 4.56 Ω at 75°C. The losses are equal to 28 W. A no-load test of the machine gave the following results: P = 215 W, I = 3.0 A and V = 380 V. Calculate the output power, the efficiency and the power factor for the given load. If the generator was Δ-connected the same data as above, what would be the difference in your calculations?

2.4 Explain the difference between rotor speed and rotor frequency. For s = –0.01 and a rotor speed of 1740 rpm what is the rotor frequency for a 60 Hz induction generator?

2.5 Explain why the efficiency of an induction generator can be smaller when working as an asynchronous motor under certain conditions.

2.6 Calculate the efficiency of a 110 V, 60 Hz induction generator connected directly to the grid with the following parameters: $R_m = 36.2$ Ω; losses equal to 0.8% of the rated power; s = –0.03; $R_1 = 0.2$ Ω; $R_2 = 0.15$ Ω; $X_1 = 0.42$ Ω; $X_2 = 0.43$ Ω.

References

1. Lawrence, R.L., *Principles of Alternating Current Machinery*, McGraw-Hill Book Co. Inc., 1953, p. 640.
2. Kostenko, M. and Piotrovsky, L., *Electrical Machines, Vol. II*, Mir Publishers, Moscow, 1969, p. 775.
3. Smith, I.R., and Sriharan, S., Transients in induction motors with terminal capacitors, *Proc. IEEE*, Vol. 115, 519–527, 1968.
4. Murthy, S.S., Bhim Singh, M. and Tandon, A. K., Dynamic models for the transient analysis of induction machines with asymmetrical winding connections, *Electric Machines and Electromechanics*, 6(6), 479–492, Nov./Dec. 1981.
5. IEEE Std. 1547, Standard for Interconnecting Distributed Resources with Electrical Power Systems.
6. IEEE Std. (Draft) P1547.1, Standard Conformance Test Procedures for Interconnecting Distributed Energy Resources with Electric Power Systems.
7. IEEE Std. (Draft) P1547.2, Application Guide for IEEE Standard 1547: Interconnecting Distributed Resources with Electric Power Systems.
8. IEEE Std. (Draft) P1547.3, Guide for Monitoring, Information Exchange, and Control of Distributed Resources Interconnected with Electric Power Systems.
9. IEEE Std. 112-1991, IEEE Standard Test Procedure for Polyphase Induction Motors and Generators.
10. Liwschitz-Gärik, M. and Whipple, C.C., *Alternating Current Machines (Máquinas de Corriente Alterna)*, Companhia Editorial Continental, authorized by D. Van Nostrand Company, Inc., 1970, p. 768.

11. Murthy, S.S., Malik, O.P. and Tandon, A.K., Analysis of self-excited induction generators, *Proc. IEEE*, 129(6), 260–265, 1982.
12. Watson, D.B. and Milner, I.P., Autonomous and parallel operation of self-excited induction generators, *Elect. Engin. Educ.*, Vol. 22, 365–374, Manchester University Press, 1985.
13. Murthy, S.S., Nagaraj, H.S., and Kuriyan, A., Design-based computational procedure for performance prediction and analysis of self-excited induction generators using motor design packages, *Proc. IEEE*, 135(1), 8–16, 1988.
14. Grantham, C., Sutanto, D., and Mismail, B., Steady-State and transient analysis of self-excited induction generators, *Proc. IEEE*, Vol. 136, Pt. B, (2), 61–68, March 1989.
15. Singh, S.P., Bhim Singh, M. and Jain, P., Performance characteristics and optimum utilization of a cage machine as capacitor excited induction generator, *IEEE Transactions on Energy Conversion*, 5(4), 679–684, Dec. 1990.
16. Hallenius, K.E., Vas, P., and Brown, J.E., The analysis of a saturated self-excited asynchronous generator, *IEEE Transactions on Energy Conversion*, 6(2), 336–345, June 1991.
17. Barbi, I., *Fundamental Theory of the Induction Motor (Teoria fundamental do motor de indução)*, Florianópolis, Brazil, Ed. da UFSC, Eletrobrás, 1985.
18. Langsdorf, A.S., *Theory of Alternating Current Machines (Teoría de Máquinas de Corriente Alterna)*, McGraw-Hill Book Co., 1977, p. 701.
19. Hancock, N.N., *Matrix Analysis of Electric Machinery*, Pergamon, Oxford, 1964, p. 55.
20. Grantham, C., Determination of induction motor parameter variations from a frequency stand still test, *Electrical Machine Power Systems*, 10, 239–248, 1985.
21. Chapman, S.J., *Electric Machinery Fundamentals*, McGraw-Hill International Edition, Third Edition, New York, 1999.

3

The Transient Model of the Induction Generator

3.1 Scope of This Chapter

This chapter will explain how steady-state terminal voltage builds up during the self-excitation process and during the recovery of voltage during perturbations across the terminal voltage and stator current due to load changes. A set of state equations will be derived to obtain the instantaneous output voltage and current in the self-excitation process. In the example in this chapter, the equations describe an induction machine with the following rated data: 220/380 V; 26/15 A; 7.5 kW; and 1.800 rpm. We will also show that the equations in this chapter can be easily used to calculate output voltage, including the variation of mutual inductance with the magnetizing current seen in Chapter 4. Furthermore, we will present a general matrix equation for simulation of the parallel aggregation of induction generators. In any case, for every machine at a given speed, we will show that there is a minimum capacitance value that causes self-excitation in agreement with the steady state case shown in Chapter 2. We will discuss rotor parameter variation and demonstrate that it has little effect on the accuracy of these calculations and that comparable results can be achieved using the standard open circuit and locked rotor parameters. Saturation effects will also be considered in this chapter by taking into account the nonlinear relationship between magnetizing reactance and the magnetizing current of a machine; using this approach, the mutual inductance, M, will be shown to vary continually.

3.2 An Induction Machine in a Transient State

During self-excitation, asynchronous or induction generators exhibit transient phenomena that are very difficult to model from an operational point of view; the impact of self-excitation is more pronounced in generators with

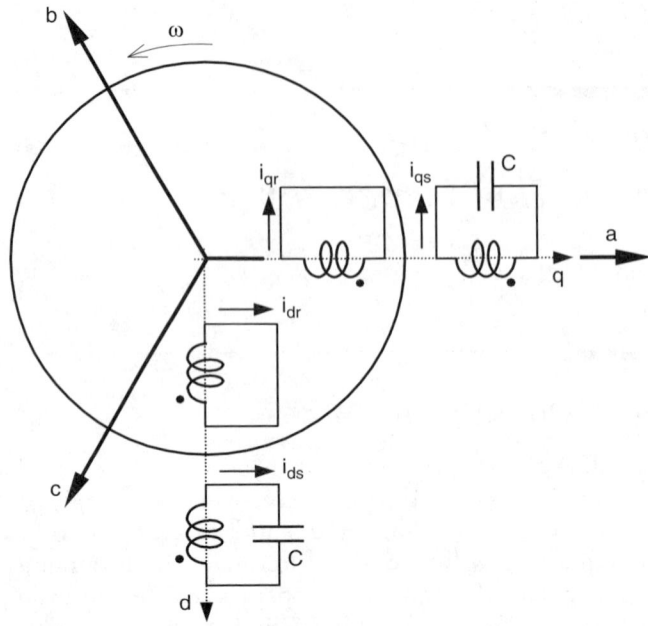

FIGURE 3.1
Equivalent circuit of a SEIG connected to an RLC load.

heavier loads.[1-3] A crucial problem to be avoided is the demagnetization of the induction generator. To analyze an induction generator in a transient state, we will apply the Park transformation (also known as the Blondel or Blondel-Park transformation), which is associated with the general theory of rotating machines used in this chapter.

The Park transformation specifies a two-phase primitive machine with fixed stator windings and rotating rotor windings to represent fixed stator windings (direct axis) and pseudo-stationary rotor windings (quadrature axis) (see Figure 3.1). Any machine can be shown to be equivalent to a primitive machine with an appropriate number of coils on each axis.[4-6]

Saturation effects are also considered in this chapter by taking into account a nonlinear relationship between the magnetizing reactance and the magnetizing current of the machine. Under this approach, the mutual inductance M varies continuously.

3.3   State Space Based Induction Generator Modeling

Figure 3.1 represents a primitive machine through a Park transformation as applied to the induction generator. Currents I_{ds} and I_{qs} refer to the stator

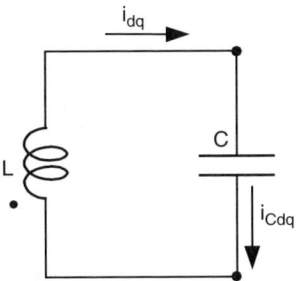

FIGURE 3.2
Representation of SEIG in d–q axis.

currents, and I_{dr} and I_{qr} to the rotor currents, in the direct and quadrature axis, respectively. The angular speed $\omega = d\theta/dt$ is the mechanical rotor speed.

As can be seen in Figure 3.1, no external voltage is applied across the rotor or the stator windings. This is the standard form of stationary reference axis used in the machine theory texts.[7-10] However, there is an additional component: the self-excitation capacitor C.

3.3.1 A No-load Induction Generator

A self-excited induction generator (SEIG) with a capacitor is initially considered to be operating at no load.[1,2] The relationships between the resulting voltages and currents, direct and quadrature, can be obtained from Figure 3.1 and Figure 3.2 and Equation 3.1. That expression represents the ordinary symmetrical three-phase machine connected to a three-phase bank of identical parallel capacitors. The reference position is put on the stator under every normal operating condition including transient ones. The subscript *s* refers to the stator while *r* refers to the rotor. The voltage drop across the capacitor, (direct axis, v_{Cd} and quadrature axis, v_{Cq}), is included in the matrix in the expression $1/pC$. The transformer ratio from the stator to the rotor is assumed to be unity and transformations should be introduced in the rotor parameters when there is need to refer them to the stator.

$$\begin{bmatrix} v_{ds} \\ v_{qs} \\ v_{dr} \\ v_{qr} \end{bmatrix} = \begin{bmatrix} R_1 + L_1 p + \dfrac{1}{pC} & 0 & Mp & 0 \\ 0 & R_1 + L_1 p + \dfrac{1}{pC} & 0 & Mp \\ Mp & \omega M & R_2 + L_2 p & \omega L_2 \\ -\omega M & Mp & -\omega L_2 & R_2 + L_2 p \end{bmatrix} * \begin{bmatrix} i_{ds} \\ i_{qs} \\ i_{dr} \\ i_{qr} \end{bmatrix} \quad (3.1)$$

The mutual inductance between stator and rotor (*M*) varies with the current through the windings and the relative positions between stator and

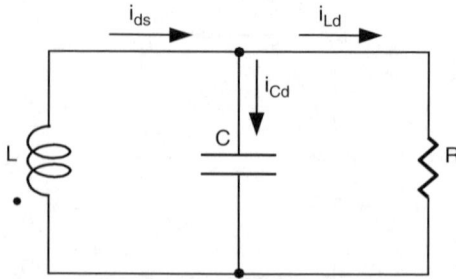

FIGURE 3.3
Stator direct axis component with an RL load.

rotor. So, the d–q coupled voltage drops are due to the relative position of the coils as well as the current variation through them. Therefore, an additional term ipM is included in the analytical solution; that is, for the mutual flux given by $\phi_{mi} = Mi$, an additional voltage drop is established as:

$$v = \frac{d\phi_m}{dt} = M\frac{di}{dt} + i\frac{dM}{dt} = Mpi + ipM \tag{3.2}$$

The first term of Equation 3.2 represents current variation due to generator stator current; the second, due to the rotation of the rotor which, taking into account the short period of the calculation step until the self-excitation reaches the steady state, can be considered either constant and equal to ω or varied and recalculated at each step according to the desired changing law from zero to the steady state value.[9–12]

3.3.2   State Equations of SEIG with Resistive Load, *R*

A resistive load, R, is added in parallel with the self-excitation capacitor. There is a voltage across this load that will be the same as the self-excitation capacitor, as illustrated in Figure 3.3. So, taking i_{LD} as the direct axis current through the load, we have:

$$v_{Ld} = V_{Cd} = Ri_{Ld} \tag{3.3}$$

Therefore:

$$i_{Cd} = Cpv_{Cd} = CpRi_{Ld}$$

As $i_{ds} = i_{Cd} + i_{Ld}$, then $i_{ds} = RCpi_{Ld} + i_{Ld}$ that, rearranged as a function of i_{Ld}, results in:

$$i_{Ld} = \frac{i_{ds}}{RCp + 1} \tag{3.4}$$

Replacing Equation 3.4 in 3.3 yields:

$$v_{Ld} = \frac{R}{RCp + 1} i_{ds} \tag{3.5}$$

Similarly, the quadrature load voltage can be obtained.

$$v_{Lq} = \frac{R}{RCp + 1} i_{qs} \tag{3.6}$$

So Equation 3.1 assumes the following form to take into account the R load:

$$
\begin{bmatrix} v_{ds} \\ v_{qs} \\ v_{dr} \\ v_{qr} \end{bmatrix} =
\begin{bmatrix}
R_1 + L_1 p + \left(\dfrac{R}{RCp + 1} \right) & 0 & Mp & 0 \\
0 & R_1 + L_1 p + \left(\dfrac{R}{RCp + 1} \right) & 0 & Mp \\
Mp & \omega M & R_2 + L_2 p & \omega L_2 \\
-\omega M & Mp & -\omega L_2 & R_2 + L_2 p
\end{bmatrix}
*
\begin{bmatrix} i_{ds} \\ i_{qs} \\ i_{dr} \\ i_{qr} \end{bmatrix}
\tag{3.7}
$$

3.3.3   State Equations of SEIG with an RLC Load

We will now consider an RLC load, which is supposed to be an RL load in parallel with a certain capacitor, in parallel with the self-excitation capacitor. Therefore, an overall value C will be considered in the following discussion. As was explained in the previous section, the voltage across the self-excitation capacitor is the same as across the RL load. Therefore, with $v_{Ld} = Ri_{Ld} + Lpi_{Ld}$, for the current through the capacitor we have

$$i_{Cd} = Cpv_{Ld} = (RCp + LCp^2)i_{Ld}$$

In the same way, as $i_d = i_{Cd} + i_{Ld}$, we have $i_d = (RCp + LCp^2)i_{Ld} + i_{Ld}$, or

$$i_{Ld} = \frac{i_{ds}}{RCp + LCp^2 + 1} \tag{3.8}$$

and therefore, $v_{Ld} = (R + Lp)i_{Ld}$ or

$$v_{Ld} = \frac{R + Lp}{RCp + LCp^2 + 1} i_{ds} \tag{3.9}$$

Similarly,

$$v_{Lq} = \frac{R + Lp}{RCp + LCp^2 + 1} i_{qs} \tag{3.10}$$

Including the load inductance effects, Equation 3.1 becomes

$$
\begin{bmatrix} v_{ds} \\ v_{qs} \\ v_{dr} \\ v_{qr} \end{bmatrix} =
\begin{bmatrix}
R_1 + L_1 p + \left(\dfrac{R + Lp}{RCp + LCp^2 + 1} \right) & 0 & Mp & 0 \\
0 & R_1 + L_1 p + \left(\dfrac{R + Lp}{RCp + LCp^2 + 1} \right) & 0 & Mp \\
Mp & \omega M & R_2 + L_2 p & \omega L_2 \\
-\omega M & Mp & -\omega L_2 & R_2 + L_2 p
\end{bmatrix}
*
\begin{bmatrix} i_{ds} \\ i_{qs} \\ i_{dr} \\ i_{qr} \end{bmatrix}
\tag{3.11}
$$

The impedance matrix obtained above is similar to that in the classic literature for induction machines, except for the expression of the voltage drop across the self-excitation capacitor (Equations 3.9 and 3.10), the voltage drop across the generator terminals, and the additional terms corresponding to each load type. Notice that if the load includes a capacitor, it can be added in parallel with the self-excitation capacitor in the matrix representation or in another form linked with the load. Therefore, independently of the load type, the following system of equations holds: $[v] = [Z][i]$. For the determination of the instantaneous currents, it is enough to obtain: $[v] = [Z]^{-1}[i]$.

To find the voltage across the self-excitation capacitor, it is enough to take the case of an SEIG feeding an RL load. Knowing that i_{Ld} and i_{Lq} correspond to the currents through the load in the direct and quadrature axis, respectively, we have

$$pv_{Ld} = \frac{i_{ds}}{C} - \frac{i_{Ld}}{C} \tag{3.12}$$

$$pv_{Lq} = \frac{i_{qs}}{C} - \frac{i_{Lq}}{C} \tag{3.13}$$

The solution to the system of equations above describes the behavior of currents and voltages during the self-excitation process and in steady operation, besides allowing the verification of load variations. In the solution of Equation 3.11, the alterations of the magnetizing reactance as a function of the magnetizing current must be taken into account through corrections to satisfy the magnetic saturation data, as discussed in Subsection 3.3.1.

For a more realistic representation of induction generators, the parameters of the machine should be obtained through conventional short circuit and open circuit tests, as discussed in Chapter 4. These tests supply the resistance values and rotor and stator reactances for a given frequency. The conventional numeric methods of solving simultaneous equation systems seem to be sufficiently appropriate.[11-13] If greater precision is needed, the resistance and reactance variations of the rotor as a function of magnetism and of frequency should be considered.

It is important to observe that for the self-excitation process to be initiated, the machine must possess a residual magnetism. If this magnetism does not exist, it must be generated by connecting a battery or a net-connected rectifier, or by using capacitors previously charged or by keeping the rotor rotating for some time with the machine in a no load condition. The effect of the residual magnetism is described in the equations deduced previously in the terms v_{dr}, v_{qr}, v_{ds} and v_{qs}. Such voltages vary according to the hysteresis curve of the motor. Therefore, they cannot have a constant value because this would generate incorrect results.

3.4 Partition of the SEIG State Matrix with an RLC Load

Our objective in this section is to find a model to separate machine parameters from load and self-excitation parameters. Equation 3.11 represents the transient electric behavior of a self-excited stand-alone induction generator, so describing the aggregation of more than one induction generator is easy.

Consider the state equations that describe the voltage across the terminals of the generator to the current through any RLC load. From Figure 3.3, we get

$$i_{Cd} = i_{ds} - i_{Ld} \tag{3.14}$$

or

$$pv_{Ld} = \frac{i_{Cd}}{C} = \frac{i_{ds}}{C} - \frac{i_{Ld}}{C} \tag{3.15}$$

Similarly, for the quadrature current, we have

$$pv_{Lq} = \frac{i_{qs}}{C} - \frac{i_{Lq}}{C}$$

(3.16)

From the load voltage $v_{Ld} = Ri_{Ld} + Lpi_{Ld}$, in Figure 3.3, we get the state values for the direct current through the load:

$$pi_{Ld} = \frac{V_{Ld}}{L} - \frac{R}{L}i_{Ld}$$

(3.17)

Similarly,

$$pi_{Lq} = \frac{V_{Lq}}{L} - \frac{R}{L}i_{Lq}$$

(3.18)

We can now establish the voltage and current relationships between the IG and the self-excitation capacitor. From matrix Equation 3.11 and Equations 3.9 and 3.10 we have, respectively,

$$v_{ds} = R_1 i_{ds} + L_1 pi_{ds} + v_{Ld} + Mpi_{dr}$$

(3.19)

$$v_{qs} = R_1 i_{qs} + L_1 pi_{qs} + v_{Lq} + Mpi_{qr}$$

(3.20)

$$v_{dr} = Mpi_{ds} + \omega Mi_{qs} + \omega L_2 i_{qr} + R_2 i_{dr} + L_2 pi_{dr}$$

(3.21)

$$v_{qr} = -\omega Mi_{ds} + Mpi_{qs} + R_2 i_{qr} + L_2 pi_{qr} - \omega L_2 i_{dr}$$

(3.22)

Isolating the differential terms of the direct and quadrature currents of the rotor in Equations 3.19 and 3.20, we get

$$pi_{dr} = \frac{1}{M}\left(v_{ds} - R_1 i_{ds} - L_1 pi_{ds} - v_{Ld}\right)$$

(3.23)

$$pi_{qr} = \frac{1}{M}\left(v_{qs} - R_1 i_{qs} - L_1 pi_{qs} - v_{Lq}\right)$$

(3.24)

Substituting Equation 3.24 in Equation 3.22, and isolating pi_{qs} we get

$$pi_{qs} = \frac{1}{M}\left[\left(v_{qr} + \omega M i_{ds} - R_2 i_{qr} + \omega L_2 i_{dr}\right) - L_2 \frac{1}{M^2}\left(v_{qs} - R_1 i_{qs} - v_{Lq} - L_1 pi_{qs}\right)\right] \quad (3.25)$$

Isolating the term pi_{qs} in Equation 3.25, we get

$$pi_{qs} = \frac{M}{M^2 - L_1 L_2}\left(v_{qr} + \omega M i_{ds} - R_2 i_{qr} + \omega L_2 i_{dr}\right) - \frac{L_2}{M^2 - L_1 L_2}\left(v_{qs} - R_1 i_{qs} - v_{Lq}\right)$$

Assuming the following leakage coefficient:

$$K = \frac{1}{M^2 - L_1 L_2}$$

Then:

$$pi_{qs} = MK(v_{qr} + \omega M i_{ds} - R_2 i_{qr} + \omega L_2 i_{dr}) - L_2 K(v_{qs} - R_1 i_{qs} - v_{Lq}) \quad (3.26)$$

Similarly, for the direct axis:

$$pi_{ds} = MK(v_{dr} - \omega M i_{qs} - R_2 i_{dr} - \omega L_2 i_{qr}) - L_2 K(v_{ds} - R_1 i_{ds} - v_{Ld}) \quad (3.27)$$

Substituting Equation 3.26 in Equation 3.24, rearranging and combining like terms, we get

$$pi_{qr} = \frac{(1 + L_1 L_2 K)}{M}(v_{qs} - R_1 i_{qs} - v_{Lq}) - L_1 K(v_{qr} + \omega M i_{ds} - R_2 i_{qr} + \omega L_2 i_{dr})$$

Notice that: $MK = \dfrac{(1 + L_1 L_2 K)}{M}$ and, therefore,

$$pi_{qr} = MK(v_{qs} - R_1 i_{qs} - v_{Lq}) - L_1 K(v_{qr} + \omega M i_{ds} - R_2 i_{qr} + \omega L_2 i_{dr}) \quad (3.28)$$

Similarly, for the direct axis,

$$pi_{dr} = MK(v_{ds} - R_1 i_{ds} - v_{Ld}) - L_1 K(v_{dr} - \omega M i_{qs} - R_2 i_{dr} - \omega L_2 i_{qr}) \quad (3.29)$$

Grouping the state Equations 3.15, 3.16, 3.26, 3.27, 3.28 and 3.29 in a matrix form for $i_{Ld} = 0$ and $i_{Lq} = 0$, we get the reduced matrix equation describing the self-excitation process at no load:

$$
p\begin{bmatrix} i_{ds} \\ i_{qs} \\ i_{dr} \\ i_{qr} \\ v_{Ld} \\ v_{Lq} \end{bmatrix} = K \left\{ \begin{bmatrix} R_1L_2 & -\omega M^2 & -R_2M & -\omega ML_2 & L_2 & 0 \\ \omega M^2 & R_1L_2 & \omega ML_2 & -R_2M & 0 & L_2 \\ -R_1M & \omega ML_1 & R_2L_1 & \omega L_1L_2 & -M & 0 \\ -\omega ML_1 & -R_1M & -\omega L_1L_2 & R_2L_1 & 0 & -M \\ 1/CK & 0 & 0 & 0 & 0 & 0 \\ 0 & 1/CK & 0 & 0 & 0 & 0 \end{bmatrix} \right.
$$

(3.30)

$$
\left. \begin{bmatrix} i_{ds} \\ i_{qs} \\ i_{dr} \\ i_{qr} \\ v_{Ld} \\ v_{Lq} \end{bmatrix} + \begin{bmatrix} -L_2 & 0 & M & 0 \\ 0 & -L_2 & 0 & M \\ M & 0 & -L_1 & 0 \\ 0 & M & 0 & -L_1 \\ 0 & 0 & 0 & 0 \\ 0 & 0 & 0 & 0 \end{bmatrix} \begin{bmatrix} v_{ds} \\ v_{qs} \\ v_{dr} \\ v_{qr} \end{bmatrix} \right\}
$$

Grouping the state Equations 3.15, 3.16, 3.17, 3.18 3.26, 3.27, 3.28 and 3.29, the full matrix equation including the representation of the load connection, is

$$
p\begin{bmatrix} i_{ds} \\ i_{qs} \\ i_{dr} \\ i_{qr} \\ v_{Ld} \\ v_{Lq} \\ i_{Ld} \\ i_{Lq} \end{bmatrix} = K \left\{ \begin{bmatrix} R_1L_2 & -\omega M^2 & -R_2M & -\omega ML_2 & L_2 & 0 & 0 & 0 \\ \omega M^2 & R_1L_2 & \omega ML_2 & -R_2M & 0 & L_2 & 0 & 0 \\ -R_1M & \omega ML_1 & R_2L_1 & \omega L_1L_2 & -M & 0 & 0 & 0 \\ -\omega ML_1 & -R_1M & -\omega L_1L_2 & R_2L_1 & 0 & -M & 0 & 0 \\ 1/CK & 0 & 0 & 0 & 0 & 0 & -1/CK & 0 \\ 0 & 1/CK & 0 & 0 & 0 & 0 & 0 & -1/CK \\ 0 & 0 & 0 & 0 & 1/LK & 0 & -R/LK & 0 \\ 0 & 0 & 0 & 0 & 0 & 1/LK & 0 & -R/LK \end{bmatrix} \right.
$$

$$
\left. \begin{bmatrix} i_{ds} \\ i_{qs} \\ i_{dr} \\ i_{qr} \\ v_{Ld} \\ v_{Lq} \\ i_{Ld} \\ i_{Lq} \end{bmatrix} + \begin{bmatrix} -L_2 & 0 & M & 0 \\ 0 & -L_2 & 0 & M \\ M & 0 & -L_1 & 0 \\ 0 & M & 0 & -L_1 \\ 0 & 0 & 0 & 0 \\ 0 & 0 & 0 & 0 \\ 0 & 0 & 0 & 0 \\ 0 & 0 & 0 & 0 \end{bmatrix} \begin{bmatrix} v_{ds} \\ v_{qs} \\ v_{dr} \\ v_{qr} \end{bmatrix} \right\}
$$

(3.31)

The induction motor might also be represented by a matrix equation that does not include the self-excitation capacitor derived from the same equation set used for SEIG. This equation is

$$
p\begin{bmatrix} i_{ds} \\ i_{qs} \\ i_{dr} \\ i_{qr} \end{bmatrix} = K\left\{ \begin{bmatrix} R_1L_2 & -\omega M^2 & -R_2M & -\omega ML_2 \\ \omega M^2 & R_1L_2 & \omega ML_2 & -R_2M \\ -R_1M & \omega ML_1 & R_2L_1 & \omega L_1L_2 \\ -\omega ML_1 & -R_1M & -\omega L_1L_2 & R_2L_1 \end{bmatrix} \cdot \begin{bmatrix} i_{ds} \\ i_{qs} \\ i_{dr} \\ i_{qr} \end{bmatrix} + \right.
$$

$$
\left. \begin{bmatrix} -L_2 & 0 & M & 0 \\ 0 & -L_2 & 0 & M \\ M & 0 & -L_1 & 0 \\ 0 & M & 0 & -L_1 \end{bmatrix} \cdot \begin{bmatrix} v_{ds} \\ v_{qs} \\ v_{dr} \\ v_{qr} \end{bmatrix} \right\}
$$

(3.32)

Equations 3.30, 3.31 and 3.32 come in the classical form of the state equations as

$$
p[x] = [A][x] + [B][u]
$$

(3.33)

that is,

$$
p\begin{bmatrix} i_G \\ v_C \\ i_L \end{bmatrix} = \begin{bmatrix} G \\ C \\ L \end{bmatrix} \bullet \begin{bmatrix} i_G \\ v_C \\ i_L \end{bmatrix} + [B][v_G]
$$

where indexes G, C and L refer, respectively, to the partitioned matrix of the induction generator, the self-excitation capacitor bank and the load. The $[x]$ vector is the matrix, $[i_G \quad v_C \quad i_L]^T$ and submatrixes $[G]$, $[C]$ and $[L]$ are defined as

$$
[G] = K\begin{bmatrix} R_1L_2 & -\omega M^2 & -R_2M & \omega ML_2 & L_2 & 0 & 0 & 0 \\ \omega M^2 & R_1L_2 & -\omega ML_2 & -R_2M & 0 & L_2 & 0 & 0 \\ -R_1M & \omega ML_1 & R_2L_1 & \omega L_1L_2 & -M & 0 & 0 & 0 \\ -\omega ML_1 & -R_1M & -\omega L_1L_2 & R_2L_1 & 0 & -M & 0 & 0 \end{bmatrix}
$$

$$
[C] = [C_G : C_T] = \begin{bmatrix} 1/C & 0 & 0 & 0 & 0 & 0 & -1/C & 0 \\ 0 & 1/C & 0 & 0 & 0 & 0 & 0 & -1/C \end{bmatrix}
$$

$$[L] = \begin{bmatrix} 0 & 0 & 0 & 0 & 1/L & 0 & -R/L & 0 \\ 0 & 0 & 0 & 0 & 0 & 1/L & 0 & -R/L \end{bmatrix}$$

The excitation vector $[u] = [v_G]$ of Equation 3.33 is premultiplied by the matrix of parameters of the excitation source, $[B]$, and it defines the voltages corresponding to the residual magnetism.

$$[B] = \begin{bmatrix} -L_2 & 0 & M & 0 \\ 0 & -L_2 & 0 & M \\ M & 0 & -L_1 & 0 \\ 0 & M & 0 & -L_1 \\ \hline 0 & 0 & 0 & 0 \\ 0 & 0 & 0 & 0 \\ \hline 0 & 0 & 0 & 0 \\ 0 & 0 & 0 & 0 \end{bmatrix}$$

A more compact way of representing an aggregation of self-excited induction generators would be to consider the self-excitation capacitor as part of the load across the machine terminals. In this case, the matrix of Equation 3.33 would be transformed as it proceeds:

$$p \begin{bmatrix} i_G \\ x_G \end{bmatrix} = \begin{bmatrix} G \\ E \end{bmatrix} \begin{bmatrix} i_G \\ x_G \end{bmatrix} + [B][v_G]$$

where

$$[x_G] = \begin{bmatrix} v_C \\ i_L \end{bmatrix}$$

$$[E] = \begin{bmatrix} 1/C & 0 & 0 & 0 & 0 & 0 & -1/C & 0 \\ 0 & 1/C & 0 & 0 & 0 & 0 & 0 & -1/C \\ \hline 0 & 0 & 0 & 0 & 1/L & 0 & -R/L & 0 \\ 0 & 0 & 0 & 0 & 0 & 1/L & 0 & -R/L \end{bmatrix}$$

Figure 3.4 shows the Excel plots of the transient self-excitation voltage process with a sudden load switching at 7.6 seconds for a stand-alone SEIG. The induction generator was operated at 1780 rpm with a DC motor as prime mover. A capacitor bank of 160 μF in a star configuration is supplying reactive power for the machine and a 120 Ω-22.5 mH star load was connected

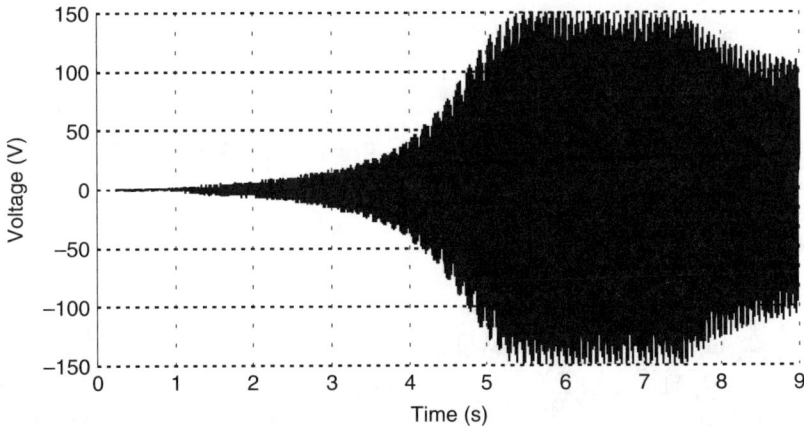

FIGURE 3.4
Voltage growth during self-excitation at no load and at load switching.

after the generator was completely excited. Remnant magnetism in the machine core is also taken into account. We see that the self-excitation follows a variation law similar to the process of magnetic saturation of the core and that a level of stable output voltage is only reached when the machine core is saturated.

3.5 Generalization of the Association of Self-Excited Generators

The method used for a stand-alone machine can be extended to multiple n generators operating in parallel. Each machine needs a self-excitation capacitor because each one has its own curve at no load and, as a consequence, a different self-excitation curve. However, for the numeric solution, there will only be a capacitor and a common load for all the generators. Equation 3.31 can be generalized even for an infinite bus bar, by the following established equation:

$$
p\begin{bmatrix} i_{G1} \\ i_{G2} \\ \vdots \\ i_{Gn} \\ v_L \\ i_L \end{bmatrix} = \begin{bmatrix} G_1 & & & & L_1 \\ & G_2 & & & L_2 \\ & & \ddots & & \vdots \\ & & & G_n & L_n \\ \hline C_G & C_G & \cdots & C_G & C_T \\ 0 & 0 & \cdots & 0 & L_T \end{bmatrix} * \begin{bmatrix} i_{G1} \\ i_{G2} \\ \vdots \\ i_{Gn} \\ v_L \\ i_L \end{bmatrix} + \begin{bmatrix} B_1 & & & \\ & B_2 & & \\ & & \ddots & \\ & & & B_n \\ \hline 0 & 0 & \cdots & 0 \end{bmatrix} \begin{bmatrix} v_{G1} \\ v_{G2} \\ \vdots \\ \cdots \\ v_{Gn} \end{bmatrix} \quad (3.34)
$$

where

$$i_{Gi} = i = \begin{bmatrix} i_{dsi} \\ i_{qsi} \\ i_{dri} \\ i_{qri} \end{bmatrix} \qquad v_{Gi} = \begin{bmatrix} v_{dsi} \\ v_{qsi} \\ v_{dri} \\ v_{qri} \end{bmatrix} \qquad v_L = \begin{bmatrix} v_{Ld} \\ v_{Lq} \end{bmatrix} \quad \text{and} \quad i_L = \begin{bmatrix} i_{Ld} \\ i_{Lq} \end{bmatrix}$$

$$G_i = K \begin{bmatrix} R_{1i}L_{2i} & -\omega M_i^2 & -R_{2i}M_i & -\omega M_i L_{2i} \\ \omega M_i^2 & R_{1i}L_{2i} & \omega M_i L_{2i} & -R_2 M_i \\ -R_{1i}M_i & \omega M_i L_{1i} & R_{2i}L_{1i} & \omega L_{1i}L_{2i} \\ -\omega M_i L_{1i} & -R_{1i}M_i & -\omega L_{1i}L_{2i} & R_{2i}L_{1i} \end{bmatrix} \quad \text{and} \quad L_i = \begin{bmatrix} L_{2i} & 0 & 0 & 0 \\ 0 & L_{2i} & 0 & 0 \\ M_i & 0 & 0 & 0 \\ 0 & M_i & 0 & 0 \end{bmatrix}$$

$$[C_G] = \begin{bmatrix} 1/C & 0 & 0 & 0 \\ 0 & 1/C & 0 & 0 \end{bmatrix} \quad \text{and} \quad [C_T] = \begin{bmatrix} 0 & 0 & -1/C & 0 \\ 0 & 0 & 0 & -1/C \end{bmatrix}$$

where $C = \sum C_k$ $(k = 1, 2, \ldots, n)$ (summation of the individual self-excitation capacitances of the n generators already connected in parallel).

That is, during simulation, the capacitance C is the summation of the self-excitation capacitances of the generators already existing in the parallel aggregation.

$$L_T = \begin{bmatrix} 1/L & 0 & -R/L & 0 \\ 0 & 1/L & 0 & -R/L \end{bmatrix}$$

$$B_i = K \begin{bmatrix} -L_{2i} & 0 & M_i & 0 \\ 0 & -L_{2i} & 0 & M_i \\ M_i & 0 & -L_{1i} & 0 \\ 0 & M_i & 0 & -L_{1i} \end{bmatrix}$$

The computer simulation models the behavior of currents and voltages of the induction generator group during the self-excitation process and in steady state, besides allowing the verification of alterations in the output voltage and variations in the load current that occur. Observe that the representation of Equation 3.31 is in the classical form of state space and, therefore, it is also adapted for the analysis of eigenvalues and eigenvectors.

The dimensions of the submatrixes given by Equation 3.34 are shown in Table 3.1.

The matrix Equation 3.34 considers that pv_{Ld} is the result of the current through the capacitor; that is, the sum of each one of the currents of the associated generators is subtracted from the current going to the load. So,

TABLE 3.1

Dimensions of the Submatrixes in Equation 3.33

Submatrix	Dimension	Submatrix	Dimension
Gi	4×4	G	$4n \times 4n$
CG	2×4	LG	$4n \times 4$
C	$2 \times 4n$	Bi	$4n \times 4n$
LT	2×4	CT	2×4
0A	$2 \times 4n$	0B	$4 \times 4n$
x	$(4n + 4) \times 1$	U	$4n \times 1$

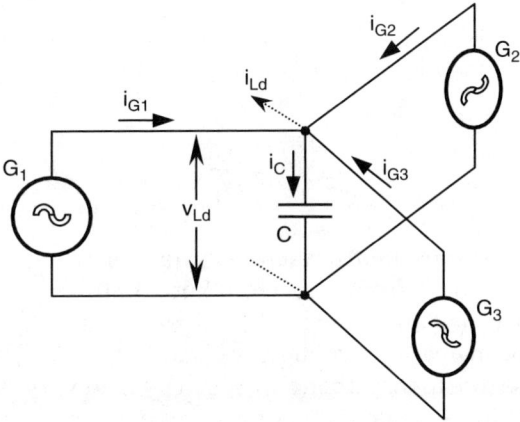

FIGURE 3.5
Current contributions through capacitor C.

$pv_{Ld} = [(\sum i_{Gk}) - i_{Ld}]/C$. For instance, in the case of three induction generators aggregated in parallel, as shown in Figure 3.5, the total current through the capacitor would be: $i_C = (i_{G1} + i_{G2} + i_{G3}) - i_{Ld}$.

Notice that the direct and quadrature axis voltages across the load terminals or across the excitation capacitor soon after the parallel connection of every new generator are given respectively by

$$V_{Ld} = \frac{C_n V_{Ld(n)} + C_{n+1} V_{Ld(n+1)}}{C_n + C_{n+1}} \qquad V_{Lq} = \frac{C_n V_{Lq(n)} + C_{n+1} V_{Lq(n+1)}}{C_n + C_{n+1}}$$

where

n refers to the number of associated generators;

$n + 1$ refers to the new generator connected to the former set of parallel generators.

Brown and Grantham demonstrated that merely taking into account the variation of the magnetization reactance is not enough if the other machine parameters are supposed to be constants.[11,12] Even for steady state analysis, unknown parameter variations of the rotor introduce considerable errors in the analysis of machine performance with rotor displacement current. In the same way, for machines not specifically designed for displacement current, better results can be obtained in behavior analysis of these machines if the variations in the machine parameters are considered.

3.6   The Relationship between Torque and Shaft Oscillation

It is important to define the torque of each machine in order to observe its oscillation with respect to the others. Torque base in p. u. is then defined as:

$$T_B = \frac{S_B}{\omega_s} \tag{3.35}$$

Analysis of the self-excited induction generator using the state variable representation starting from the well known theory of the induction machines produces instantaneous values for direct and quadrature current axis to analyze the increases in voltages and currents during the self-excitation process and those due to disturbances caused by load variations. If necessary, it is also possible to calculate the net transient electric torque. This torque is discussed in the literature[11,13-15] as

$$T = \frac{Power}{\omega_r} = -\left(\frac{3}{2}\right)\left(\frac{p}{2}\right)Mi_{dr}i_{qs} + \left(\frac{3}{2}\right)\left(\frac{p}{2}\right)Mi_{qr}i_{ds} - 2H\frac{p}{\omega_b}\omega_r \tag{3.36}$$

where
   H is the rotor inertia constant
   p is the number of stator poles
   ω_r is the angular speed of the rotor
   ω_b is the base speed, commonly specified as the one obtained from the parameter tests

3.6.1   The Oscillation Equation

To determine the oscillation or the angular displacement among the aggregated machines in a group during transient outcomes, it is necessary to solve the differential equations describing the rotor movements of the machines. The electric torque acting on the rotor of only one machine, following the mechanical laws related to rotating masses, is

$$T = \frac{M_J}{g} \alpha \qquad (3.37)$$

where

T = algebraic sum of all applied torques to the axis in Nm
M_j = moment of inertia in Nm²
g = acceleration of the gravity (= 9,81 m/s²)
α = angular mechanical acceleration in rd/s²

The electrical angle θ_e is related to the mechanical angle, θ_m by $\theta_e = (p/2)\theta_m$, or, using the relationship $f = pn/120$ (being n in rpm), we get

$$\theta_e = \frac{60f\theta_m}{n} \qquad (3.38)$$

The angular displacement δ of the rotor with respect to the rotating synchronous reference axis is

$$\delta = \theta_e - \omega_s t$$

The angular variation with respect to the reference axis is established by

$$\frac{d\delta}{dt} = \frac{d\theta_e}{dt} - \omega_s = \omega - \omega_s \qquad (3.39)$$

and the angular acceleration is

$$\frac{d^2\delta}{dt^2} = \frac{d^2\theta_e}{dt^2}$$

Combining this equality with the second derivative of Equation 3.38 with respect to time, we get

$$\frac{d^2\delta}{dt^2} = \frac{60f}{n} \frac{d^2\theta_m}{dt^2} \qquad (3.40)$$

where

$\frac{d^2\theta_m}{dt^2} = \alpha$ is the mechanical acceleration

Therefore, substituting Equation 3.40 in Equation 3.37, we get the net torque:

$$T = \frac{M_J}{g} \frac{n}{60f} \frac{d^2\delta}{dt^2} \tag{3.41}$$

It is convenient to express the torque in p. u. The torque base T_B is defined as the necessary torque to supply the rated power at the rated speed, that is,

$$T_B = \frac{S_B}{\omega_s} \qquad\qquad \text{Nm}$$

where S_B is the rated apparent power base in kVA.

Therefore, a convenient form of the torque in p. u. is

$$T(pu) = \frac{T}{T_B} = \frac{\dfrac{M_J}{g} \dfrac{p\pi}{f} \left(\dfrac{n}{60}\right)^2}{S_B} \frac{d^2\delta}{dt^2} \tag{3.42}$$

The inertia constant H of a machine is defined as the kinetic energy at the rated speed in kW–s/kVA, this last one given by

$$E_c = \frac{1}{2} \frac{M_J}{g} \omega_s^2 = HS_B$$

where

$$\omega_s = \frac{p\pi n_s}{60} \text{ at the nominal speed}$$

As a result, we have

$$H = \frac{\dfrac{1}{2} \dfrac{M_J}{g} (p\pi)^2 \left(\dfrac{n}{60}\right)^2}{S_B}$$

that, substituted in Equation 3.42, gives

$$T(pu) = \frac{2H}{p\pi f} \frac{d^2\delta}{dt^2} \tag{3.43}$$

The sum of all torques acting on the rotor of a generator includes the input mechanical torque of the primary machine, the rotating torque losses (friction, air opposition and copper plus core losses), electrical output torques, and damping torques from the primary machine, the generator, and the interconnected electric system. The electric and mechanical torques acting on the rotor of a motor are of opposite polarities, and they are the result of the electrical and mechanical loads. Despite the rotating and damping losses, the accelerating torque T_a is given by the difference between the mechanical and electrical torques: $T_a = T_m - T_e$. The electrical torque is the net torque given by Equation 3.43. Therefore,

$$T_m - T_e = T(pu) = \frac{2H}{p\pi f} \frac{d^2\delta}{dt^2}$$
(3.44)

As for small variations of speed, torque, and power (mechanical and electrical), they are the same in p. u., Equation 3.44 becomes

$$\frac{d^2\delta}{dt^2} = \frac{p\pi f}{2H}(P_m - P_e)$$
(3.45)

For numeric solutions, it is interesting to separate this equation in two simultaneous equations of first order (Equations 3.39 and 3.45):

$$\frac{d\omega}{dt} = \frac{d^2\delta}{dt^2} = \frac{p\pi f}{2H}(P_m - P_e)$$
(3.46)

$$\frac{d\delta}{dt} = \omega - 2\pi f$$
(3.47)

If the transient values of torque and power are known during the numeric solution process, they can be used to correct the variations of angular speed using the relationship $P = T\omega$.

3.7 Transient Simulation of Induction Generators

Nowadays, there are several computer packages used in simulation studies of electric machines and particularly in the transient modeling of induction generators as presented in this chapter. MatLab® was used to implement the differential matrix Equation 3.31 as discussed in Chapter 12. Such an equation describes the behavior of currents and voltages of an induction generator during the self-excitation process and in steady operation. It also allows verification of the alterations that happen in the output voltage and variations in the load current.

We must emphasize, at this point, that to begin the process of self-excitation of an induction generator, there must be a small amount of residual magnetism present. This needs to be taken into account in computer simulation of the self-excitation process; otherwise it is not possible to start the numerical method of integration. At the beginning of the integration process of Equation 3.31, a pulse function has to be present whose value fades away as soon as the first iterative step has started. This observation is very important in terms of the dynamic understanding of the self-excitation phenomenon, because it could be used during the occurrence of fortuitous core demagnetizing of the machine, recommending a small source of applied voltage to the real machine for recovery of its active state.

The beginning of the self-excitation process can be slow or fast, depending on the construction of the machine, the way the self-excitation process is done, and the level of residual magnetism. It will be fast if the machine is made to rotate at its rated rotation speed and the self-excitation capacitors are then connected. The transient process in this case takes, typically, about one second. For small machines this process can be shorter. If the machine is started with the self-excitation capacitor already connected, the process will take, typically, about 10 seconds.[13,15]

Once obtained, the state variables should be returned to the phase sequence a-b-c in order to obtain the results for a three-phase generator and not a two-phase as given by the Park transformation. For this, the voltages across each phase of the stator are established as:[8–10]

$$v_A = v_{qs}$$

$$v_B = -\frac{1}{2}v_{qs} - \frac{\sqrt{3}}{2}v_{ds} \tag{3.48}$$

$$v_C = -\frac{1}{2}v_{qs} + \frac{\sqrt{3}}{2}v_{ds}$$

Similarly, the voltages across the rotor winding are as follows.

$$v_a = v_{qr}$$

$$v_b = -\frac{1}{2}v_{qr} - \frac{\sqrt{3}}{2}v_{dr} \tag{3.49}$$

$$v_c = -\frac{1}{2}v_{qr} + \frac{\sqrt{3}}{2}v_{dr}$$

TABLE 3.2

Rated Parameters of Sample Induction Generators G1 and G2

Generator G1		Generator G2	
P = 1 HP	$R_s = 0.32\ \Omega$	P = 1 HP	$R_s = 0.39\ \Omega$
1750 rpm, 60Hz	$X_{ls} = 0.8\ \Omega$	1750 rpm, 60Hz	$X_{ls} = 0.93\ \Omega$
V = 120/220V	$R_r = 0.41\ \Omega$	V = 120/220V	$R_r = 0.44\ \Omega$
I = 11.5/5.5A	$X_{lr} = 0.8\ \Omega$	I = 11.5/5.5A	$X_{lr} = 0.93\ \Omega$

Source: CSM Laboratories, Golden, Colorado.

When connecting several SEIGs in parallel, the common transient voltage established across their terminals could be seen as the voltage of a parallel connection between one generator and a group of several generators whose power could be a lot larger than the one just connected. In these cases, a program for simulation of only two generators of distinct characteristics could be a good approximation. In the more general case, the solution presented in this chapter is quite reasonable for most of the practical problems.

3.7.1 An Example of the Transient Model of the Induction Generator

In order to validate the self-excitation process, an induction machine G1 was used with the nameplate data shown in Table 3.2. The iterative solution of Equation 3.11 was used to simulate the induction generator from the beginning of the self-excitation process at 0.0 ms up to 2.0 ms, using the Colorado School of Mines power laboratory values provided above for the generator.

3.7.2 Effect of RLC Load Connection

The RLC load can be seen as a combination of R and L in parallel with the self-excitation capacitance. Figure 3.4 displays the output voltage of the generator for the beginning of the self-excitation process and switching of the load impedance at 7.6 seconds. It is possible to observe the voltage drop caused at the instant of the connection traduced by a decrease of the load impedance, in other words, a power increase.

As a consequence of Ohm's Law, there is a reduction in the load current, as shown in Figure 3.4. No decaying component appears except at the very beginning of the self-excitation process, due to the residual magnetism, as discussed before.

3.7.3 Loss of Excitation

Figure 3.6 displays the loss of excitation of the generator caused by an excessive increase of the load current without the corresponding voltage increase at t = 2.6 s. Generator G1 was self-excited with 160 μF capacitance bank, and a 45 Ω-45.5 mH star load was applied after the generator was fully

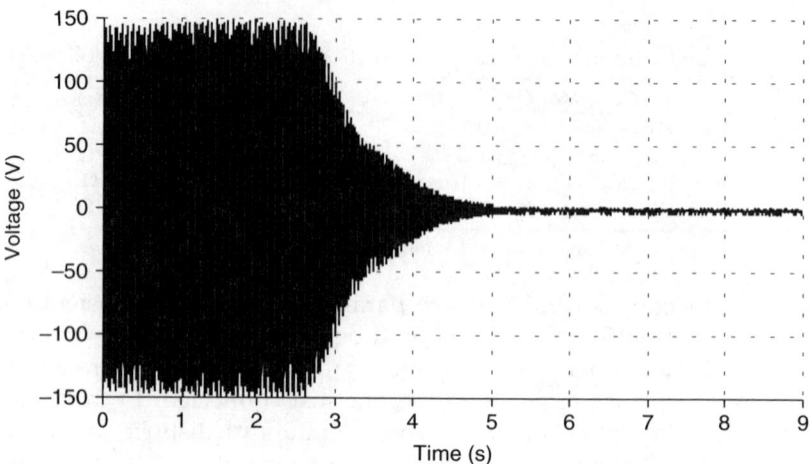

FIGURE 3.6
Terminal voltage collapse due to an excessive increase of the load current.

excited. It took about 2.5 seconds for the voltage to collapse completely. The current increase could be compensated for by an increase of reactive power or an increase in the rotor speed, which did not happen in this example. However, it is known that the voltage tends to increase, even inside of the saturation band of the iron, as the terminal capacitance increases. Therefore, to avoid core demagnetization of the generator, there should be a settled control strategy adopted to allow variation of either the excitation capacitance or the rotor speed according to changes in the load current.

3.7.4   Parallel Connection of Induction Generators[14,15]

Table 3.2 displays the experimental data of the two induction generators G1 and G2 used in this example. Figure 3.7 shows the transient process of load connection and generator parallel switching surges of two SEIGs. Generator G1 was self-excited with 160 μF capacitance at 1800 rpm. A 130 Ω-22.5 mH star connected load was applied at 2 s from the beginning of the simulation time. Generator G2 was previously self-excited with 160 μF at 1800 rpm and connected in parallel at t = 3.75 s. The sudden and brief collapse in voltage is due to differences in phase voltage between the two generators at the paralleling instant. Full common voltage was recovered at about t = 6.0 s. The rotor speed variation of generators G1 and G2 during parallel operation were carefully observed in the laboratory and incorporated in the simulation illustrated in Figure 3.7(a). At exactly the instant of parallel connection of the two generators, a heavy dip in the overall speed of the machines may happen.

Three points in time in those graphs should be observed. The first is the voltage reduction across generator G1 terminals when the R-L load was

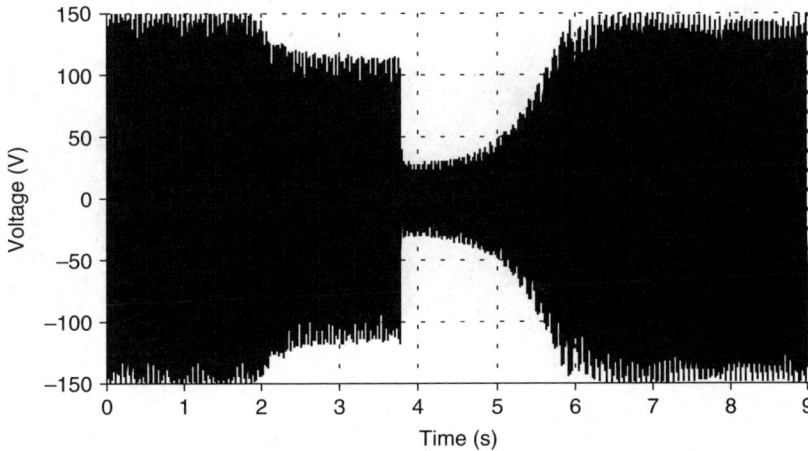

FIGURE 3.7
Transient switching voltage for a parallel connection of SEIGs operating at distinct voltage levels.

switched on. The second is the partial voltage collapse at t = 3.75 s and its recovery up to the steady state parallel common voltage level at about 6.0 s. The third is the recovered voltage of generators G1 and G2 with respect to the no-load voltage level of G2 alone representing an appreciable voltage difference at the instant of parallel connecting. These three points can be clearly and closely observed in both theoretical and experimental setups.

Figure 3.8 shows the simulated plot of the transient switching process of the generator when parallel connection was made between SEIGs operating at identical voltage levels. Generator G1 was self-excited with 180 μF capacitance, and a 130 Ω-22.5 mH star load was applied at t = 1.2 s. Generator G2 was already self-excited with 160 μF and was connected in parallel at t = 4.4 s. Full common voltage was recovered at t = 5.2 s. As the voltages were similar, there was no collapse of the kind observed in the previous case, and the voltage and speed dips were not very pronounced.[7]

3.8 Using This Chapter to Solve Typical Problems

This chapter developed a transient modeling framework. We saw the growth of voltage and current during the self-excitation process and its stabilization in steady state. The models enable visualization of the transient disturbances that the stator output voltage and current suffer when a load is connected to the terminals. Connection of parallel generators was considered as a load connection across each induction generator terminals.

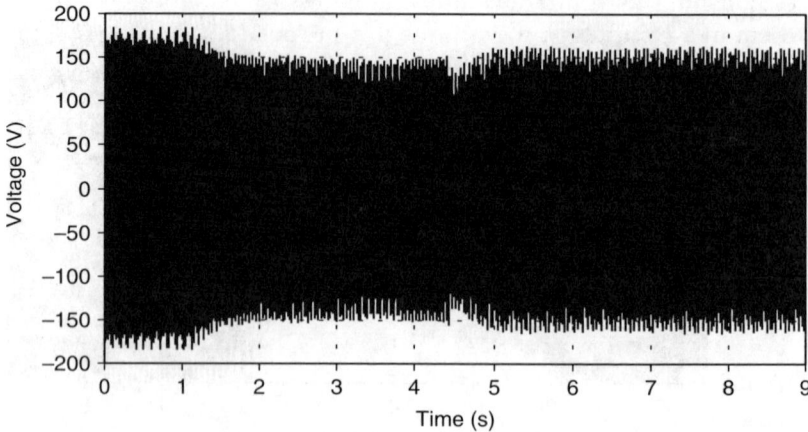

FIGURE 3.8
Transient switching voltage for a parallel connection of SEIGs operating at identical voltage levels.

An induction generator simulation takes a very short time. It must be emphasized that the machine needs a residual magnetism so that the self-excitation process can be started; it cannot be zero at the beginning of the process. During calculations, the parameter variation should follow the curve of the machine hysteresis.

The value of the mutual inductance M should vary continually, following a non-linear relationship between the gap voltage and the terminal magnetizing current in such a way that correct results are obtained. For even more precise results, it is advisable that the variation of the frequency be included during the simulation. A more detailed analysis, including the variation of frequency is presented in Chapter 5.

This description of the behavior of induction generators provides the basis for simulations, taking into consideration the core demagnetization of the generator and transient phenomena. This description in more detail is of fundamental importance so that asynchronous micropower plants can operate with higher safety margins.

3.9 Problems

3.1 Why is the equivalent p. u. circuit model of the induction generator, whose stator and rotor p. u. frequencies are denoted by F and v, respectively, inadequate to represent the transient state of the capacitor self-excited induction generator?

3.2 Develop a form to represent the variation of the mutual inductance M during the process of self-excitation of the induction generator, establishing its limitations and the necessary equations to include them in a computer program for this purpose.

3.3 Given the instantaneous stator phase voltages $v_A = V_m \sin(\omega t)$, $v_B = V_m \sin(\omega t - 120°)$ and $v_C = V_m \sin(\omega t + 120°)$ and the rotor instantaneous phase voltages $v_a = V_m \sin(\omega t)$, $v_b = V_m \sin(\omega t - 120°)$ and $v_c = V_m \sin(\omega t + 120°)$, of an induction machine, for $V_m = 110\sqrt{2}V$ e $f = 60$ Hz, calculate $[v_{ds}v_{qs}v_{dr}v_{qr}]$.

3.4 Based on the data in Table 3.2, calculate the values of matrix $[v_{ds}v_{qs}v_{dr}v_{qr}]$ assuming a squirrel cage rotor.

3.5 Calculate the eigenvalues and eigenvectors of Equation 3.11 to establish the oscillating modes of operation of the induction generator.

3.6 Using MatLab, write a computer program to represent the self-excited induction generator in its transient state.

3.7 Explain why an induction generator cannot self-excite under the following conditions: a) no residual magnetism; b) too much inductive load; c) overcurrent.

3.8 Discuss the four best-known practical methods of generating residual magnetism.

3.9 Discuss the advantages of a matrix partition to represent the transient state of an aggregation of self-excited induction generators.

3.10 Using the state variable theory for generalization of an aggregation of self-excited induction generators, write a matrix system for an aggregation of three induction generators with distinct machine parameters. What are the dimensions of the necessary submatrixes for this?

3.11 Describe an induction motor using the parameters of the partitioned matrix of Equation 3.30 in the classical state equation form.

References

1. Smith, I.R., and Sriharan, S., Transients in induction motors with terminal capacitors, *Proc. IEEE*, Vol. 115, 519–527, 1968.
2. Murthy, S.S., Bhim Singh, M. and Tandon, A.K., Dynamic models for the transient analysis of induction machines with asymmetrical winding connections, in *Electric machines and Electro-mechanics*, 6(6), 479–92, 1981.
3. Murthy, S.S., Malik, O.P. and Tandon, A.K., Analysis of self-excited induction generators, *Proc. IEEE*, 129(6), 260–265, 1982.
4. Watson, D. B. and Milner, I. P., Autonomous and parallel operation of self-excited induction generators, in *Elect. Engineering Education*, Vol. 22, 365–374, Manchester University Press, 1985.
5. Murthy, S.S., Nagaraj, H.S. and Kuriyan, A., Design-based computational procedure for performance prediction and analysis of self-excited induction generators using motor design packages, *Proc. IEEE*, 135(1), 8–16, 1988.

6. Grantham, C., Sutanto, D. and Mismail, B., Steady State and transient analysis of self-excited induction generators, *Proc. IEEE*, 136(2), 61–68, 1989.
7. Barbi, I., *Fundamental Theory of the Induction Motor (Teoria Fundamental do Motor de Indução)*, Florianópolis, Brazil, UFSC Press, Eletrobrás, 1985.
8. Langsdorf, A.S., *Theory of the alternating current machines (Teoría de Máquinas de Corriente Alterna)*, McGraw-Hill Book Co., 1977, p. 701.
9. Singh, S.P., Bhim Singh, M. and Jain, P., Performance characteristics and optimum utilization of a cage machine as capacitor excited induction generator, in *IEEE Transactions on Energy Conversion*, 5(4), 679–684, 1990.
10. Hallenius, K.E., Vas, P. and Brown, J. E., The analysis of a saturated self-excited asynchronous generator, in *IEEE Transactions on Energy Conversion*, 6(2), 1991.
11. Brown, J.E. and Grantham, C., Determination of induction motor parameter and parameter variations of a 3-phase induction motor having a current-displacement rotor, in *Proc. IEEE*, 122(9), 919–921, 1975.
12. Grantham, C., Determination of induction motor parameter variations from a variable frequency standstill test, in *Electrical Mach. Power Systems, United States*, 10, 239–248, 1985.
13. Li Wang and Ching-Huei Lee, A novel analysis on the performance of an isolated self-excited induction generator. in *IEEE Transactions on Energy Conversion*, 12(2), 109–117, 1997.
14. Farret, F.A., Schramm, D.S. and Neves, L.S., Modeling of the self-excited induction generator: transient and steady state (Modelagem de geradores de indução autoexcitados: estados transitório and permanente), in *Tecnologia Magazine, UFSM Press, Santa Maria, RS*, 16(1–2), 49–59, 1995.
15. Hancock, N.N., *Matrix Analysis of Electrical Machinery: Second Edition*, Pergamon Press, Oxford, 1974.

4

The Self-Excited Induction Generator

4.1 Scope of This Chapter

As previously discussed in Chapter 2, the induction generator has a serious limitation: an inherent need for reactive power. It consumes reactive power when connected to the distribution network and, as a matter of fact, it needs an external reactive source permanently connected to its stator windings to provide output voltage control. This source of reactive power keeps the current going through the machine windings needed to generate the variable magnetic field that, with the movement of the rotor conductors, causes the voltage across the induction generator terminals. Another way of inducing this voltage is to use the residual magnetism associated with an external capacitor that generates current by rotor movement inside of this magnetic field so as to establish an induced voltage as described below.

The advantages of the induction generator will be enhanced under optimized conditions of performance, as discussed in Chapter 10.

4.2 Performance of the Self-Excited Induction Generator

Interaction among the operating states of the primary source of energy, the induction generator, the self-excitation process, and the load state will define the global performance of the power plant.[1,2] Performance is greatly affected by the random character of many of the variables involved related to the availability of primary energy and to the way consumers use the load. So, the performance of induction generators depends on appropriate power plant specifications at the design stage; in particular, the following items:

- Parameters of the induction machine
  - Operating voltage
  - Rated power

— Rated frequency used in the parameter measurements

— Power factor of the machine

— Rotor speed

— Capacity for acceleration

— Isolation class

— Operating temperature

— Carcass type

— Ventilation system

— Service factor

— Noise

- Load parameters

— Power factor

— Starting torque and current

— Maximum torque and current

— Generated harmonics

— Form of connection to the load: directly to the distribution network or through converters

— Load type: resistive; inductive or capacitive; constant or variable; passive or active

— Evolution of the load over time

- Self-exciting process

— Degree of iron saturation of the generator caused by the choice of capacitor

— Fixed or controlled self-excitation capacitor

— Speed control

- Type of primary source: hydro, wind, biomass or combinations.

Nonlinear loads like electronic power converters generate many harmonics, and they can have a variable power factor if some control techniques are not adopted. The harmonics can be minimized with the installation of filters or the utilization of supplementary signals. In general, the cost of passive filters is relatively small with respect to the cost of the speed control in the conventional power plants exerted by the electronic variation of frequency. The self-excitation capacitor in stand-alone power plants or with electric or electronic control of the load, contributes favorably in these cases. Supplementary signals for harmonic minimization, in spite of being a relatively recent technology and not very well established yet, seem to be more promising techniques for such situations. The IEEE Std. 519 establishes the limits of 2% of harmonic content for single- and three-phase induction motors

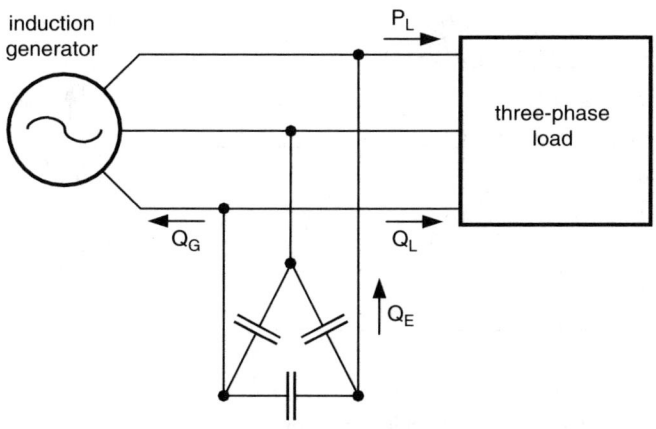

P_L - Active power to the load
Q_L - Reactive power to the load
Q_G - Reactive power to the generator
Q_E - Reactive power to the excitation

FIGURE 4.1
Capacitor self-excited induction generator.

(except category N, that is, conventional) and 3% for high efficiency (H and D). The other standard is IEC1000-2-2.

In the case of stand-alone operation of power plants, the connection of a capacitor bank across the terminals of the induction generator is necessary, as displayed in Figure 4.1, to supply its need for reactive power. Notice that in practical schemes it is advisable to connect each excitation capacitor across each motor winding phase, in either $\Delta - \Delta$ or $Y - Y$. The necessary capacitance of the bank will depend on the primary energy and on the instantaneous load as discussed in the following sections.

For a self-excited induction generator (SEIG), an equivalent circuit in per unit values (p. u.) shown in Figure 4.2 can be used to represent a more generic form of power plant.[2] The frequency effect on the reactance should be considered if it is used at different frequencies from the base frequency in Hertz f_b at which the parameters of the machine were measured. For this purpose, if F is the p. u. frequency, a relationship can be defined between the self-excitation frequency f_{exc} and the base frequency f_b (usually 60 Hz):[3-5]

$$F = \frac{f_{exc}}{f_b} = \frac{\omega_{exc}}{\omega_b}$$

In a more generic way the inductive reactance parameters can be defined for the base frequency as $X = F\omega L$. Figure 4.2 displays the generic equivalent circuit in steady state per phase of the self-excited induction generator with all circuit parameters divided by F, making the source voltage equal to V_{ph}/F. From the definition of the secondary resistance (rotor resistance) shown in

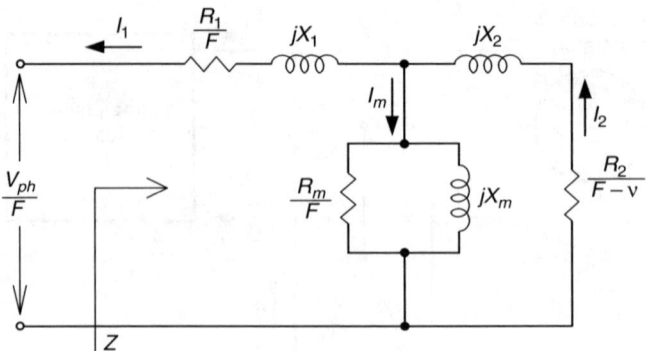

FIGURE 4.2
Generic model per phase of the induction generator.

Figure 2.3 and from Equation 2.11, the following modification can be used to correct R_2/s to take into account changes in the stator and rotor p. u. frequencies:

$$\frac{R_2}{Fs} = \frac{R_2}{F\left(1 - \frac{n_r}{n_s}\right)} = \frac{R_2}{F - v}$$

where n_r is the rotor speed in p. u. referred to the test speed used in the rotor.

Although taking into account the variation of the magnetizing reactance due to the magnetic saturation in transient studies of induction generators, the other parameters are not considered constant.[4,5] The use of uncorrected parameters may lead to rough mistakes in the representation of the machine rotor with current changes. In Chapter 3 the general characteristics of the induction generator under the transient state is discussed in more detail.

4.3   Voltage Regulation

In Figure 4.2, the equivalent impedance seen by the voltage across the terminals of the stator at the synchronous speed ($F = 1$) is[2]

$$Z = R_1 + jX_1 + \cfrac{1}{\cfrac{1}{jX_m} + \cfrac{1}{\cfrac{R_2}{s} + jX_2}} \tag{4.1}$$

where

$$s = 1 - \frac{n_r}{n_s} = 1 - \frac{\omega_r}{\omega_s}, \text{ that is, } n_r = (1 - s)n_s$$

Sometimes, it can be more convenient to enter the value of the electric frequency supplied to the load first and then obtain the mechanical rotation. Then, the equation used for calculation of these variables should be changed conveniently to:[3]

$$\omega_r = \frac{p}{2}\left(2\pi\frac{n_r}{60}\right) = \frac{p}{2}\omega_r'$$

where
n_r is the rotor mechanical speed (rpm)
$\omega_r' = 2\pi\frac{n_r}{60}$ is the angular mechanical speed of the rotor (*rad/s*)
p is the number of poles of the machine

The complex terms of Equation 4.1 can be separated into real and imaginary parts as

$$Z = \left(R_1 + \frac{a}{a^2 + b^2}\right) + j\left(X_1 + \frac{b}{a^2 + b^2}\right) \tag{4.2}$$

whose polar form can be given by:

$$Z = |Z|\underline{\theta} \tag{4.3}$$

where

$$a = \frac{\cos(\theta_2)}{|Z_2|}$$

$$b = \frac{1}{X_m} + \frac{\sin(\theta_2)}{|Z_2|}$$

$$|Z_2| = \sqrt{\left(\frac{R_2}{s}\right)^2 + X_2^2}$$

$$\theta_2 = \tan^{-1}\left(\frac{sX_2}{R_2}\right)$$

$$|Z| = \sqrt{\left(R_1 + \frac{a}{a^2 + b^2}\right)^2 + \left(X_1 + \frac{b}{a^2 + b^2}\right)^2}$$

$$\theta = \tan^{-1}\left[\frac{\left(a^2 + b^2\right)X_1 + b}{\left(a^2 + b^2\right)R_1 + a}\right]$$

The impedance given by Equation 4.2 is applicable to the self-excited generator as well as to a generator interconnected to the public network.

4.4 Magnetizing Curves and Self-Excitation

The magnetizing curve, also known as the saturation or excitation curve, is related directly to the quality of the iron, core dimensions, overall geometry, and coil windings. In other words, the induction generator characteristics determine the terminal voltage for a given magnetizing current through the windings (see Figure 4.3). As discussed in Section 5.5, the magnetizing curve is customarily represented either by a polynomial or a nonlinear expression, like this:

$$L_m = a_0 + a_1 V_{ph} + a_2 V_{ph}^2 + a_3 V_{ph}^3 + a_4 V_{ph}^4$$

$$X_m = F\omega_s L_m = \frac{V_g}{I_m} = F\left(K_1 e^{K_2 I_m^2} + K_3\right)$$

The operation of the induction generator as a function of the terminal voltage (at a given frequency) will need the magnetizing current I_m to describe the operation. That is determined by feeding the machine as an induction motor without load and measuring the current as a function of the terminal voltage variation. This curve starts at the value of the residual magnetism (zero current) existing before the beginning of the test in the iron hysteresis curve of the machine.

It is interesting at this point to compare the induction generator and the DC generator; the exciting capacitor of the induction generator takes the place of the field resistor of the DC generator. When the shunt DC generator starts, the residual magnetism in the field generates an initial voltage which

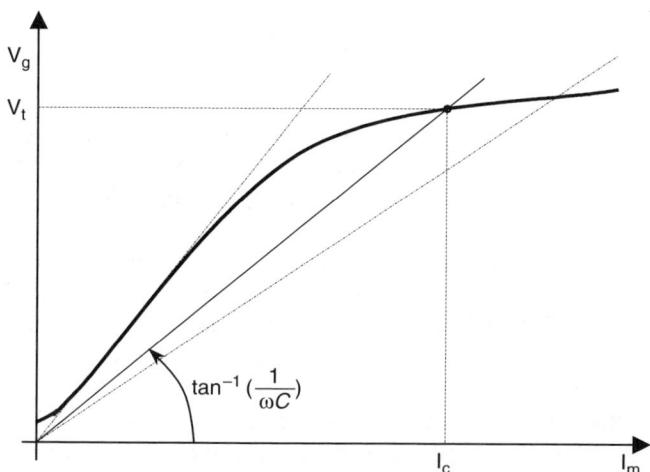

FIGURE 4.3
Magnetizing curve of the induction generator.

produces a field current that in turn produces more field voltage and so on. The process goes until the iron is saturated. This is a self-excitation process very similar to what happens with the induction generator, as we will see further on. In this case, the residual magnetism in the iron produces a small voltage that produces a capacitive current (a delayed current). This current produces an increased voltage that provokes a higher increase of the capacitive current, and so on, until the iron saturation of the magnetic field. Without residual magnetism, neither the DC generator nor the induction generator can produce any voltage.

It is important to emphasize that it is very difficult to lose residual magnetism completely; a minimum value always remains. In the case of complete loss, however, there are four commonly used techniques for recovering it: (1) Make the machine rotate at no load and a high speed until the residual magnetism is recomposed; in extreme cases this is not possible. (2) Use a battery to cause a current surge in one of the machine windings. (3) Maintain a charged high-capacity capacitor — it can be one of the electrolytic types — to cause a current surge as in the previous method. (4) Use a rectifier fed from the network to substitute for the battery in the second method.

A way of obtaining a more coherent magnetizing curve with the frequency-dependent nature of the induction motor parameters is to use a secondary driving machine coupled to its shaft. With this, a constant rotation is always guaranteed during the tests, once a good terminal voltage control may produce a wide speed and frequency variation of the rotor.

For operation as a stand-alone generator, it should be connected to a three-phase bank of capacitors. Like the excitation curve, the capacitive reactance (Figure 4.3) will be a straight line passing through zero whose slope is

$$X_C = \frac{1}{\omega C}$$

As has already been said above, the magnetizing current lags behind the terminal voltage by about 90°, depending on the losses of the motor in a no load condition while the current through the capacitor is approximately 90° ahead.

The value of C can be chosen for a given rotation in such a way that the straight line of the capacitive reactance intercepts the magnetizing curve at the point of the desired rated voltage (V_{rated}). This means that the intersection of these two lines is the point at which the necessary reactive power of the generator is supplied only by capacitors (the resonant point). Therefore, to have the excitation process, this value should be between the straight line slope passing through the origin and the tangent to the most sloped part of the excitation curve (the air gap line), also passing through the origin. The current corresponding to the interception point should not be much above the rated current of the machine. In the first case, the output voltage will be unstable, once there are infinite common points between the two lines. The second extreme end is justified by the maximum current capacity that the motor windings can support.

Notice in Figure 4.3 that the magnetizing curve does not pass through zero, which is very typical of the laboratory tests as the core is rarely fully demagnetized in such tests. This displacement from the zero is due to the residual magnetism that should be corrected to compensate for the distortion.

The compensation for the residual magnetism is according to the Equation below; that is, there is an ordinate change suitable for the dashed line of Figure 4.4 where

$$\frac{y_2}{y_1} = \frac{x_2 + x_0}{x_1 + x_0}$$

This expression can be put in a more directly useful form:

$$x_0 = \frac{x_1 y_2 - x_2 y_1}{y_1 - y_2}$$

Similarly, in Figure 4.4 we see that one can obtain x_0 from the values of the residual magnetism and from the first point obtained as being

$$x_0 = \frac{x_1 y_0}{y_1 - y_0}$$

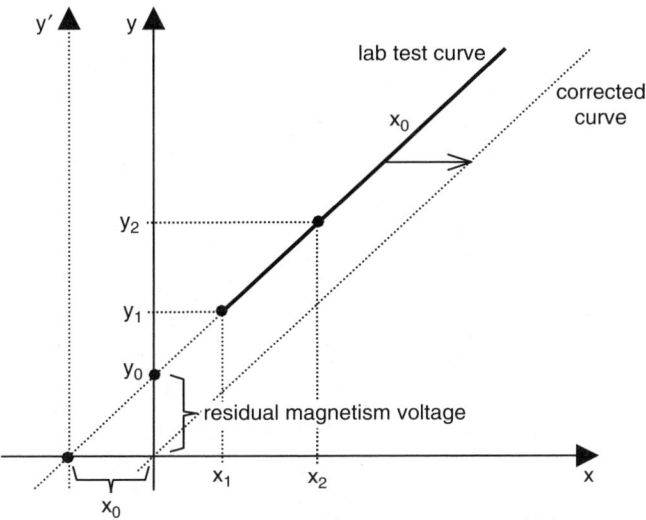

FIGURE 4.4
Compensation for the residual magnetism.

We see that the mutual inductance is essentially nonlinear. For numerical computation it should be determined for each value of the instantaneous magnetizing current I_m. To solve this problem it is suggested that the no-load curve of the machine can be initially obtained and then, the terminal voltage found is divided by the product of the magnetizing current times $2\pi f_{base}$.

4.5 Mathematical Description of the Self-Excitation Process

The mathematical description of the self-excitation process can be based on the Doxey model, which comes from the classical steady state model of the induction machine presented in Figure 2.7, adapted for the induction generator as discussed in this Section. As the excitation impedance is much larger than any of the winding impedances, for simplicity, it is common to represent the induction machine with separate resistances of the losses and including the load as in Figure 4.5.

In the specific case of the self-excited induction generator, the circuit of Figure 4.5 combines machine resistances and reactances separate from the parallel load circuit as in Figure 4.6.[2]

R_m represents core losses; it can be determined by making the machine rotate experimentally at no load and measuring the active power per phase P_0, the average voltage per phase V_0, and the average current per phase I_0. As the machine is rotating at no load, the only loads to be gained by the

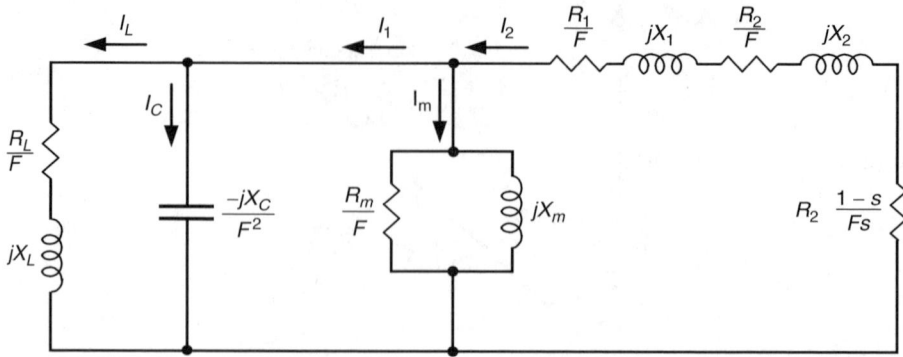

FIGURE 4.5
Simplified classic model of the self-excited induction generator.

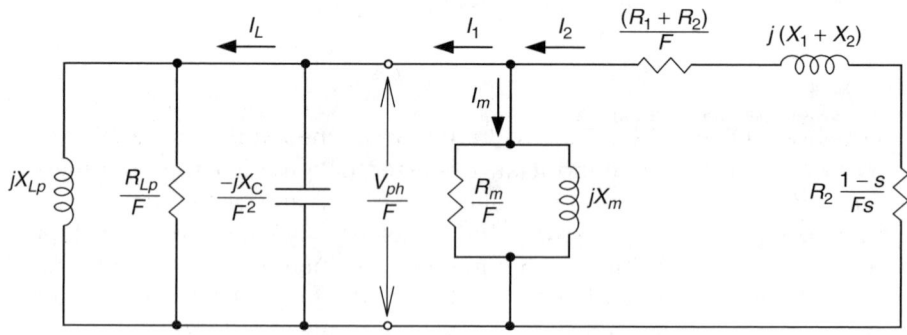

FIGURE 4.6
Compact equivalent circuit of the parallel loaded induction generator.

primary machine are the losses by hysteresis, parasite currents in the core, friction, air resistance, and other spurious losses. So, with the help of another external machine, the induction machine is made to turn at the synchronous rotation ($s = 0$), to allow separation of the mechanical losses from the total losses. Then, the losses by hysteresis and parasites currents in the core can be represented by:

$$R_m = \frac{V_0^2}{P_0 - I_0^2 R_1} \tag{4.4}$$

Such losses are dependent on the temperature, voltage levels used, and, above all, on the frequency. However, for a rough estimate like the one we are making here, the value of R_m as given by Equation 4.4 is acceptable.[6–10]

Taking into consideration that any load impedance Z can be transformed from a series to a parallel configuration based on the classical concepts of circuit:

$$Z = R_s + jX_s = \frac{jX_pR_p}{R_p + jX_p} = \frac{R_pX_p^2 + jX_pR_p^2}{R_p^2 + X_p^2} \tag{4.5}$$

where

$$R_p = \frac{R_s^2 + X_s^2}{R_s} = \frac{Z_s^2}{R_s} \text{ and } X_p = \frac{R_s^2 + X_s^2}{X_s} = \frac{Z_s^2}{X_s}$$

For $\dfrac{Z_L}{F}$ we have

$$\frac{Z_L}{F} = \frac{R_L}{F} + jX_L = \frac{jX_{Lp}\dfrac{R_{Lp}}{F}}{\dfrac{R_{Lp}}{F} + jX_{Lp}} \tag{4.6}$$

Equations 4.5 and 4.6 enable us to determine the load power angle as

$$\theta_L = \tan^{-1}\left(\frac{FX_L}{R_L}\right) = \tan^{-1}\left(\frac{R_{Lp}}{FX_{Lp}}\right) \tag{4.7}$$

Also, from Equations 4.5 and 4.6, it can be easily shown that

$$R_{Lp} = \frac{Z_L^2}{FR_L} \qquad X_{Lp} = \frac{Z_L^2}{F^2X_L} \qquad P_L = \frac{V_{ph}^2}{FR_{Lp}} = \frac{V_{ph}^2}{Z_L^2}R_L \tag{4.8}$$

An even simpler model of the self-excited induction generator can be obtained making the circuit of Figure 4.7 the equivalent of that of Figure 4.6.

Notice in Figure 4.7 that the influence of an inductive load on the self-excitation capacitance is a decrease in its effective value. In a quantitative form, it can bring down this value from the load parallel inductance X_{Lp} in parallel with X_c to obtain the effective self-exciting capacitance:

$$C_{rms} = C - \frac{L}{R_L^2 + (\omega_sL)^2} \tag{4.9}$$

Equation 4.9 leads to the conclusion that for small values of L the inductive reactance of the load does not have much influence on C_{rms}. However, for high values of L this inductance can have a decisive effect on the output voltage because it approximates the straight line of the capacitive reactance slope to the linear portion of the excitation curve slope (the air gap line),

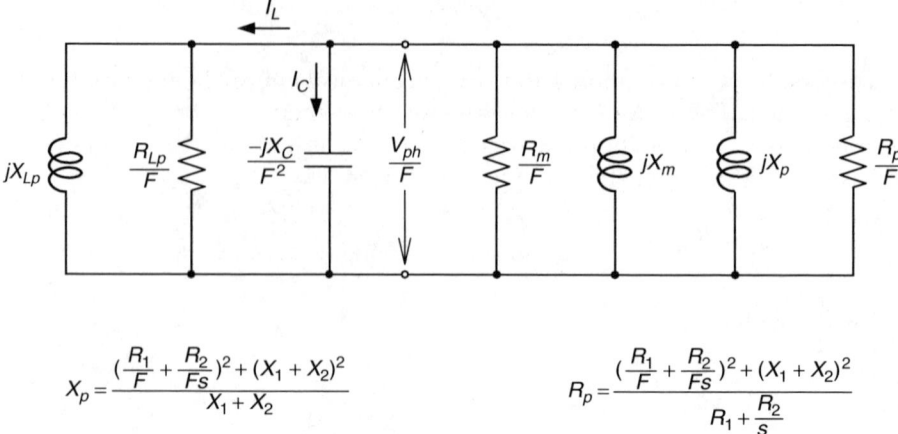

$$X_p = \frac{(\frac{R_1}{F} + \frac{R_2}{Fs})^2 + (X_1 + X_2)^2}{X_1 + X_2}$$

$$R_p = \frac{(\frac{R_1}{F} + \frac{R_2}{Fs})^2 + (X_1 + X_2)^2}{R_1 + \frac{R_2}{s}}$$

FIGURE 4.7
Doxey simplified model for the induction generator.

reducing drastically the terminal voltage until total collapse. The positive side of this matter is that when the load resistance is too small, it will help to discharge the self-excitation capacitor more quickly, taking the generator to the deexcitation process. This is a natural protection against high currents and short-circuits. On the other hand, the increase of the self-excitation capacitance is limited by the heavy iron saturation when the intersection point on the straight line of the capacitive reactance enters into the saturation portion of the machine for values beyond the rated current value of the generator. The winding heats up quickly, which can cause permanent damage in the isolations and in the magnetic properties of the iron.

To evaluate the performance of the induction generator it should be taken into account that the only power entering in the circuit is from the primary machine in the form of active power given by Equation 2.23. The reactive power should also be balanced. For calculation effects, the active power balance can be obtained from Figure 4.6 and Figure 4.7 for the active and reactive powers, respectively, by

$$I_2^2 R_2 \frac{1-s}{s} + I_2^2 (R_1 + R_2) + \frac{V_{ph}^2}{R_m} + \frac{V_{ph}^2}{R_{Lp}} = \sum P = 0 \tag{4.10}$$

$$\frac{V_{ph}^2}{X_p} + \frac{V_{ph}^2}{X_m} - \frac{V_{ph}^2}{X_c} + \frac{V_{ph}^2}{X_{Lp}} = \sum Q = 0 \tag{4.11}$$

where

$$X_m = \omega_s L_m$$

The first term of Equation 4.10 is the mechanical energy being supplied to the generator. The sign of this term is inverted when $s > 1$ or $s < 0$; in other words, when the induction machine is working as a brake or as a generator.

Dividing Equation 4.10 by I_2^2 and simplifying we get

$$\frac{R_2}{s} + R_1 + \frac{V_{ph}^2}{I_2^2} \frac{1}{R_{mL}} = 0 \tag{4.12}$$

where

$$\frac{1}{R_{mL}} = \frac{1}{R_m} + \frac{1}{R_{Lp}}$$

From Figure 4.6 we get

$$\frac{V_{ph}^2}{F^2 I_2^2} = \left(\frac{R_2}{Fs} + \frac{R_1}{F} \right)^2 + (X_1 + X_2)^2 \tag{4.13}$$

and, with that, Equation 4.12 becomes

$$\frac{R_2}{s} + R_1 + \left[\left(\frac{R_2}{s} + R_1 \right)^2 + F^2 (X_1 + X_2)^2 \right] \frac{1}{R_{mL}} = 0$$

or

$$\left(\frac{R_2}{s} + R_1 \right) R_{mL} + \left(\frac{R_2}{s} + R_1 \right)^2 + F^2 (X_1 + X_2)^2 = 0 \tag{4.14}$$

Equation 4.14 is an equation of the second order, and if the other parameters of the machine are constant, it can be used to determine the corresponding value of s for the conditions of primary power and load given by

$$s = \frac{2R_2}{-2R_1 - R_{mL} \pm \sqrt{R_{mL}^2 - 4F^2 (X_1 + X_2)^2}} \tag{4.15}$$

The radical in the denominator of Equation 4.15 will always be negative to satisfy the practical condition that if $X_1 + X_2$ were very small, the load R_{mL}

would have almost no influence on s, which cannot be the case. So, it is convenient sometimes to approximate Equation 4.15 associated to 4.12 by the following expression:

$$s \cong -\frac{R_2}{R_1 + R_{mL}} = -\frac{R_2(R_m + R_{Lp})}{R_1 R_m + R_1 R_{Lp} + R_m R_{Lp}} \tag{4.16}$$

A numerical process may be necessary when using Equation 4.15 instead of Equation 4.16 because the calculation of F depends on ω_s, which depends on s which, in turn, depends on F.

The voltage regulation also depends on the variation of X_m, which can be determined by solving Equation 4.11 as:

$$X_m = \frac{1}{F^2 \omega_s C - \dfrac{1}{X_p} - \dfrac{1}{X_{Lp}}} \tag{4.17}$$

The efficiency can be estimated by

$$\eta = \frac{P_{out}}{P_{in}} = \frac{P_{in} - P_{losses}}{P_{in}} \tag{4.18}$$

The input power is the mechanical power should be subtracted from the copper losses $(R_1 + R_2)$ and the losses represented by the resistance R_m. Using Equation 4.10 once again for the three-phase case without the portion corresponding to the load power $(R_{Lp} \rightarrow \infty)$ together with Equation 4.18 and remembering what we said about s in Equation 4.10 we get

$$P_{out} = P_{in} - P_{losses} = 3I_2^2 R_2 \frac{1-s}{s} + 3I_2^2 (R_1 + R_2) + \frac{3V_{ph}^2}{R_m} = -\frac{3V_{ph}^2}{R_{Lp}}$$

where

$$P_{in} = 3I_2^2 R_2 \frac{1-s}{s}$$

Therefore, using these values in Equation 4.18 and simplifying, we get

$$\eta = \frac{I_2^2 R_2 \dfrac{1-s}{s} + I_2^2 (R_1 + R_2) + \dfrac{V_{ph}^2}{R_m}}{I_2^2 R_2 \dfrac{1-s}{s}} \tag{4.19}$$

Dividing the numerator and the denominator of Equation 4.19 by I_2^2 and using Equation 4.13 again, we get

$$\eta = \frac{\left(\dfrac{R_2}{s} + R_1\right) + \dfrac{1}{R_m}\left[\left(\dfrac{R_2}{s} + R_1\right)^2 + F^2(X_1 + X_2)^2\right]}{R_2 \dfrac{1-s}{s}} \qquad (4.20)$$

Besides the generator parameters being studied, the complete solution for the performance of the induction generator depends on the self-excitation capacitance C, on the angular mechanic frequency of the rotor ω_r, and on the load impedance Z_L. The block diagram of Figure 4.8 illustrates the sequence of calculations required to obtain the load current, the output power, and the overall efficiency of the generator.

Furthermore, the magnetizing reactance can be determined point by point in the laboratory instead of using a nonlinear equation adjusted for losses that can enter into the calculations under a table form.

4.6 Series Capacitors and Composed Excitation of the Induction Generator

Another analogy between the DC generator and the induction generator can be seen with the connection of capacitors in series across the generator terminals instead of in parallel in a form of composed excitation (Figure 4.9).[11]

The voltage-current characteristic results from the loaded generator with a constant lagging power factor. The capacitive reactive power increases with the load increase, partially compensating the reactive power demanded by the load as sketched in Figure 4.10.

From everything we have seen in this chapter, the biggest advantage of the induction generator is its simplicity in not needing a separate field circuit as the synchronous generator and not needing to work at a constant speed. It is sufficient that it works at a speed above the synchronous speed, which is the stator magnetic field speed. The higher the torque is, the higher the output power and, therefore, the speed. The distribution network, as cited previously, controls the voltage, and the power factor should be compensated by capacitors. These factors make the induction generator advisable for wind energy, heat recovery, and other small sources of electrical energy. The great majority of the wind and hydro turbines up to 500 kVA use this type of generator.

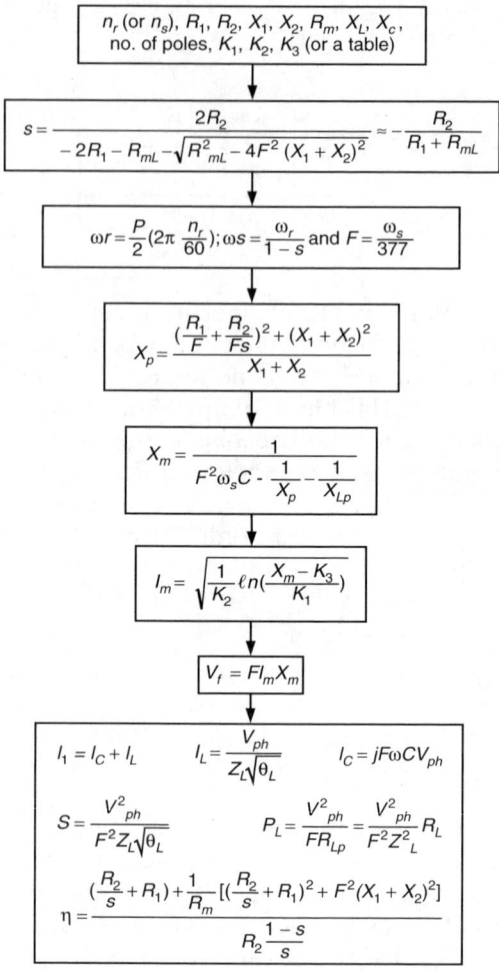

FIGURE 4.8
Calculation flowchart of performance of the self-excited induction generators.

4.7   Problems

4.1 How can harmonic voltage and current affect the power factor of an induction generator? What are the usual measures to sort out generating harmonics?

4.2 Based on Section 4.2, suggest some performance values for an induction generator being considered for supplying a small load of 100 kW for a small group of residential houses.

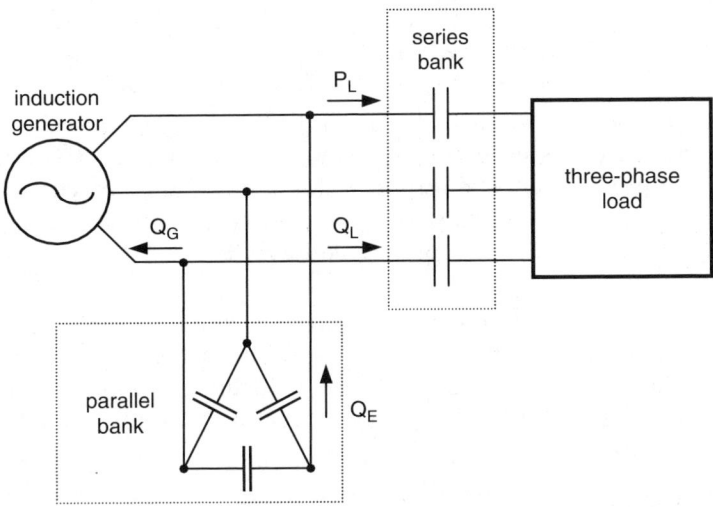

FIGURE 4.9
Influence of the power factor on the relationship between terminal voltage and load.

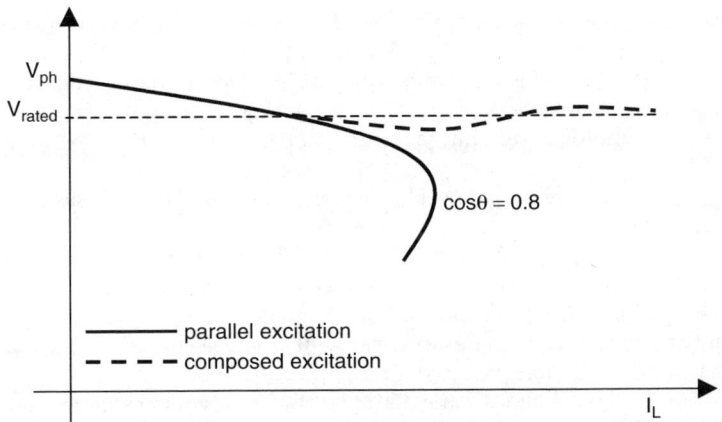

FIGURE 4.10
Influence of the self-excitation capacitance on the relationship between terminal voltage and load.

4.3 Specify the self-exciting capacitance for a three-phase, 4-pole, 1700-rpm, Y-stator connection induction generator supplying 110 V, 1.0 kW at 60 Hz, whose equivalent circuit is given as: $R_1 = 0.20\ \Omega$, $R_2 = 0.42\ \Omega$, $X_1 = 0.15\ \Omega$ and $X_2 = 0.43\ \Omega$; R_2 and X_2 parameters refer to the stator. The excitation current is 8.1 A and is assumed to be constant at the rated values.

4.4 Give the effective self-exciting capacitance at 1900 rpm and 60 Hz of an induction generator having a rated self-exciting capacitance of 180 μF and a load connected across its terminals of $R_L = 50\ \Omega$ and

L = 0.05 H. What would the effective capacitance be if the frequency went up to 65 Hz? What would the new value of the real capacitance be to keep the self-exciting point at its original value?

4.5 Using the data given in Tables 5.2 and 5.3, calculate the self-exciting capacitance of the induction generator to give the nominal electrical values for the specified machine. If this self-exciting capacitance was electronically controlled, what should its maximum and minimum values be to keep the voltage regulation within a ±10% range at the rated speed?

4.6 Calculate the induction generator's total internal current to feed the load specified in Exercise 4.4, separating the self-exciting current to the capacitor and the current to the load itself. What is the active power actually supplied by the generator in comparison to that supplied to the load?

References

1. Doxey, B.C., Theory and application of the capacitor-excited induction generator, *The Engineer Magazine*, 893–897, Nov. 1963.
2. Murthy, S.S., Nagaraj, H.S., and Kuriyan, A., Design-based computer procedures for performance prediction and analysis of self-excited induction generators using motor design packages, *IEEE Proc.*, Vol. 135, Pt. B, No. 1, 8–16, Jan. 1988.
3. Stagg, G.W. and El-Abiad, A.H., *Computer methods in power systems analysis*, McGraw-Hill Book Company, New York, 1968.
4. Tabosa, R.P., Soares, G.A., and Shindo, R., *The high efficiency motor (Motor de alto rendimento), technical handbook of the PROCEL program*, edited by Eletrobrás/Procel/CEPEL, Rio de Janeiro, Brazil, Aug. 1998.
5. Grantham, C., Steady-state and transient analysis of self-excited induction generators, *IEEE Proc.*, Vol. 136, Pt. B, No. 2, 61–68, March 1989.
6. Seyoum, D., Grantham, C. and Rahman, F., The dynamics of an isolated self-excited induction generator driven by a wind turbine, The 27th annual conference of the IEEE Industrial Electronics Society, IECON'01, Denver, Colorado, USA, 1364–1369, Dec. 2001.
7. Singh, S.P., Bhim Singh, M., and Jain, P., Performance characteristics and optimum utilization of a cage machine as capacitor excited induction generator, *IEEE Transactions on energy conversion*, 5(4), 679–684, Dec. 1990.
8. Hallenius, K.E., Vas, P., and Brown, J.E., The analysis of a saturated self-excited asynchronous generator, *IEEE Transactions on energy conversion*, 6(2), 336–345, June 1991.
9. Barbi, I., *Fundamental Theory of the Induction Motor (Teoria fundamental do motor de indução)*, UFSC University press-Eletrobrás, Florianópolis, Brazil, 1985.
10. Langsdorf, A.S., *Theory of the Alternating Current Machines (Teoría de máquinas de corriente alterna)*, McGraw-Hill Book Co., 1977, p. 701.
11. Chapman, S.J., *Electric Machinery Fundamentals, Third Edition*, McGraw-Hill International Edition, New York, 1999.

12. Hancock, N.N., *Matrix Analysis of Electric Machinery*, Pergamon, 1964, p. 55.
13. Grantham, C., Steady-state and transient analysis of self-excited induction generators, *IEEE Proc.*, Vol. 136, Pt. B, No. 2, March 1989.
14. Seyoum, D., Grantham, C., and Rahman, F., The dynamics of an isolated self-excited induction generator driven by a wind turbine, The 27th annual conference of the IEEE Industrial Electronics Society, IECON'01, Denver, Colorado, USA, 1364–1369, Dec. 2001.
15. Muljadi, E., Sallan, J., Sanz, M., and Butterfield, C.P., Investigation of self-excited induction generators for wind turbine applications, *IEEE Proc.*, 1, 509–515, 1999.
16. Lee, C.H. and Wang, L., A novel analysis of parallel operated self-excited induction generators, *IEEE Transactions on Energy Conversion*, 13(2), June 1998.
17. Murthy, S.S., Bhim Singh, M., and Tandon, A.K., Dynamic models for the transient analysis of induction machines with asymmetrical winding connections, *Electric Machines and Electromechanics*, 6(6), 479–492, Nov./Dec. 1981.

5

General Characteristics
of the Induction Generator

5.1 Scope of this Chapter

Chapters 2, 3, and 4 encompassed an overall discussion of the behavior of the induction machine as a generator. Based on standard machine models and on circuit equations relating voltage and current many characteristics can be derived. In this chapter, mechanical and electrical characteristics are combined in such a way as to give a clear picture of the induction generator as one of the most important tools for mechanical conversion of renewable and alternative energies into electrical energy. The relationships between load voltage and frequency that result from this electromechanical interaction, are made clear as the basic parameters of the turbine as a prime mover and the induction machine as a generator.

5.2 Torque–Speed Characteristics of the Induction Generator

Figure 2.3 can be used to determine the torque–speed characteristics of the induction generator by neglecting the magnetizing loss resistance R_m. It is repeated here as Figure 5.1 for convenience. From this figure, and from Equations 2.19, 2.23 and 2.26, neglecting R_m, we can obtain the expressions for air-gap power, converted power, and converted torque, which are

$$P_{airgap} = \frac{3I_2^2 R_2}{s} = \frac{3V_{ph}^2 R_2/s}{\left(R_1 + R_2/s\right)^2 + (X_1 + X_2)^2} \qquad (5.1)$$

$$P_{converted} = (1-s)P_{airgap} \qquad (5.2)$$

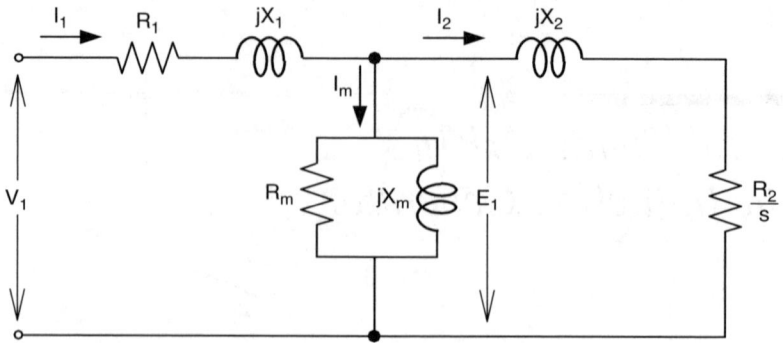

FIGURE 5.1
Stator equivalent circuit of the induction generator.

$$T_{converted} = \frac{P_{airgap}}{\omega_s} \qquad (5.3)$$

Notice that the precision of R_1 and X_1 can be improved by replacing them with the values taken from the Thévenin equivalent in Figure 5.1. This subject is left for the student as an exercise.

Output power may be obtained from Equation 2.33 if there is a way of determining the stray losses, friction, and air resistance as being

$$P_{out} = P_{mec} - P_{losses} = -3I_2^2 R_2 \frac{1-s}{s} - 3I_1^2 R_1 - \frac{3E_1^2}{R_m} - 3I_2^2 R_2 - P_{frict+air} - P_{stray} \qquad (5.4)$$

or

$$P_{out} = P_{airgap} - P_{losses} = -3I_2^2 R_2 \frac{1}{s} - 3I_1^2 R_1 - \frac{3E_1^2}{R_m} - P_{frict+air} - P_{stray} \qquad (5.5)$$

Substituting Equation 5.1 into Equation 5.3, we get

$$T_{converted} = \frac{3V_{ph}^2 R_2/s}{\omega_s \left[\left(R_1 + R_2/s\right)^2 + \left(X_1 + X_2\right)^2 \right]} \qquad (5.6)$$

where $s = 1 - \dfrac{n_r}{n_s}$.

Equations 5.5 and 5.6 are plotted in Figure 5.2. It can be observed that: (a) there is no torque at the synchronous speed; (b) both the torque–speed

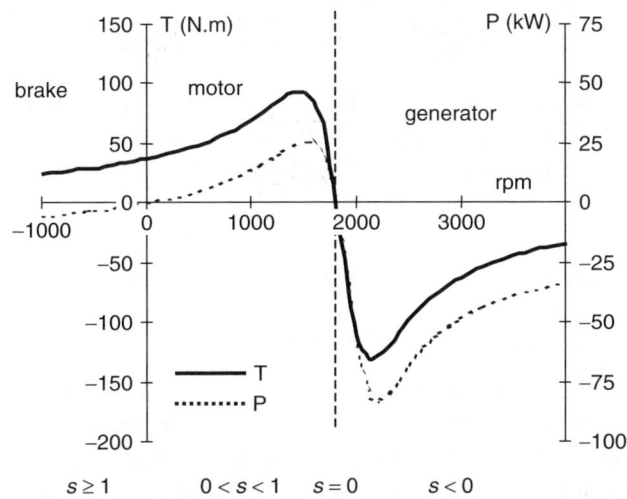

FIGURE 5.2
Torque-power–speed characteristic of the induction machine working as a brake, motor or generator.

and the power–speed curves are almost linear since from no load to full load the machine's rotor resistance is much larger than its reactance; (c) as resistance is predominant in this range, current and the rotor field as well as the induced torque increase almost linearly with the increase of the slip factor s; (d) the rotor torque varies as the square of the voltage across the terminals of the generator; (e) if the speed slows down close to the synchronous speed, the generator motorizes, that is, it works as a motor; (f) as we will show, the generated power has a maximum value for a given current drained from the generator; (g) in the same way, there is a maximum possible induced generator torque called pullout or breakdown torque, and from this torque value on there will be overspeed.

As can be observed in Figure 5.2, the peak power supplied by the induction generator happens at a speed slightly different from the maximum torque and, naturally, no electric power is converted into mechanical power when the rotor is at rest (zero speed). In the same way, in spite of the same rotation, the frequency of the induction generator varies with the load variation. But as the torque–speed characteristic is very accentuated in the normal range of operation, the total frequency variation should be limited to not much more than 5%. Such variation is acceptable for most of the loads in stand-alone power plants.[1-3]

In Figure 5.2 it can be observed that, for $0 < s < 1$ (or $0 < n_r < 1800$ rpm), the machine absorbs electric power in the rotor to convert it to a positive mechanical torque. If $s = 0$ (or $n_r = n_s$) there is no electromotive force induced in the rotor and, therefore, there is no power transfer from the stator to the rotor, and there is no mechanical torque. If $s < 0$ (or $n_r > 1800$ rpm), the mechanical power is converted into electric power by the external application of torque in the shaft to turn the rotor, and the machine works as a generator. If $s \geq 1$ (or $n_r \leq 0$), the machine works as a brake, absorbing mechanical power, which acts negatively on its shaft ($P_{mec} < 0$).

In the characteristic Txn_r, the current I_2 assumes values that depend on the rotation assumed for the rotor. In the characteristic $P_{mec}xI_2$, in contrast, when establishing the values of I_2, the slip factor s varies according to it.

5.3   Power Versus Current Characteristics

Chapter 10 presents optimized methods of control for induction generators. Thus it is important to know the characteristics of mechanical power versus load current (PxI). This relationship is showed in Figure 2.7 and obtained from the definition of mechanical power in the rotor. The power dissipation in the rotor is expressed by Equation 2.23 and repeated here for convenience.

$$P_{mec} = 3I_2^2 \left(\frac{R_2}{s} - R_2 \right) \tag{5.7}$$

Figure 5.1 can be used to make Equation 5.7 independent of slip s and only a function of the rotor current. To do this, I_2 may be given as:

$$I_2 = \frac{V_{ph}}{\sqrt{\left(R_1 + R_2/s \right)^2 + \left(X_1 + X_2 \right)^2}} \tag{5.8}$$

from which the term R_2/s can be isolated to give

$$\frac{R_2}{s} = \sqrt{\left(\frac{V_{ph}}{I_2} \right)^2 - \left(X_1 + X_2 \right)^2} - R_1 \tag{5.9}$$

Substituting Equation 5.9 into Equation 5.7 we get, finally,

$$P_{mec} = 3I_2^2 \left[\sqrt{\left(\frac{V_{ph}}{I_2} \right)^2 - \left(X_1 + X_2 \right)^2} - \left(R_1 + R_2 \right) \right] \tag{5.10}$$

Notice in Equation 5.10 that necessarily

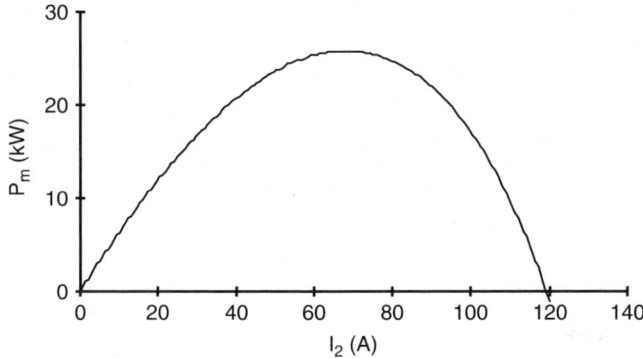

FIGURE 5.3
PxI characteristic of the induction generator (V_{ph} = 220 V).

$$\left(\frac{V_{ph}}{I_2}\right)^2 \geq (X_1 + X_2)^2$$

or

$$I_2 \leq \frac{V_{ph}}{X_1 + X_2} \tag{5.11}$$

Maximum power is obtained by substituting the inequality 5.11 in 5.10 to give

$$P_{mec(max)} = -I_2^2 (R_1 + R_2) \tag{5.12}$$

Notice that $I_2 \approx I_1$ when we neglect the excitation admittance. The negative sign in Equation 5.12 is due to reversion of the power flow through the terminals of the induction machine. Equation 5.10 is plotted in Figure 5.3, assuming a constant output voltage control.

Knowledge of the characteristic shown in Figure 5.3 is particularly useful in studies of the types of hill-climbing and fuzzy control (peak power tracking) discussed in Chapter 10. In this figure, we see that when the load current of the induction generator increases, two limiting values of current stand out for which the power is null: when there is no load across its terminals (I_{2min} = 0.0) and when there is an excess of current over the maximum defined by P_{mec} = 0.0 in Equation 5.10. Therefore,

$$I_{2max} = \frac{V_{ph}}{\sqrt{(R_1 + R_2)^2 + (X_1 + X_2)^2}} \tag{5.13}$$

On the other hand, there is also a value for generator current at which generated power is maximized. This value is obtained by the Theorem of the Average Values of Maxima and Minima of Equation 5.10, which is

$$I_{p\max} = \frac{V_{ph}}{\sqrt{2}(X_1 + R_2)}\left[1 - \frac{R_1 + R_2}{\sqrt{(R_1 + R_2)^2 + (X_1 + R_2)^2}}\right] \tag{5.14}$$

5.4   The Rotor Power Factor in Rotation

The power factor of an induction generator is the cosine of the angle represented by the impedance argument seen across the output terminals of the equivalent circuit whose vector diagram is represented in Figure 2.6.[4,5] The angle δ between the stator and rotor magnetic fields can express the rotor power factor by addition of an angle of 90°, that is, $\delta = 90° + \theta_r$. As a consequence, $\cos\theta_r = \sin\delta$. In terms of the rotor parameters, the power factor can be expressed from Equation 2.7 as

$$FP = \cos\left(\arctan\frac{sX_{r0}}{R_r}\right) \tag{5.15}$$

The plot of Equation 5.15 is shown in Figure 5.4 for an induction machine operating as motor and generator. So, for low power factor and high currents, the voltage regulation is deeply affected.

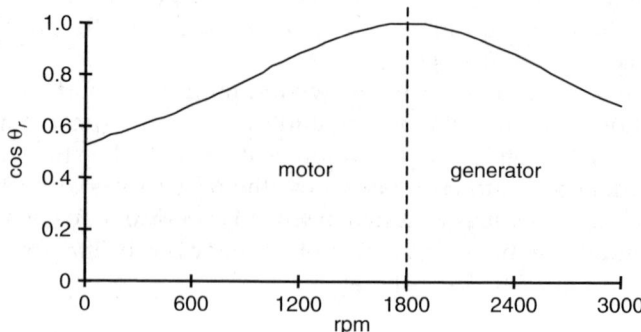

FIGURE 5.4
Rotor power factor as a function of rotation.

5.5 The Nonlinear Relationship Between Air-Gap Voltage V_g and Magnetizing Current I_m

Variation of the magnetizing inductance is the main factor in the process of voltage buildup and stabilization in operating SEIGs. Since the magnetization inductance L_m at a rated voltage is derived from the relationship between V_g and I_m, the result is a highly nonlinear function. For better results in terms of representation, a table of experimental values usually gives the magnetization inductance used in laboratory setups as a polynomial series or an analytically adjusted expression.[6–9]

The polynomial representation of the magnetizing inductance taken from experimental results can be reasonably given for practical purposes by a fourth order curve fit:

$$L_m = a_0 + a_1 V_{ph} + a_2 V_{ph}^2 + a_3 V_{ph}^3 + a_4 V_{ph}^4$$

Another way of representing the magnetization curve is a linear piecewise association within certain ranges of the magnetizing current. The magnetizing reactance $X_m = \omega L_m$ may appear as

$$X_m = \begin{cases} a_0 & \text{for} \quad 0 \le I_m < I_1 \\[2mm] \dfrac{a_1}{(I_m + b_1)} & \text{for} \quad I_1 \le I_m < I_2 \\[2mm] \dfrac{a_2}{(I_m + b_2)} & \text{for} \quad I_2 \le I_m < I_3 \\[2mm] \dfrac{a_3}{(I_m + b_3)} & \text{for} \quad I_3 \le I_m < I_4 \\[2mm] \dfrac{a_4}{(I_m + b_4)} & \text{for} \quad I_4 \le I_m \end{cases} \tag{5.16}$$

The technical literature has reported on several possible ways to relate the air-gap voltage to the magnetizing current,[2–4] showing that the relationship between V_g and I_m can be established through the following nonlinear equation.

$$V_g = F I_m \left(K_1 e^{K_2 I_m^2} + K_3 \right) \tag{5.17}$$

where

K_1, K_2, and K_3 are constants to be determined

V_g is the air-gap voltage across the magnetizing reactance (without external access)

F is the frequency in p. u., defined as

$$F = \frac{f}{f_{base}} \tag{5.18}$$

where

f is rotor frequency

f_{base} is the reference frequency used in the tests to obtain the excitation curve (usually, the 60 Hz from the distribution network)

In this case, the magnetizing reactance can be obtained directly from Equation 5.19 as

$$X_m = \omega L_m = \frac{V_g}{I_m} = F\left(K_1 e^{K_2 I_m^2} + K_3\right) \tag{5.19}$$

The reason for using the magnetizing curve[6] is that the representation of the table points from the lab magnetizing curve test should be adjusted in such a way that it passes through all points. Because of its computational advantages of easy and compact representation it has been adopted in the discussions in this book.

Another observation that can be made regarding Equation 5.19 is that it can be used to establish the variation of the generator magnetizing reactance, $x_m = dV_g/dI_m$, as well as its boundary limits of validations. For this, taking the derivative of Equation 5.19 with respect to I_m, gives

$$\frac{dV_g}{dI_m} = x_m = K_3 + \left(1 + 2K_2 I_m^2\right)K_1 e^{K_2 I_m^2} \tag{5.20}$$

After some algebraic manipulation from Equations 5.19 and 5.20, we get

$$x_m = \frac{dV_g}{dI_m} = \left(1 + 2K_2 I_m^2\right)\frac{V_g}{I_m} - 2K_2 K_3 I_m^2 \tag{5.21}$$

At the origin ($I_m = 0$), Equation 5.19 yields

$$X_{m0} = (K_1 + K_3) \tag{5.22}$$

The theoretical maximum of the saturation curve would occur when dV_g/dI_m tends to a minimum value and with Equation 5.21 we get

$$\frac{V_g}{I_m} = \frac{2K_2K_3I_m^2}{1+2K_2I_m^2} = \frac{K_3}{1+\dfrac{1}{2K_2I_m^2}} \tag{5.23}$$

As a matter of fact, for the size of induction motor used in asynchronous generation, the magnetizing current, in the saturation region is very high, so

$$1 >> \frac{1}{2K_2I_m^2}$$

and as a result, for Equation 5.23, we get

$$X_m = \frac{V_g}{I_m} \approx K_3 \tag{5.24}$$

The limit conditions established by Equations 5.22 and 5.24 can be gathered together as follows.

$$K_3 < X_m < K_1 + K_3$$

With the test data obtained from a machine at no load, a graph can be drawn to describe the behavior of the magnetizing reactance with the increase of current I_m. It can be concluded that the value of M decreases steeply as the machine reaches the saturating point as in Figure 5.5. The larger the I_m, the larger the variation of X_m.

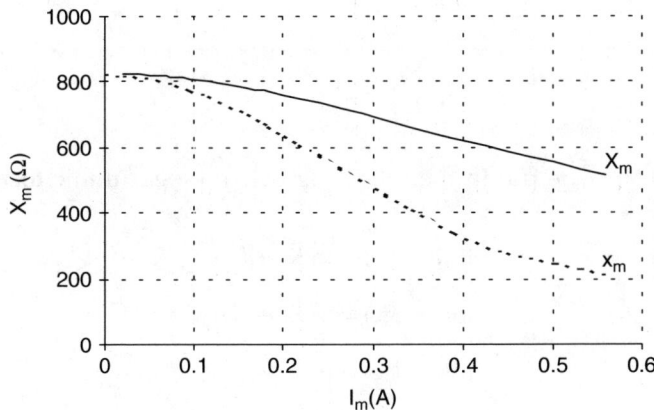

FIGURE 5.5
Variation of the magnetizing reactance with the saturation current.

5.5.1 Minimization of Laboratory Tests

The number of points needed during laboratory tests to obtain the constants K_1, K_2 and K_3, and, therefore, the magnetization curve, can be minimized if it can be taken into account that there are only three of these constants. Therefore, the minimum number of tests necessary to obtain them is three. As quadratic exponents are used to establish these constants, good precision in measuring is fundamental; otherwise, the results can lead to incorrect conclusions.

The theoretical basis for selecting the correct current and voltage values to use in the laboratory tests for minimization of the number of measurements, comes from Equation 5.17 for $F = 1$ (test frequency). Assume that the following three points were already fitted into Equation 5.19 from the lab measurements and corrected for the residual magnetism according to Equation 5.16:

$$V_{g1} = \left(K_1 e^{K_2 I_{m1}^2} + K_3 \right) I_{m1}$$

$$V_{g2} = \left(K_1 e^{K_2 I_{m2}^2} + K_3 \right) I_{m2} \qquad (5.25)$$

$$V_{g3} = \left(K_1 e^{K_2 I_{m3}^2} + K_3 \right) I_{m3}$$

or, more specifically,

$$a = \frac{V_{g1}}{I_{m1}} = \left(K_1 e^{K_2 I_{m1}^2} + K_3 \right)$$

$$b = \frac{V_{g2}}{I_{m2}} = \left(K_1 e^{K_2 I_{m2}^2} + K_3 \right) \qquad (5.26)$$

$$c = \frac{V_{g3}}{I_{m3}} = \left(K_1 e^{K_2 I_{m3}^2} + K_3 \right)$$

that can, in turn, be put in the following general logarithmic form:

$$\ln(a - K_3) = \ln K_1 + K_2 I_{m1}^2$$

$$\ln(b - K_3) = \ln K_1 + K_2 I_{m2}^2 \qquad (5.27)$$

$$\ln(c - K_3) = \ln K_1 + K_2 I_{m3}^2$$

Subtracting the first equation in 5.27 from the second, we get

$$K_2\left(I_{m2}^2 - I_{m1}^2\right) = \ln\left(\frac{b - K_3}{a - K_3}\right) \tag{5.28}$$

Similarly, subtracting the first equation in 5.28 from the third, we get

$$K_2\left(I_{m3}^2 - I_{m1}^2\right) = \ln\left(\frac{c - K_3}{a - K_3}\right) \tag{5.29}$$

Dividing 5.29 by 5.28 we get

$$k = \frac{I_{m3}^2 - I_{m1}^2}{I_{m2}^2 - I_{m1}^2} = \frac{\ln\left(\dfrac{c - K_3}{a - K_3}\right)}{\ln\left(\dfrac{b - K_3}{a - K_3}\right)}$$

or

$$\frac{c - K_3}{a - K_3} = \left(\frac{b - K_3}{a - K_3}\right)^k \tag{5.30}$$

To make Equation 5.30 more useful, the value of k should be any real value not equal to 1. However, to establish a sequence of multiple integer values easy to memorize and to minimize the current to be used in laboratory tests, it is done as follows.

$$k = \frac{I_{m3}^2 - I_{m1}^2}{I_{m2}^2 - I_{m1}^2} = 2$$

Therefore,

$$I_{m3}^2 - I_{m1}^2 = 2\left(I_{m2}^2 - I_{m1}^2\right)$$

or

$$I_{m2}^2 = \left(I_{m1}^2 + I_{m3}^2\right)\big/2 \tag{5.31}$$

If it is done in Equation 5.31, $I_{m3} = nI_{m1}$, and n is an integer, it we get

$$I_{m2} = I_{m1}\sqrt{(n^2 + 1)\big/2}$$

To have a minimum sequence of integer numbers it is important to choose the smallest convenient value of n; in this case, it could be $n = 7$. So,

$$I_{m3} = 7I_{m1}$$

Also, for $n = 7$ in Equation 3.1, I_{m2} becomes equal to $5I_{m1}$.

Notice that the advisable value for I_{m3} is that of the motor rated current so as not to overload it during tests.

Equation 5.30 has an infinite set of solutions for integers $k > 1$. Therefore, except for trivial solutions, the next smallest integer is $k = 2$, the simplest solution. So,

$$(c - K_3)(a - K_3) = (b - K_3)^2$$

or, simplifying,

$$b^2 - 2bK_3 + K_3(a + c) - ac = 0$$

Isolating K_3 we get

$$K_3 = \frac{b^2 - ac}{2b - (a + c)} \tag{5.32}$$

From Equations 5.28 and 5.29 we see that is convenient to obtain the K_i's as functions of K_3 since this value is present in both of these equations. So, from Equation 5.26 we get

$$a - K_3 = K_1 e^{K_2 I_{m1}^2}$$
$$c - K_3 = K_1 e^{K_2 I_{m3}^2} \tag{5.33}$$

Dividing the second equation by the first in 5.33, applying the logarithm and isolating K_2, we get

$$K_2 = \frac{\ln\left(\dfrac{c - K_3}{a - K_3}\right)}{I_{m3}^2 - I_{m1}^2} \tag{5.34}$$

Furthermore, the second equation in 5.33 gives

$$K_1 = \frac{c - K_3}{e^{K_2 I_{m3}^2}} \tag{5.35}$$

Notice that, to obtain higher numerical precision in the results it is good to use the laboratory values measured directly so as to avoid the accumulation of errors with the calculation of each parameter as a function of another parameter. In the same way, starting calculations with the rated current of the machine I_{m3} also helps with the precision of K_1. Equation 5.36 already is in this form.

For K_1 and K_2, the following derivations can be used.

1. If Equation 5.32 is replaced in expressions $a - K_3$ and $c - K_3$ we get

$$a - K_3 = \frac{(a-b)^2}{a+c-2b} \qquad (5.36)$$

$$c - K_3 = \frac{(c-b)^2}{a+c-2b} \qquad (5.37)$$

Replacing Equations 5.36 and 5.37 in Equation 5.34 and, taking into account that $I_{m1} = I_{m3}/7$, we get

$$K_2 = \frac{49}{24} \frac{\ell n\left(\frac{b-c}{a-b}\right)}{I_{m3}^2}$$

Observe that a, b, and c represent the slope of the curve at points 1, 2 and 3 respectively of the straight line parallel to the saturation curve tangent. The highest slope in decreasing order will always be a, b, and c and the negative sign of the squared argument in the logarithmic expression of K_2 must be rectified.

2. If we multiply the above expression of K_2 by I_{m3}^2 and substitute the results in Equation 5.35 (recalling that $e^{\ell nx} = x$), we get

$$K_1 = \frac{c - K_3}{e^{K_2 I_{m3}^2}} = (c - K_3)e^{-\ell n\left(\frac{b-c}{a-b}\right)^{49/24}}$$

or, after some algebraic transformations,

$$K_1 = (c - K_3)\left(\frac{a-b}{b-c}\right)^{49/24} = \frac{(b-c)^2}{a+c-2b}\left(\frac{a-b}{b-c}\right)^{49/24}$$

However, this is a very complex form in which to express K_1 which introduces high sensitivity to the values obtained in the lab measurements. Table 5.1 gathers the approaches for best numerical results.

TABLE 5.1

Table of Measurements for Determination of $K_i's$

Order	Measured Current	Measured Voltages	X_m	Formulas for $K_i's$
1	$I_{m1} = I_{m1}$	V_{g1}	$a = V_{g1}/I_{m1}$	$K_1 = \left(c - K_3\right)\left(\dfrac{a-b}{b-c}\right)^{49/24}$
2	$I_{m2} = 5I_{m1}$	V_{g2}	$b = V_{g2}/I_{m2}$	$K_2 = \dfrac{49}{24}\dfrac{\ell n\left(\dfrac{b-c}{a-b}\right)}{I_{m3}^2}$
3	$I_{m3} = 7I_{m1}$	V_{g3}	$c = V_{g3}/I_{m3}$	$K_3 = \dfrac{b^2 - ac}{2b - (a+c)}$

Notice that the excitation curve represented by Equation 5.17 can be used to predict the terminal voltage of induction generators as a function of the excitation capacitance. For that, the point of intersection between the straight line $V_t = I_t/F\omega C$ and the excitation curve can be determined by substituting I_m in Equation 5.17 by $I_t = F\omega C V_t$ to give

$$\frac{1}{\omega C} = F^2\left(K_1 e^{K_2(\omega CFV_t)^2} + K_3\right)$$

(5.38)

Expressing Equation 5.38 in terms of V_t, we get

$$V_t = \frac{1}{\omega CF\sqrt{K_2}}\sqrt{\ell n\left(\frac{1}{\omega CF^2 K_1} - \frac{K_3}{K_1}\right)}$$

(5.39)

Equation 5.39 imposes limits to the existence of V_t, that is,

$$\frac{1}{\omega CF^2 K_1} > \frac{K_3}{K_1} \quad \text{or} \quad C < \frac{1}{\omega F^2 K_3}$$

5.6 An Example of Determination of the Magnetizing Curve and the Magnetizing Reactance

In Table 5.2 are the plate data of an induction motor used for obtaining some lab results with formulas from the previous section. The asynchronous motor was made to rotate at no load by a DC motor with the current progressively increased by an increase of the voltage across its terminals. The corresponding

TABLE 5.2

Plate Data of the Tested Motor

Power	10 hp	Δ connection	220V; 26 A
Isolation	B. IP54	Y connection	380V; 15 A
Regime	S1	frequency	60 Hz
I_p/I_n	8.6	rotation	1765 RPM
Category	N	FS	1.0

TABLE 5.3

Excitation Curves

Measured Current (a)	Exciting Voltage (v)	Theoretical Voltage (v)
0.000	0	0.0000
0.020	15	16.4538
0.060	45	49.0341
0.080	65	65.0013
0.115	90	92.14222
0.160	115	125.0500
0.185	140	142.1579
0.220	165	164.5150
0.265	190	190.3397
0.315	215	215.0967
0.375	240	239.6143
0.450	265	263.5572
0.560	290	289.9996

Lab Tests	V (V)	I_m (A)
1	65.00	0.08
2	248.33	0.40
3	290.00	0.56

voltage and current values were measured in the three phases of the motor and the average of these values is shown in Table 5.3. Eventual differences among the excitation voltage measured in the laboratory and the theoretical values obtained by Equation 5.17 are attributed to imprecision in the reading since no theory predicts such oscillations on the magnetizing curve.

To appreciate the precision of the values given by Equation 5.17 and of the method suggested to obtain the magnetizing curve in the laboratory, the data presented in Table 5.3 can be used. The maximum current of the table was taken as the reference, being the largest possible value of current without putting the integrity of the motor in danger. The other values were obtained by interpolation to satisfy the relationship 1:5:7 foreseen in the method described in Section 5.5 for obtaining K_1, K_2, and K_3.

The values for the formulas in Table 5.1 are filled out in Table 5.4. The no-load curves foreseen by Equation 5.17 for $F = 1$ and the values obtained in laboratory tests (marked with an "x") are plotted in Figure 5.6.

TABLE 5.4

Test Data of the Motor Described in Table 5.2

Order	Measured Current	Measured Voltages	X_m	Formulas for K_i's
1	$I_{m1} = 0.08$	65.00	$a = V_{g1}/I_{m1}$ $= 812.5$	$K_1 = (c - K_3)\left(\dfrac{a-b}{b-c}\right)^{49/24}$ $= 425.05$
2	$I_{m2} = 0.40$	248.33	$b = V_{g2}/I_{m2}$ $= 620.8$	$K_2 = \dfrac{49}{24}\dfrac{\ell n\left(\dfrac{b-c}{a-b}\right)}{I_{m3}^2}$ $= -4.0455$
3	$I_{m3} = 0.56$	290.00	$c = V_{g3}/I_{m3}$ $= 517.9$	$K_3 = \dfrac{b^2 - ac}{2b - (a+c)}$ $= 398.33$

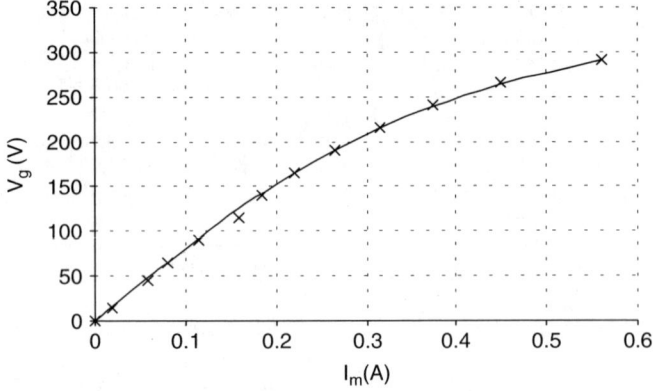

FIGURE 5.6
Practical and theoretical magnetizing curve.

5.7 Voltage Regulation

The type of load connected to the self-excited induction generator is the most serious problem for voltage regulation because highly resistive loads (high current), and inductive loads in general, can vary the terminal voltage over a very wide range. Figure 5.7 displays the regulation curves where the dashed line suggests the desired values of rated voltage for a purely resistive load. On top of the secondary effect of the load resistance on the generator voltage drops, the effect of an inductive load in parallel with the excitation capacitor is to reduce its resulting effective value as given by

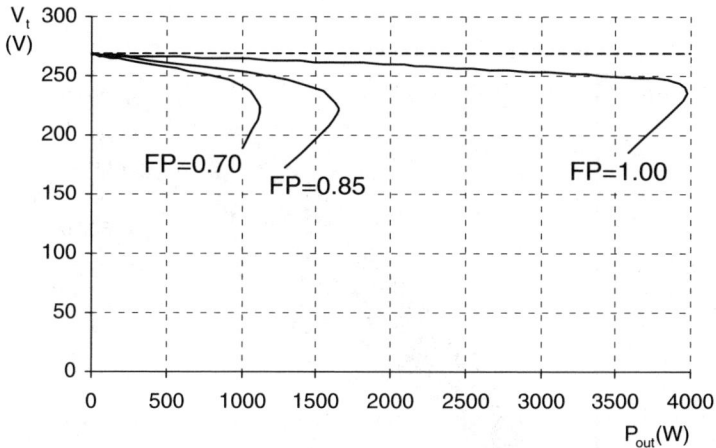

FIGURE 5.7
Influence of the load power factor on the voltage regulation of the induction generator.

$$Z_{eff} = R + j\left(\omega L - \frac{1}{\omega C}\right)$$

This change in the effective self-excitation capacitance increases the slope of the straight line of the capacitive reactance, reducing the terminal voltage. This phenomenon is even more drastic the higher the load inductance is. A capacitance for individual compensation can be connected across the terminals of each inductive load in such a way that the induction generator sees the load as almost resistive. This capacitance should be added or removed naturally with the connection or disconnection of the load. This solution is particularly useful when loads like refrigerators, whose thermostats turn their compressors on and off countless times to maintain their cold temperature at a certain level, are connected to micropower plants.

Figure 5.8 illustrates the influence of the exciting frequency on the air-gap voltage, which almost doubles when the frequency changes from 45 to 75 Hz. An interesting characteristic of the induction generator is the variation of the stator frequency with the load, which is remarkably affected by the power factor, shown in Figure 5.9.

5.8 Characteristics of Rotation

Rotation also influences the terminal voltage of the induction generator, as would be expected.[9] The faster the rotation, the higher the induced voltage. However, it should not move too far away from the machine rated limits,

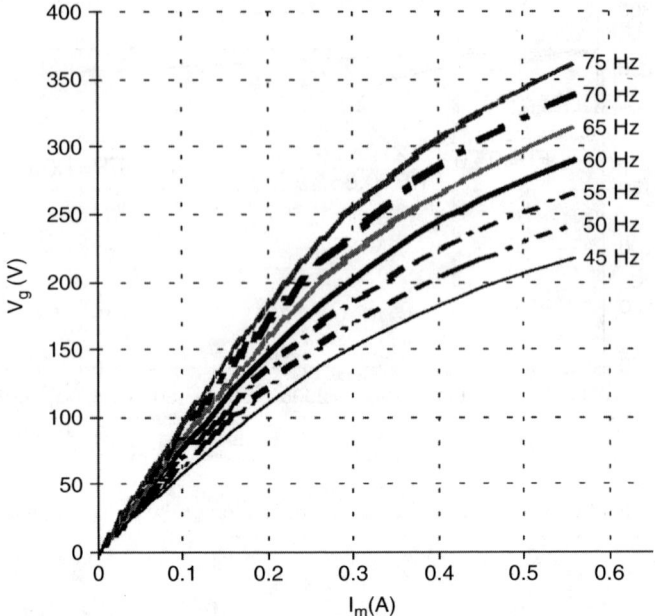

FIGURE 5.8
Influence of the exciting frequency on the air-gap voltage.

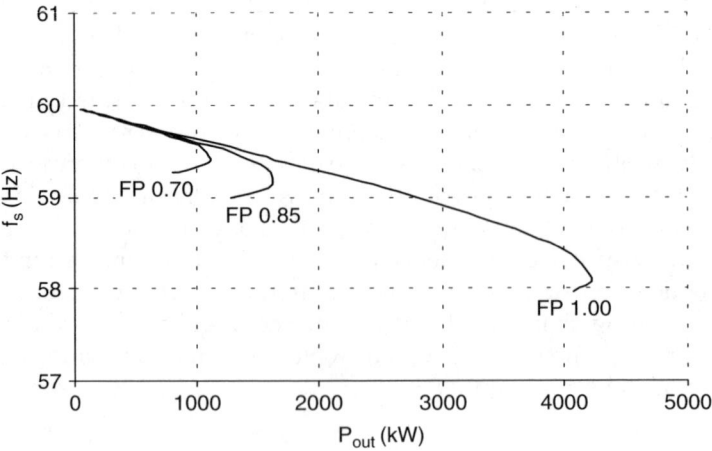

FIGURE 5.9
Frequency variation with the load (kW).

not only by mechanical limitations, but also by the low terminal voltage induced, as can be observed in Figure 5.10 for a 4-pole induction generator. From Equations 5.1 and 5.2 we get

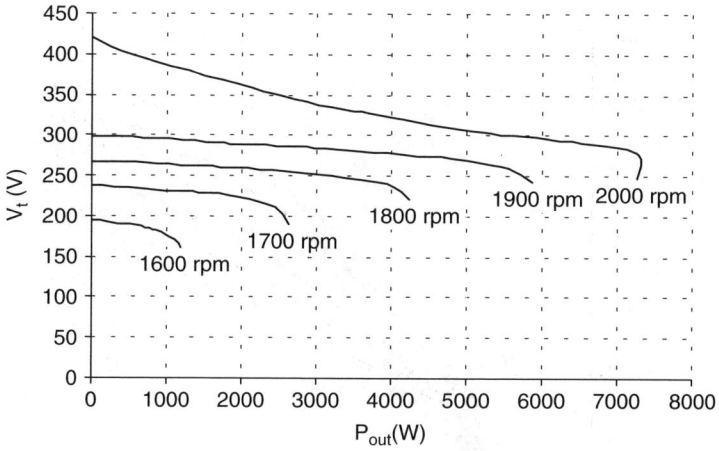

FIGURE 5.10
Voltage–rotation characteristics.

$$P_{conv} = (1-s)P_{airgap} = \frac{3V_{ph}^2 R_2 \dfrac{1-s}{s}}{\left(R_1 + R_2/s\right)^2 + (X_1 + X_2)^2} \quad (5.40)$$

Isolating V_{ph} from the rest, we get

$$V_{ph} = \sqrt{\frac{P_{conv}\left[\left(R_1 + \dfrac{R_2}{s}\right)^2 + (X_1 + X_2)^2\right]}{3R_2 \dfrac{1-s}{s}}} \quad (5.41)$$

Finally, substituting Equation 2.2 into Equation 5.41 after some simplification, we get

$$V_{ph} = \sqrt{\frac{P_{conv}\left(\dfrac{n_s}{n_r} - 1\right)\left[\left(R_1 + \dfrac{R_2}{1 - \dfrac{n_r}{n_s}}\right)^2 + (X_1 + X_2)^2\right]}{3R_2}} \quad (5.42)$$

Equation 5.42 should be conveniently interpreted; it does not have a finite solution at $n_r = n_s$. It is important to emphasize that in Equation 5.42 when the rotor is turning at a speed faster than synchronous, there must be a sign change on the converted power. Equation 5.42 is plotted in Figure 5.10

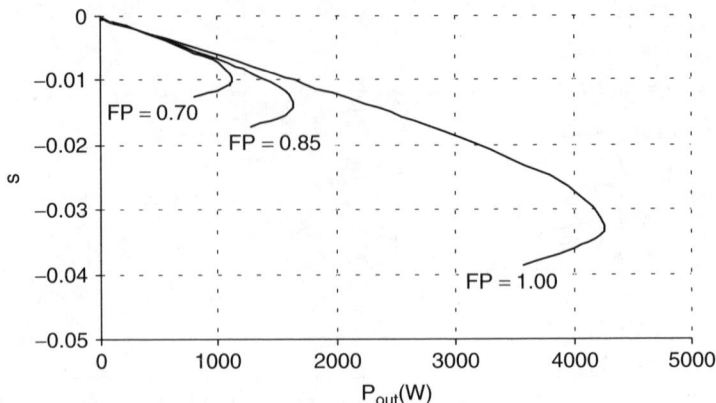

FIGURE 5.11
Slip factor–power factor.

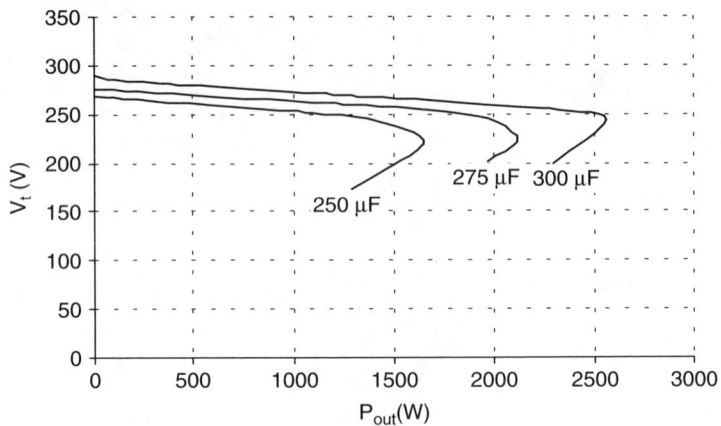

FIGURE 5.12
Self-exciting capacitance–terminal voltage.

illustrating the influence of rotor speed on terminal voltage. As expected, the induced voltage is approximately proportional to the speed. Rotor speed also affects frequency and, as a consequence, terminal voltage, as depicted in the same Figure 5.10. The slip factor can also be displayed with the power factor, as in Figure 5.11.

Another decisive factor affecting frequency is the generator's load, even at constant speed, as in Figure 5.11 and Figure 5.12, which show the influence of the self-exciting capacitance on the terminal voltage. The voltage increases with the amount of reactive power being supplied to the terminals.

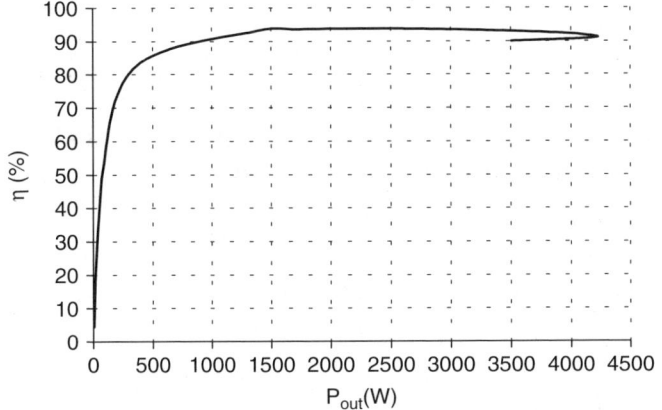

FIGURE 5.13
Relationships between efficiency and load for n_r = 1700 rpm.

On the other hand, the efficiency of the induction generator varies widely for small loads, reaching close to 90% for larger loads not very far from the rated value, as one can see in Figure 5.13. Rotation has little influence on efficiency.

As one can easily observe it is very difficult to start up an induction motor in a network supplied only by induction generators; special techniques must be used to temporarily increase the effective capacitance at start-up time, later reducing it gradually to normal. With small loads — for example, the motor impedance of a refrigerator — one of these techniques is the simple connection of a small reactor to limit surge currents during the start-up of the machine (soft start). For heavier loads, modern static reactive compensators are a good practical solution.

5.9 Problems

5.1 Using the experimental values of the magnetization curve obtained in laboratory tests for an induction generator,

a. Estimate the values K_1, K_2, and K_3 (interpolation) of the non-linear relationship for L_m given in equation 5.17.

b. Plot the magnetization curve with the parameters obtained in (a) and with the values from the lab tests. What is the maximum percentage difference?

c. Calculate the excitation capacitance to have as output terminal voltage the following values: 110 V, 127 V, and 220 V.

    d. With the capacitors calculated in item (c), calculate the voltage regulation for the rated power.

    e. What are the respective efficiencies of the induction generator in item (d)?

5.2 Using Excel or a similar program, plot the following relationships of the induction generator you have used in the lab tests:

    a. Output power x efficiency

    b. Output power x power factor; c) rotor speed x load frequency

    d. Voltage and frequency regulation for 10%, 50% and the rated load (100%)

5.3 Calculate the power injected by your induction generator into the grid for speeds 5%, 10%, and 15% above synchronous speed. What are the respective efficiencies in each case?

5.4 Explain why in cases of loads sensitive to frequency variation it is not recommended to connect an induction generator in parallel with a synchronous generator of comparable rated output power.

5.5 Calculate the necessary torque of a primary machine and the maximum rated power of an induction generator fed at 6.0 kV, 60 Hz, 4 poles, $R_1 = 0.054\ \Omega$, $R_2 = 0.004\ \Omega$, $X_1 = 0.852\ \Omega$ and $X_2 = 0.430\ \Omega$. What is the rated mechanical power? Calculate the distribution transformer ratings if the subtransmission voltage is 69 kV.

5.6 Using the data given in Problem 5.5, estimate the difference in your calculations if the Thévenin equivalent circuit was used rather than the more approximate model given by the straight values when $R_m = 29.82\ \Omega$ and $X_m = 52.0\ \Omega$. Express the final differences in percentage terms.

5.7 Using Excel or any other automatic calculator, plot the turbine torque and the corresponding induction generator output power according to the mechanical speed variation from 0.0 rpm to 2000 rpm and the data given in Exercise 5.5.

5.8 A small three-phase induction generator has the following nameplate data: 110 V, 1.140 rpm, 60 Hz, Y connection, and 1,250 watts. The blocked rotor test at 60 Hz yielded the following: I = 69.4 A, P = 4,200 W. The turn ratio between stator and rotor is 1.13 and the rotor resistance referred to the stator is 4.2 Ω. What are the full load torque and air-gap power?

5.9 In a small generating system there are two generators in parallel. One of them is a synchronous generator and the other, an induction generator. The common load is 3.0 kW with a power factor of 0.92. If the induction generator supplies only 1.0 kW of the total load at a power factor of 0.85, at what power factor would the synchronous generator be running?

5.10 There is a three-phase induction generator of 110 V, 60 Hz, 4 poles, 1900 rpm, Y stator connection, and an equivalent circuit given as $R_1 = 0.20\ \Omega$, $R_2 = 0.42\ \Omega$, $X_1 = 0.15\ \Omega$, and $X_2 = 0.43\ \Omega$; R_2 and X_2 refer to the stator. The excitation current is 8.1 A and is assumed constant at the rated values. Due to a reduced flux density, the excitation current is neglected at starting conditions. Evaluate:

  a. The maximum torque in Nm

  b. The slip factor at the maximum torque

  c. The torque to generate the rated voltage

  d. Full load torque

References

1. Lawrence, R.R., *Principles of Alternating Current Machinery*, McGraw-Hill Book Co. Inc., 1953, p. 640.
2. Kostenko, M. and Piotrovsky, L., *Electrical Machines, Vol. II*, Mir Publishers, Moscow, 1969, p. 775.
3. Smith, I.R. and Sriharan, S., Transients in induction motors with terminal capacitors, *Proc. IEEE*, Vol. 115, 519–527, 1968.
4. Chapman, S.J., *Electric Machinery Fundamentals*, McGraw-Hill International Edition, Third Edition, New York, 1999.
5. Liwschitz-Gärik, M. and Whipple, C.C., *Alternating Current Machines (Maquinas de corriente alterna)*, Companhia Editorial Continental (authorized by D. Van Nostrand Company, Inc.), 1970, p. 768.
6. Murthy, S.S., Malik, O.P., and Tandon, A.K., Analysis of self-excited induction generators, *Proc. IEEE*, 129(6), 260–265, 1982.
7. Watson, D.B. and Milner, I.P., Autonomous and parallel operation of self-excited induction generators, *Elect. Engin. Educ.*, Vol. 22, 365–374, Manchester U. P., 1985.
8. Murthy, S.S., Nagaraj, H.S., and Kuriyan, A., Design-based computational procedure for performance prediction and analysis of self-excited induction generators using motor design packages, *Proc. IEEE*, 135(1), 8–16, Jan. 1988.
9. Grantham, C., Sutanto, D., and Mismail, B., Steady-state and transient analysis of self-excited induction generators, *Proc. IEEE*, 136(2), 61–68, 1989.

6

Construction Features
of the Induction Generator

6.1 Scope of This Chapter

Induction machines are of robust construction and relative low manufacturing cost. Induction machines are more economical than synchronous machines. For high power applications, this difference is less perceptible because those units are custom made. However, for medium and small sizes, the difference in price is dramatic, as much as 80%. For the same kVA rating, though, induction machines are larger than synchronous machines because their magnetizing current circulates through the stator winding. Induction generators connected to the grid need no voltage regulation, less constraints on the turbine speed control, and practically no maintenance. These advantages plus the fact that induction machines are readily available from several manufacturers, make them very competitive for just-in-time installation. The U.S. National Electrical Manufacturers Association (NEMA) has standardized the variations in torque–speed characteristics and frame sizes, assuring physical interchangeability between motors of competing manufacturers, thereby making them a commercial success and available for integral horsepower ratings with typical voltages ranging from 110 to 4160 V. As discussed in Chapter 2, the only perceived weaknesses of induction machines are lower efficiency — the rotor dissipates power — and the need for reactive power in the stator. This chapter will discuss the physical construction of various types of induction machines, emphasizing their features in generator mode.

6.2 Electromechanical Considerations

An induction generator is made up of two major components: (1) the stator, which consists of steel laminations mounted on a frame so that slots are formed on the inside diameter of the assembly as in a synchronous machine,

FIGURE 6.1
Exploded view of induction generator major parts (Courtesy of WEG Motores, Brazil).

and (2) the rotor, which consists of a structure of steel laminations mounted on a shaft with two possible configurations: wound rotor or cage rotor. Figure 6.1 shows an exploded view of all the major parts that make up an induction generator. Figure 6.2 shows a schematic cut along the longitudinal axis of a typical induction machine. Figure 6.2(a) shows the external case with the stator yoke internally providing the magnetic path for the three-phase stator circuits. Bearings provide mechanical support for the shaft clearance (the air gap) between the rotor and stator cores. For a wound rotor, a group of brush holders and carbon brushes, indicated on the left side of Figure 6.2(a) allow for connection to the rotor windings. A schematic diagram of a wound rotor is shown in Figure 6.2(b). The winding of the wound rotor is of the three-phase type with the same number of poles as the stator, generally connected in Y. Three terminal leads are connected to the slip rings by means of carbon brushes.

Wound rotors are usually available for very large power machines. External converters in the rotor circuit, rated with slip power, control the secondary currents providing the rated frequency at the stator. For most medium power applications squirrel cage rotors as in Figure 6.2(c) are used. Squirrel cage rotor windings consist of solid bars of conducting material embedded in the rotor slots and shorted at the two ends by conducting rings. In large machines the rotor bars may be of copper alloy brazed to the end rings. Rotors sized up to about 20 inches in diameter are usually stacked in a mold made by aluminum casting, enabling a very economical structure combining the rotor bars, end rings, and cooling fan as indicated in Figure 6.3. Such

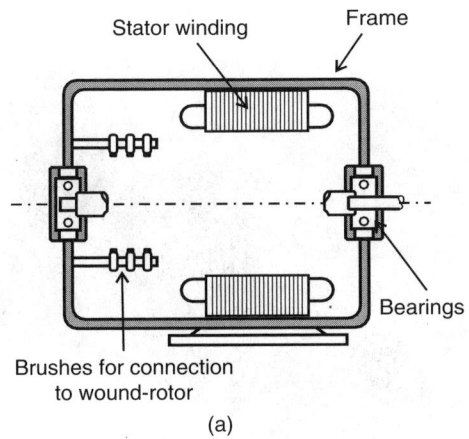

Stator winding   Frame

Brushes for connection
to wound-rotor

Bearings

(a)

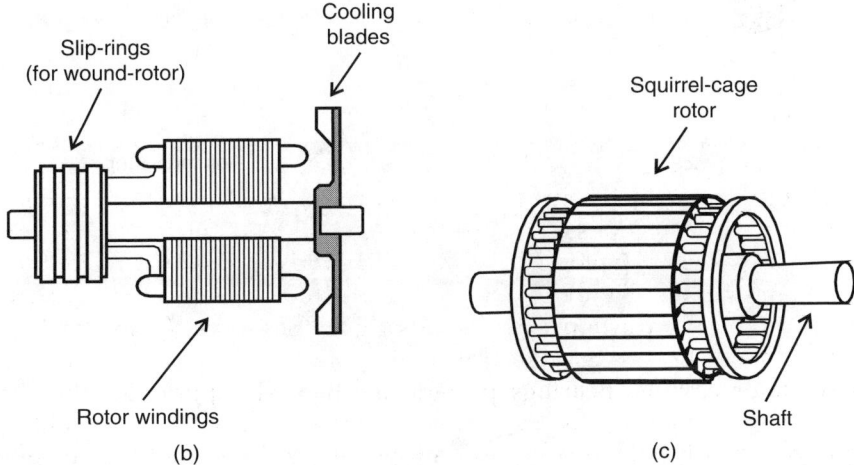

Slip-rings
(for wound-rotor)

Cooling
blades

Squirrel-cage
rotor

Rotor windings

Shaft

(b)   (c)

FIGURE 6.2
Induction machine longitudinal cut, (a) stator, (b) wound rotor, and (c) cage rotor.

induction machines are very simple and rugged with very low manufacturing cost. Some variations on the rotor design are used to alter the torque–speed features.

Figure 6.4 shows a cross-sectional cut indicating the distributed windings for three-phase stator excitation. Each winding (a, b or c) occupies the contiguous slots within a 120° spatial distribution. The stator or stationary core is built up from silicon steel laminations punched and assembled so that it has a number of uniformly spaced identical slots, integral multiples of six (such as 48 or 72 slots), roughly parallel to the machine shaft. Sometimes the slots are slightly twisted or skewed in relation to the longitudinal axis, so as to reduce cogging torque, noise, and vibration, and smooth up the generated voltage. Machines up to a few hundreds of kW rating and low voltage

FIGURE 6.3
Finished squirrel cage for large power induction motor (Courtesy of Equacional Motores, Brazil).

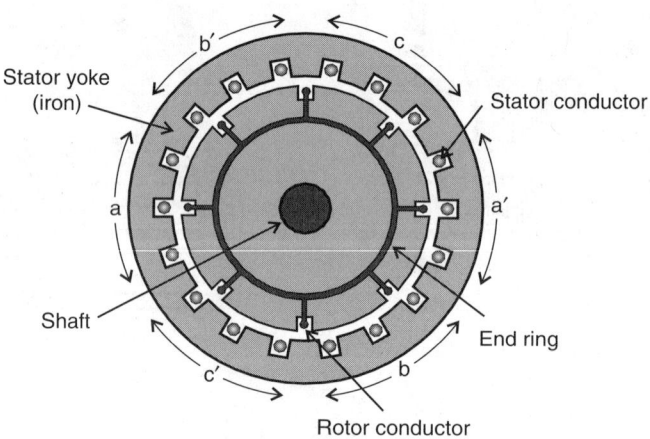

FIGURE 6.4
Cross section cut for an induction machine.

have semi-closed slots while larger machines with medium voltage have open slots. The coils are usually preformed in a six-sided diamond shape as shown in Figure 6.5, where Figure 6.5(a) indicates a random-wound coil with several turns of wire and Figure 6.5(b) is a preformed-wound coil.

The conventional winding arrangement has as many coils as there are slots, with each slot containing the left-hand side of one coil and the right-hand side of another coil. Such top and bottom coils may belong to different phases with relatively high voltage between them requiring reinforced insulation. Those coils and insulation elements are inserted into the slots, and the finished cored and winding assembly are dipped in insulating varnish and baked, making a solid structure.

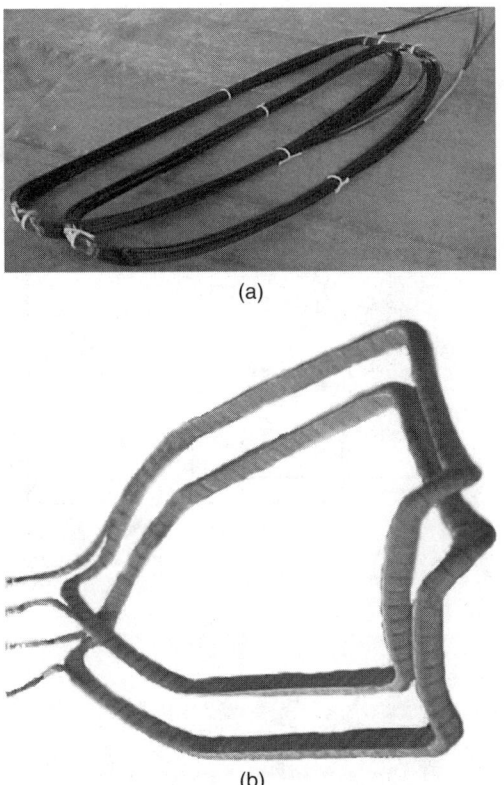

(a)

(b)

FIGURE 6.5
Stator coils, (a) form-wound coil and (b) random-wound coil.

The number of poles defines how the magnetic flux is distributed along the stator core. An elementary two-pole stator winding is depicted in Figure 6.6, where the pole created by one phase winding links half of the other phases. Two-pole machines are usually constructed only for motoring applications, despite the fact they have large stator conductors with considerable return paths at the end of the windings resulting in high leakage reactance and more core material to withstand higher flux per pole. For generator operation, a large number of poles is most common, because the prime movers to which they are coupled are usually low speed turbines.

A four-pole machine winding diagram is outlined in Figure 6.7. In order to obtain a rotating field with p number of poles, each phase of the stator winding must create those magnetic poles by splitting the coils into 3p groups, known as pole-phase groups or phase belts and connected in series with their mmf's adding to each other. Sometimes in large machines the pole-phase groups are also connected in parallel circuits. The objective is to divide the phase current in several paths as a means of limiting the coil wire diameter, as well as to adjust the required flux per pole for a given voltage

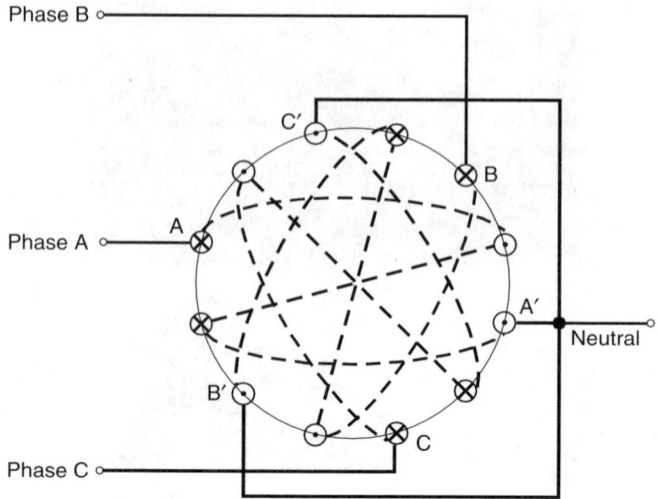

FIGURE 6.6
Concept of a two-pole stator winding.

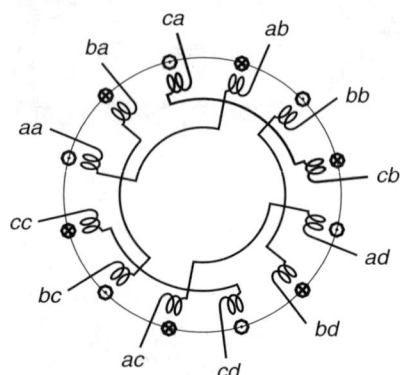

FIGURE 6.7
A four-pole stator winding.

and number of turns. Another feature of the parallel connection of phase groups is the control of asymmetric magnetic pull between stator and rotor cores, limiting vibration and noise of electromagnetic origin, particularly in large machines with a great number of poles. Figure 6.8 shows that several different connections of the phase groups are possible. Star and delta connections with series and parallel combinations offer flexibility in altering the rated voltages for the generator. No matter how the phase belts are distributed, the adjacent pole-phase groups will have opposite magnetic polarities. Eventually, the adjacent phase belts may have equal polarities in small consecutive-pole machines or machines with very large number of poles and a limited number of stator slots.

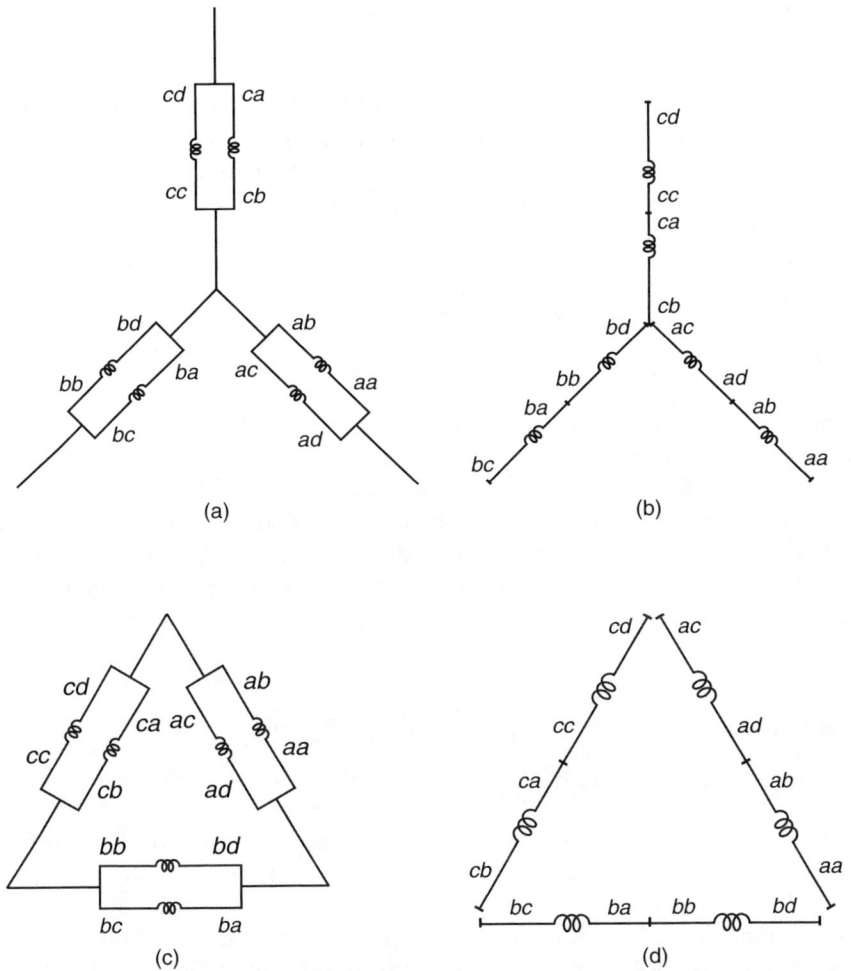

FIGURE 6.8
Possible stator winding connections: (a) star for high current, (b) star for high voltage, (c) Y for high current, and (d) Y for high voltage.

6.3 Optimization of the Manufacturing Process

There are several factors to be considered in the design of an induction generator. There must be a minimum of materials in the machine with optimized manufacturing costs and the design should be compatible with machining and assembly equipment. Materials should be available without time consuming and costly delays. The cost of losses and maintenance over the life of the machine must also be minimal, and there must be a reasonable trade-off between manufacturing and efficient operating costs. Total costs include material costs, fabrication and selling costs, capitalization of losses over generator life plus maintenance costs.

The design of electrical machines, particularly induction generators, is closely constrained by the performance, dimensions, limits, and technical and economic properties of materials. Progress in magnetic-steel laminations, in copper and aluminum for windings, and in mechanical devices has been continuously affecting the design of machines. Insulating materials in particular play a very important role in electrical machines and have been continually developed, looking for ever higher maximum operating temperature limits. Such developments have permitted better utilization of the materials, reducing the size and cost of induction generators. Metallurgical technologies in casting, molding, plastic forming, cutting, and surfacing will strongly affect the design. In recent years, several advances in efficient higher-flux amorphous nanocrystalline materials promise, in the near future, new improvements in the performance and cost of electrical machinery. Maximum ambient temperature, altitude, and cooling systems, especially for high-rated machines, must be considered in the design and optimization of induction generators. These factors strongly affect the size of the machine and consequently the cost of the machine itself and its associated systems.

6.4　Types of Design

Rotor lamination design is particularly important in defining the torque–speed characteristic of an induction machine, in both motor and generator mode. Figure 6.9 shows a cross section of a typical rotor with distinct features. The equivalent rotor leakage reactance defines the rotor flux lines that do not couple with the stator windings. If the bars of a squirrel cage rotor are placed near the surface, as in Figure 6.9(a), there is a small reactance leakage and the pull-out torque will be high and closer to the synchronous speed. This occurs at the expense of a small amount of torque for low speed and high starting current in motoring operation and a hard self-excitation process in generator mode. On the other hand, if the bars of a squirrel cage are deeper in the rotor, as in Figure 6.9(e), the leakage reactance is higher, the torque curve is not so steep close to the synchronous speed, and the maximum torque is not so high. There is a lower starting torque with a better self-excitation process and easier closed loop current control. However, the leakage flux at the air gap is very significant, and the output voltage of the generator fluctuates more when operating as an isolated machine or off-grid generator. Double-cage rotors, shown in Figure 6.9(c), or deep-bar with skin effect rotor, Figure 6.9(b), combining both features, lead to better starting characteristics that are more appropriate for motoring operation. They are the most expensive.

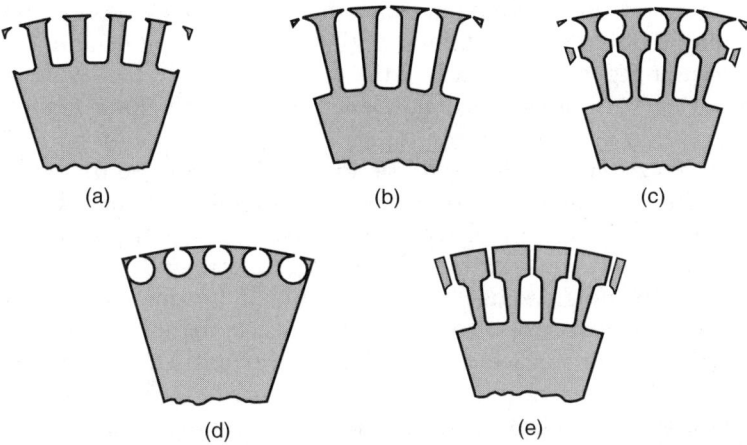

FIGURE 6.9
Laminations for squirrel cage induction generators: (a) class A, (b) class B, (c) class C, (d) class D, and (e) deep bars.

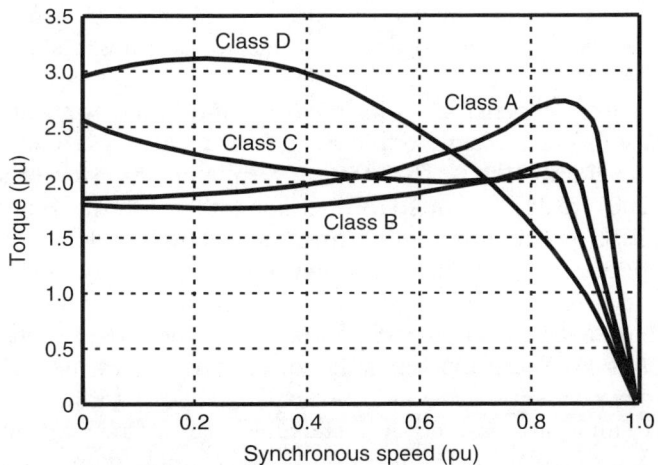

FIGURE 6.10
Torque–speed characteristics for NEMA standard machines.

The design, performance, and materials used in squirrel cage rotors are subject to standards issued by technical or standardization organizations such as IEEE, IEC, and the National Electrical Manufacturers Association (NEMA). These organizations have set a plethora of standards — for example, torque, slip, and current requirements to meet design classifications are defined by the NEMA Standard MG-1. Figure 6.10 shows typical torque–speed characteristics for a squirrel cage induction machine operating in the motor range. For the generator range, that is, rotating above synchronous speed, the characteristics are essentially the same, with torque in the negative quadrant and breakdown values slightly higher.

The following general characteristics, presented here for reference only, are those of the various classes of machines operating in motor mode defined by NEMA:

Design Class A. The full-load slip of a Class A machine must be less than 5%, lower than that of a Class B machine of equivalent rating. The push-out torque is greater than that of Class B. Specified starting torques are the same as those for Class B. The starting current is greater than that of Class B and may be more than eight times the rated full-load current. Class A machines have single-cage, low-resistance rotors. In Class A there are large rotor bars placed near the rotor surface, leading to little slip at full load, high efficiency, and high starting current.

Design Class B. This class provides a full-load slip of less than 5% (usually 2% to 3%). The breakdown torque is at least 200% of full-load torque (175% for motors of 250 hp and above). The starting current usually does not exceed five times rated current. Starting torque is at least as great as full-load torque for motors of 200 hp or less and is more than twice the rated torque for four-pole machines of less than 3 hp. These characteristics may be achieved by either deep-bar with skin effect or double-cage rotors. Class B machines have large narrow rotor bars placed deep in the rotor and lower starting current than Class A. They became the standard for industrial applications and probably the least expensive choice for rural and home applications.

Design Class C. These machines have a full-load slip of less than 5%, usually slightly larger than Class B. The starting torque is larger than that of Class B, running 200% to 250% of rated torque. The pull-out torque is a few percent less than for Class B, and the starting current is about the same. These excellent overall characteristics are achieved by means of a double-cage rotor with large low resistance bars buried deep in the rotor and small high-resistance bars near the rotor surface. Class C is the most expensive.

Design Class D. These machines are designed so that maximum torque occurs near zero speed; so they have a very high starting torque (about 275% of full-load value) at relatively low starting current (the starting current is limited to that of Class B). The term pull-out or breakdown torque does not apply to this design, since there may be no maximum in the torque–speed curve between zero and synchronous speed. For this reason, these are often called "high-slip" motors. Class D motors have a single, high-resistance rotor cage. Their full-load efficiency is less than that of the other classes. These motors are used in applications where it is important not to stall under heavy overload, like loads employing flywheel energy storage or lifting applications.

Insulation is very important in defining a motor for overall electrical insulation and resistance to environmental contamination. Motors are built with a full Class F insulation system using nonhygroscopic materials throughout to resist moisture and extend thermal endurance. Air-gap surfaces are protected against corrosion and windings are impregnated with polyester resin to provide mechanical strength.

Operating as a generator, an induction machine may exhibit some different and important characteristics. As a rule, induction generators do not have to start directly on line, like motors, but are accelerated by means of the prime movers they are coupled to. For this reason, starting torque and starting current are of minor importance. Efficiency is a very important issue in induction generators; keeping losses as low as possible implies a low slip machine. Moreover, to improve dynamic stability, a large maximum transient power is desired, especially when on-grid induction generators are used. To achieve this characteristic, a high breakdown torque is important, which implies a small leakage reactance rotor design. In the same way, to minimize the amount of reactive power needed for the excitation of induction generators, small leakage and small magnetizing reactance are required. In a general way, to achieve the characteristics mentioned above, a cage rotor resembling that of Design Class A is usually the most appropriate for an induction machine designed for generator operation.

As already stated, a rotor design with low leakage reactance, presents a steeper torque–slip characteristic near synchronous speed. So, a relatively precise and fast responding speed regulator is required for the prime mover. When this is not possible or desirable, as in induction generators used with wind turbines or small low-cost hydropower plants, a squirrel cage rotor induction machine is not the best choice. In that case, a possible solution is the use of a wound rotor induction generator, associated to a secondary converter. This electronic device, rated for slip power only, injects into the rotor circuit current with the slip frequency, enabling the generator to be connected directly to the grid, although the rotor speed may vary through a relatively wide range.

In some applications special considerations may override some preconceived issues. For example, aircraft generators require minimum weight and maximum reliability, while military generators and motors require reliability and ease of servicing. Large hydropower generators require better starting torque per ampere, induction generators for renewable energy applications require good steady and dynamic active and reactive capabilities and better efficiency. But if they work self-excited, a trade-off with a consistent saturation curve might be important.

6.5 Sizing the Machine

When designing an induction generator, the developed torque depends on the following equation.

$$T = \frac{\pi}{4} D^2 l_a J_{sm} B_{rm} \sin(\phi) \qquad (6.1)$$

where
   D is the stator bore diameter
   l_a is the stator core length
   T is the output torque
   ϕ is the rotor power factor

For a given air-gap volume and the stator winding J_{sm} current density, as well as the assumed air gap B_{rm} flux density, the following is true.

$$\frac{D^2 l_a \phi}{T} \cong constant \tag{6.2}$$

Since power is the product of torque and speed, the following coefficient (C_0) can be used for sizing AC generators.

$$C_0 = \frac{kVA}{D^2 l_a \omega} \tag{6.3}$$

In order to optimize size, the values of current and flux densities are maintained at the limits allowed by cooling capability and saturation of the ferromagnetic material utilized in the laminations. The output sizing coefficient C_0 is calculated from those considerations and put into Equation 6.3 to support the decision on diameter and stack length with respect to output power and rated speed. A fairly complex design procedure is undertaken to decide the ratio of core length to pole pitch.

The major issues in designing induction generators are in five areas: electrical, magnetic, insulation, thermal and mechanical.

Electrical. Electrical design involves the selection of the supply voltage, frequency, number of phases, minimum power factor, and efficiency; and in a second stage, the phase connection (delta or Y), winding type, number of poles, slot number, winding factors, and current densities. Based on the output coefficient, power, speed, number of poles, and type of cooling, the rotor diameter is calculated. Then, based on a specific current loading and air-gap flux density, the stack length is determined.

Magnetic. Magnetic design implies establishing the flux densities in various parts of the magnetic path and calculating the magnetomotive force requirements. Slots are sized to optimize stack length and find a suitable arrangement of the winding conductors. While aiming to design machines with large diameters to benefit from increased output coefficients, the designers of induction generators need to ensure that the machine operates with a satisfactory power factor. The magnetizing current decreases (and the power factor improves) as the diameter grows, but it becomes larger as the number of poles is increased. A generator having a large pole pitch and a small number of poles will have a higher power factor than one with a small pole pitch. The diameters and lengths are selected for minimum cost, while these values are proportioned to provide a good power factor at reasonable cost.

Insulation. Insulation design is concerned with determining the composition of slot and coil insulating materials and their thickness for a given voltage level. End connection insulation and terminal leads depend upon voltage, insulation type, and the environment in which the generator operates.

Thermal. Thermal design depends on the calculation of losses and temperature distribution in order to define the best cooling system to keep the windings, core, and frame temperatures within safe limits.

Mechanical. Mechanical design refers to selection of bearings, inertia calculations, definition of critical rotating speed, noise, and vibration modes that may cause stresses and deformations. Forces on the windings during current transients must be calculated and centrifugal stresses on the rotor must not surpass safe limits. Some isolated generator systems require pole changing winding stator procedures and the right parallel-star or series-star connection will influence the design.

The design of induction generators is generally divided into two distinct levels: less than and greater than 100 kW. In general, below-100-kW generators have a single winding stator built for low voltages and a rotor with an aluminum cast squirrel cage. Induction generators above 100 kW are built for 460 V/60 Hz or even higher voltages, 2.4 to 6 kV, and sometimes 12 kV. For such high voltages the design is more constrained with regard to insulation issues and more appropriate for air or water cooling techniques. At such high power levels, wound rotor generators are frequently used. The configuration and construction of wound-rotor machines are different from squirrel-cage ones, basically in the design of the rotor. They incorporate brushes and slip rings, and skin effect must be considered for higher power generators. Unlike cage rotors, wound rotors need an insulating system for coils similar to their stators, and also require banding tapes or retaining rings to support the end windings against centrifugal forces. Figure 6.11 shows one typical flowchart used in designing an induction generator comprising stator, rotor, and mechanical design. More details on the design equations can be found in references 4 and 5.

6.6 Efficiency Issues

An induction generator is a long-term investment. An appropriate design should consider maximizing machine efficiency. There are four different kinds of losses that occur in a generator: electrical losses, magnetic losses, mechanical losses, and stray load losses. Those losses can be reduced by using better quality materials, as well as by optimizing the design. Electrical losses increase with generator load. Increasing the cross section of the stator and rotor conductors can reduce them, but more copper is required, at a

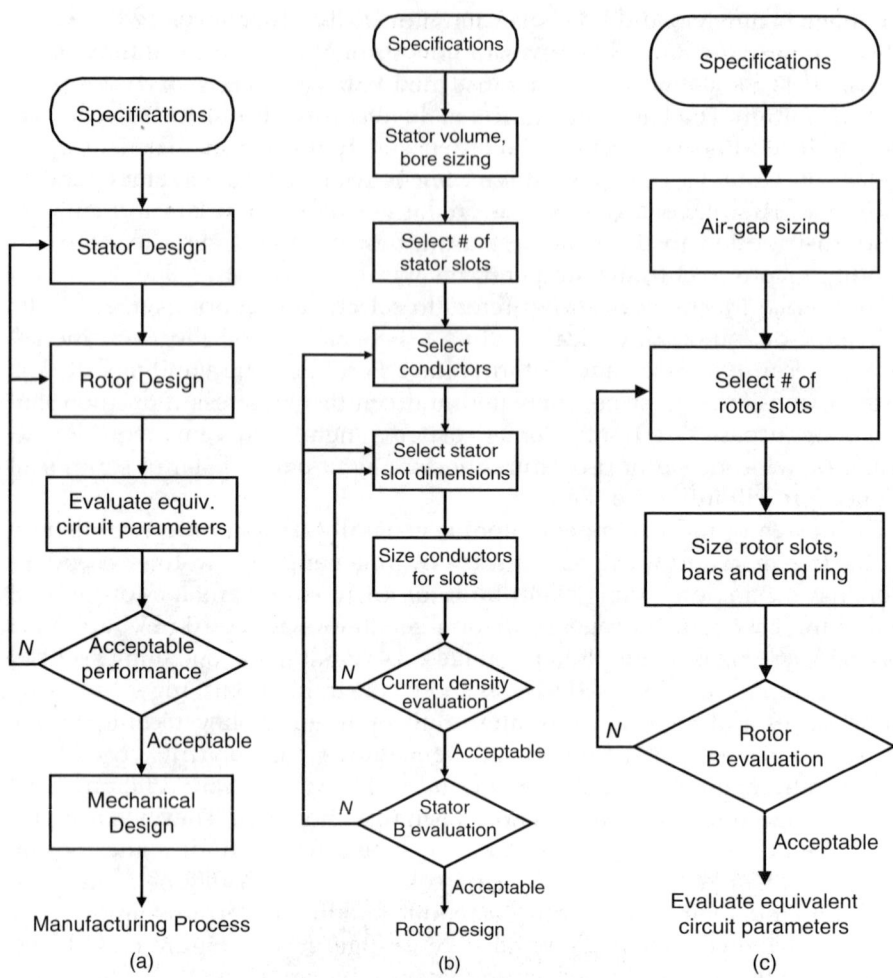

FIGURE 6.11
Flowchart for design: (a) overall process, (b) stator design, and (c) rotor design.

higher cost. Magnetic losses occur in the steel laminations of the stator and at a much lower level in the rotor. They are due to hysteresis and eddy currents and vary with the flux density and the frequency. These losses can be reduced by increasing the cross-sectional area of the iron paths mainly in the stator. They can also be limited by using thinner laminations and improved magnetic materials. If rotor and stator core length is increased, the magnetic flux density in the air gap of the machine is reduced, limiting the magnetic saturation and core losses, again at a higher cost.

Mechanical losses are due to friction in the bearings, power consumption in the ventilating fans, and windage losses. They can be decreased by using more expensive bearings and a more elaborate cooling system. The magnetic field inside a generator is not completely uniform. Thus as the rotor turns, a voltage is developed on the shaft longitudinally (directly along the shaft). This voltage causes microcurrents to flow through the lubricant film on the bearings. These currents, in turn, cause minor arcing, heating, and eventually bearing failure. The larger the machine, the worse the problem. To overcome this problem, the rotor side of the bearing body is often insulated from the stator side. In most instances, at least one bearing will be insulated, usually the one farthest from the prime mover for generators and farthest from the load for motors. Sometimes, both bearings are insulated.

Efficient generators also need improved fan design and more steel used in the stator of the machine. This enables a greater amount of heat to be transferred out of the machine, reducing its operating temperature. The fan and the rotor surface are then redesigned to reduce windage losses. Stray load losses are due to leakage flux, nonuniform current distribution, mechanical imperfections in stator and rotor surfaces, and irregularities in the air-gap flux density. Contributing to these losses are the skin effect in stator and rotor conductors, flux pulsation due to slotting, and harmonic currents. Stray load losses can be reduced by optimization of the electromagnetic design and careful manufacturing, eventually leading to a larger volume machine.

Figure 6.12 to Figure 6.17 portray some pictures of actual medium- and large-size machines. Figure 6.12 shows a medium-sized eight-pole induction machine, with wound rotor coils, Figure 6.13 is a squirrel cage rotor showing the brazed bars to the end rings. Figure 6.14 depicts the stator and rotor punchings for a large induction machine. Figure 6.15 shows the stator core of a medium-sized induction machine prepared for winding. Figure 6.16 shows a medium-sized squirrel cage rotor. Figure 6.17 portrays a random stator winding of a small induction machine.

FIGURE 6.12
Medium sized eight-pole induction machine, with form wound stator coils. (Courtesy of Equacional Motores, Brazil.)

FIGURE 6.13
Squirrel cage rotor for a large induction machine showing the brazed bars to the end rings (Courtesy of Equacional Motores, Brazil).

FIGURE 6.14
Stator and rotor punchings for a large induction machine (Courtesy of Equacional Motores, Brazil).

6.7   Problems

6.1 Torque production by an induction generator depends on the air-gap volume. Write a summary of the practical limitations on magnetic flux and conductor current densities that constrain this physical fact.

6.2 Since most renewable energy prime movers are very low speed, discuss the design characteristics of an induction generator for those applications.

FIGURE 6.15
Stator core of a medium-sized induction machine prepared for winding (Courtesy of Equacional Motores, Brazil).

FIGURE 6.16
Medium-sized squirrel cage rotor winding (Courtesy of Equacional Motores, Brazil).

6.3 Gas turbine and compressed air systems will serve as prime movers at very high speed. Discuss the design characteristics of an induction generator for those applications.

6.4 Explain in detail why induction generators of the same rating as synchronous generators are larger in size.

6.5 What are the differences between a high standard induction generator and the corresponding induction motor?

FIGURE 6.17
Random stator winding of a small induction machine (Courtesy of Equacional Motores, Brazil).

References

1. Nasar, S.A., *Handbook of Electric Machines*, McGraw-Hill, New York, 1987.
2. Say, M.G., *Alternating Current Machines*, John Wiley & Sons, New York, 1976.
3. Boldea, I. and Nasar, S.A, *The Induction Machine Handbook*, CRC Press, Boca Raton, 2001.
4. Hamdi, H.S. and Hamdi, E.S., *Design of Small Electrical Machines*, John Wiley & Sons, New York, 1994.
5. Cathey, J.J., *Electric Machines: Analysis and Design Applying MATLAB*, McGraw-Hill, New York, 2000.
6. Salon, S.J. and Lipo, T., *Finite Element Analysis of Electrical Machines*, Kluwer Academic Publishers, Dordrecht, the Netherlands, 2002.

7

Power Electronics for Interfacing
Induction Generators

7.1 Scope of This Chapter

Power electronics is the branch of electronics that studies application systems ranging from less than a few watts to more than 2 gigawatts. These systems encompass the entire field of power engineering, from generation to transmission and distribution of electricity and its industrial use, as well as transportation, storage systems, and domestic services. The progress of power electronics has generally followed the microelectronic device evolution and influenced the current technological status of machine drives. The amount of power produced by renewable energy devices like photovoltaic cells and wind turbines varies significantly on an hourly, daily, and seasonal basis due to variations in the availability of the sun, the wind, and other renewable resources. This variation means that sometimes power is not available when it is required, and sometimes there is excess power.

Figure 7.1 shows how power electronics interfaces energy sources with loads. The variable output from renewable energy devices also means that power conditioning and control equipment is required to transform this output into a form (voltage, current, and frequency) that can be used by electrical appliances. This chapter will present power electronic semiconductor devices and their requirements for interfacing with renewable energy systems: AC-DC, DC-DC, DC-AC and AC-AC conversion topologies as they apply to the control of induction machines used for motoring and generation purposes.

7.2 Power Semiconductor Devices

After the transistor was invented, the emergence of thyristors initiated the first generation of power electronics. In the 1960s, inverter-grade thyristors

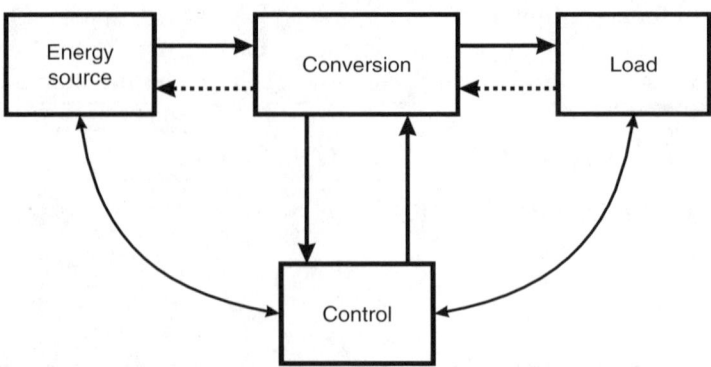

FIGURE 7.1
Power electronics interface energy sources with load.

enabled the introduction of force-commutated thyristor inverters like the McMurray inverter, the McMurray-Bedford inverter, the Verhoef inverter, the AC-switched inverter, and the DC-side commutated inverter. Many other commutation techniques were discussed in the literature. This class of inverters gradually faced obsolescence because of the advent of self-commutated gate turn-off thyristors (GTOs) and improved bipolar transistors, ending the first generation of power electronics by the middle of the 1970s.

Thyristors and GTOs continue to grow in power rating (most recently 6000 V, 6000 A) for multimegawatt voltage-fed and current-fed converter applications (with reverse blocking devices). The device's slow switching, which causes large switching losses, keeps its switching frequency low (a few hundred Hz) in high power applications. Large dissipative snubbers are essential for GTO converters. It is important to note that as the evolution of the new and advanced devices continued, the voltage and current ratings and electrical characteristics of those devices dramatically improved the performance of power electronic circuits, making possible a widespread deployment of new applications. Even today, thyristors are indispensable for handling very high power at low frequency for applications such as HVDC converters, phase-control type static VAR compensators, cycloconverters and load-commutated inverters. It appears that the dominance of thyristors in high power handling will not be challenged, at least in the near future.

The introduction of large Darlington and Ziklai BJTs in the 1970s brought great expectations in power electronics. However, at the same time power MOSFETs were introduced. Although the conduction drop is large for higher voltage devices, the switching loss is small. MOSFET is the most popular device for low-voltage high-frequency applications, like switching mode power supplies (SMPS) and other battery-operated power electronic apparatuses. Bipolar junction transistors have been totally ousted in new applications in favor of power MOSFETs (higher frequency) and IGBTs (higher voltage).

The introduction of the IGBT in the 1980s was an important milestone in the history of power semiconductor devices, marking the second generation of power electronics. Its switching frequency is much higher than that of the BJT, and its square safe operating area (SOA) permits easy snubberless operation. The power rating (currently 3500 V, 1200 A) and electrical characteristics of the device are continuously improving. IGBT intelligent power modules (IPM) are available with built-in gate drivers and control and protection for up to several hundred kW power ratings. The present IGBT technology has conduction drops slightly larger than that of a diode and a much higher switching speed. MCT is another MOS-gated device, which was commercially introduced in 1992. The present MCT (P-type with 1200 V, 500 A), however, has limited reverse-biased SOA (RBSOA), and its switching speed is much lower than that of IGBTs. MCTs are being promoted for soft-switched converter applications where these limitations are not barriers.

The integrated gate-commutated thyristor (IGCT) is the newest member of the power conductor family at this time. It was introduced by ABB in 1997. It is a high voltage, high-power, hard-driven, asymmetric-blocking GTO with unity turn-off current gain (currently 4500 V, 3000 A), turning on with a positive gate current pulse and turning off with a large negative gate current pulse of about 1μs. It is fully integrated with a fiber-optic-based gate drive. Basically, by means of a "hard-drive" gate control with a unity gain turn-off, the element changes directly from thyristor mode to transistor mode during turn-off. This allows operation without any snubber. The gate driver required is part of the IGCT, and no special gating requirements are needed. The claimed advantages over GTO are smaller conduction drop, faster switching, and a monolithic bypass diode that combines the excellent forward characteristics of the thyristor and the switching performance of the bipolar transistor. It is claimed that it acts as a short circuit in failure mode (advantageous in converter series operation) while IGBT modules usually act as an open circuit.

The third generation of power electronics is currently under way through the integration of power devices with intelligent circuits. Isolation, protection, and interfacing are already incorporated in several commercial modules. Digital computation-intensive embedded systems are making it easier to integrate PWM techniques, signal estimation, and complex algorithms for control of machines. Although silicon has been the basic raw material for power semiconductor devices for a long time, several other raw materials, like silicon carbide and diamond, are showing promise. These materials have a large band gap, high carrier mobility, high electrical and thermal conductivity, and strong radiation hardness. Therefore, devices can be built for higher voltage, higher temperature, higher frequency, and lower conduction drop and are likely to be commercially available soon. Devices capable of withstanding voltages and currents in both directions (AC-switches) are also under development and will make possible the implementation of direct frequency changers (matrix converters) in the fourth generation of power electronics. Table 7.1 depicts the most important power electronic devices.

TABLE 7.1

Power Electronic Device Capabilities

Device	Symbol	Applications	Ratings
SCR	Anode A $\quad \downarrow i_A$ G $\quad + v_{AK}$ Gate $\quad -$ K Cathode	Very high power, multi-megawatt power systems	100 MW to 1 GW, frequency < 100 Hz
GTO	Anode Gate Cathode	Very high power, multi-megawatt traction and controlled systems	1 MW to 100 MW, frequency < 500 Hz
IGCT	Anode Gate Cathode	High voltage/high current megawatt systems	1 MW to 100 MW, frequency < 500 Hz
TRIAC	MT2 Gate MT1	Low power household AC-ac control	Up to 250 W, 60 Hz
Power MOSFET	Drain D $\quad \downarrow i_D$ Gate G $\quad + v_{DS}$ $\quad -$ S Source	Switching mode power supplies and small power actuators/drives	Up to 10 kW, frequency < 1 MHz

TABLE 7.1 (continued)

Power Electronic Device Capabilities

Device	Symbol	Applications	Ratings
IGBT	Collector C Gate G E Emitter	Medium power industrial drives, machine control, inverters, converters and active filter	Up to 500 kW, frequency < 100 kHz
MCT	Anode A Gate G K Cathode	Resonant based topologies	Up to 10 kW, frequency < 100 kHz

7.3 Power Electronics and Converter Circuits

There are different circuit topologies that can be used to handle variations in supply or change the electrical current into a form that can be used by industrial, rural, or household loads. The following classes of power electronic equipment are frequently used in renewable energy systems: regulators for battery charge controllers, inverters for interfacing DC to AC, and protection/monitoring units.

7.3.1 Regulators

A battery charge controller or regulator should be used to protect the battery bank from overcharging and overdischarging. The simplest method of charge control is to turn off the energy source as the battery voltage reaches a maximum and turn off the load when the battery voltage reaches a preset minimum. There are three main types of regulators: shunt, series, and chopper.

- Shunt regulators. Once the batteries are fully charged, the power from the renewable source is dissipated across a dump load. These are commonly used with wind turbines.

- Series regulators. Once the batteries are fully charged, the power from the renewable source is switched off in the simplest series regulator.
- Chopper regulators. These regulators use a high-frequency switching technique. The regulator switches the control device on and off quickly. When the batteries are discharged the unit will be fully switched on. As the battery reaches a fully charged state, the unit will start switching the control device on and off in proportion to the level of charging required. When the battery is fully charged no current will be allowed to flow to the battery.

7.3.2 Inverters

Renewable energy systems often provide low voltage, direct current (DC) from batteries, solar panels, or wind generators with rectification. An inverter is an electrical device that changes direct current (DC) into alternating current (AC). The inverter enables the use of standard household appliances. Inverters often incorporate extra electronic circuits that control battery charging and load management. If inverters are used only to supply power from a generator to a load, they are unidirectional. On the other hand, if they need to transfer power to an electrical machine to operate in motoring mode (such as for start-up, braking, or pumping), they need to be bidirectional. Inverters must produce power of a similar quality to that in the main electricity grids. Therefore, power quality and compliance with harmonic distortion standards are very important. Inverters can operate in stand-alone mode or connected to the main grid. In the latter case, interconnection guidelines must be negotiated with the local public services company.

7.3.3 Protection and Monitoring Units

In systems with a number of power sources, sophisticated system controllers are required. These controllers are usually computer controlled, with inputs indicating the state of the system being fed into the controller. The microprocessor makes changes to the system operation if necessary. System controllers can measure energy demanded in a house and communicate with the utility company to achieve contracted load management — for example, water heaters and furnaces, or to dispatch to a local generator. The functions performed by system controllers include:

- Disconnecting or reconnecting renewable energy sources
- Disconnecting or reconnecting loads
- Implementing a load management strategy
- Starting diesel generators when battery voltage is too low or if the load becomes too heavy

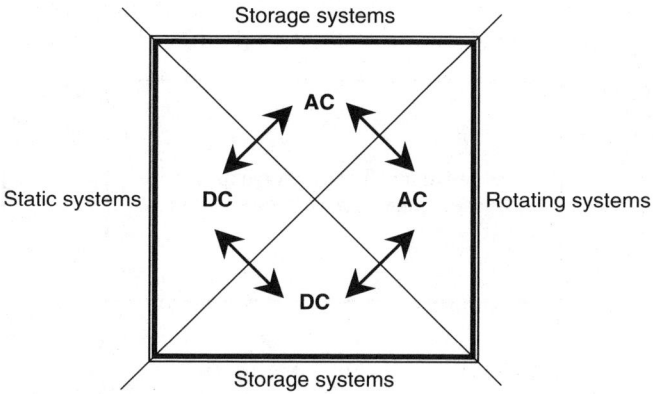

FIGURE 7.2
Power electronic needs for renewable energy conversion systems.

- Synchronizing AC power sources (e.g., inverters and diesel generators)
- Shutting systems down if overload conditions occur
- Monitoring and recording of key system parameters

Renewable energy systems can be classified as stationary and rotatory. Stationary systems usually provide direct current; photovoltaic (PV) arrays and fuel cells (FC) are the main renewable energy sources in this group. Rotatory systems usually provide alternating current; induction, synchronous, and permanent magnet generators are the main drivers for hydropower, wind, and gas turbine energy sources. Of course, DC machines are of the rotatory type, but not usually employed due to their high cost and maintenance needs.

Renewable energy sources in the stationary group do not absorb any power; energy only flows outwards to the load. Sources in the rotatory group, on the other hand, require bidirectional power flows, either to rotate in motoring mode (for pumping, braking, or starting up) or to absorb reactive power (for induction generators). The storage systems can also be either stationary (batteries, super-capacitors, magnetic) or rotatory (connection of generators to the grid, flywheels, hydro pumping). A generalized concept of power electronic needs for renewable energy systems is portrayed in Figure 7.2, embracing three categories:

- Static systems with a DC input converter to a DC or AC output and unidirectional power flow (no moving parts, lower time constant, silent, no vibrations, minimally pollutant, lower power up to now)

- Rotating systems with an AC input converter to a DC or AC output and bidirectional power flow (angular speed, noisy, mechanical inertia, weight and volume, inrush characteristics)

- Storage systems—necessarily bi-directional with requirements of AC and DC conversion (dependent on the application)

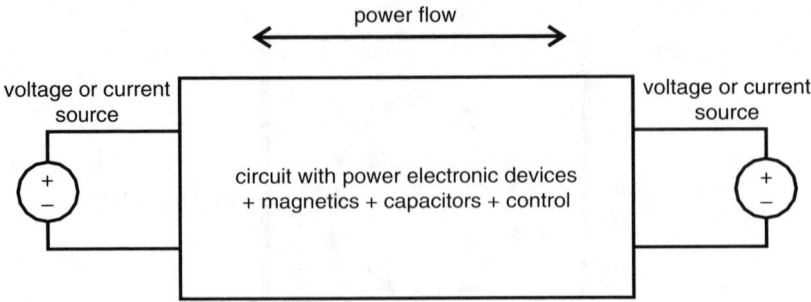

FIGURE 7.3
Power electronic converter for connection of two sources.

A converter uses a matrix of power semiconductor switches to convert electrical power at high efficiency. The standard definition of a converter assumes a source to be connected to a load as indicated in Figure 7.1. The inherent regenerative capabilities of renewable energy systems require less restrictive terms as input/output or supply/load when referencing to the connection of two sources. The links between a power converter and the outside world can be perceived as having the possibility of power reversal, where the sink can be considered as a negative source, in accordance to Figure 7.3. Therefore, it is a better fit to define the terms "voltage source" (VS) and "current source" (CS) when describing the immediate connection of converters to its entry and exit ports. The voltage source maintains a prescribed voltage across its terminals, irrespective of the magnitude or polarity of the current flowing through the source. The prescribed voltage may be constant DC, sinusoidal AC, or a pulse train. The current source maintains a prescribed current flowing between its terminals, irrespective of the magnitude or polarity of the voltage applied across these terminals. Again, the prescribed current may be constant DC, sinusoidal AC, or a pulse train.

Current source inverters (CSI) have several distinct advantages over variable voltage inverters. They provide protection against short circuits in the output stage; they can handle oversized motors. In addition, CSIs have relatively simple control circuits and good efficiency. Their disadvantages are that CSIs produce torque pulsations at low speed. They cannot handle undersized motors, and they are large and heavy. The phase-controlled bridge rectifier CSI is less noisy than its chopper-controlled counterpart. It does not need high-speed switching devices, but it cannot operate from direct DC voltage. The chopper-controlled CSI can operate from batteries and produces more noise as a result of its need for high-speed switching devices. With the advent of fast high-power controlled devices, voltage source inverters became very popular. Table 7.2 summarizes a comparison between VS and CS inverters. The following discussion in this chapter covers only voltage-source inverters.

TABLE 7.2

Summarized Comparison Between VS Inverters and CS Inverters

VSI	CSI
Output is constrained voltage	Output is constrained current
DC bus is dominated by shunt capacitor	DC bus dominated by series inductor
DC bus current proportional to motor/ generator power and hence dependent on motor/generator power factor	DC bus voltage proportional to motor/ generator power and hence dependent on motor/generator power factor
Output contains voltage harmonics varying inversely as harmonic order	Output contains current harmonics varying inversely as harmonic order
Prefers motor/generator with larger leakage reactance	Prefers motor/generator with lower leakage reactance
Can handle motor/generator smaller that inverter rating	Can handle motor/generator larger than inverter rating
DC bus current reverses regeneration	DC bus voltage reverses in regeneration
Immune to open circuit	Immune to short circuit

7.4 DC to DC Conversion

Electrical power in DC form is readily available from batteries and other inputs like photovoltaic arrays and fuel cells. Batteries are a common energy storage device for portable applications ranging from handheld drills and portable computers to electric vehicles. Another typical DC source is the output of rectifiers. When using power electronic devices to synthesize AC or variable DC waveforms, it is usually easier to begin with a known, fixed source rather than a varying source.

A fundamental form of DC-DC converter is the buck converter, sometimes called a chopper, shown in Figure 7.4. A buck converter can create an average DC output less than or equal to its DC source. The switch used in a buck converter must be fully controllable, that is, a MOSFET or an IGBT, so that the switch can forcibly turn off its current. On the other hand, it is rarely desirable to rapidly extinguish the current in an inductor. In fact, one would like the inductor to store the excess power from the source when the switch is off, then surrender the stored energy to the load while the switch is on. For the inductor to power the load while the main switch is open, the designer must add a second switch to provide another current path.

Pulse width modulation (PWM) is a technique of digitally encoding analog signal levels. Power circuits need to be turned on and off in order to modulate a voltage or current level. PWM can be implemented through old-fashioned operational amplifier-based analog circuits or, recently, through timers available in most microcontrollers, including on-chip PWM firmware. Figure 7.4 shows the principle of PWM, where by turning on and off the switches, a variable average voltage can be synthesized at the output.

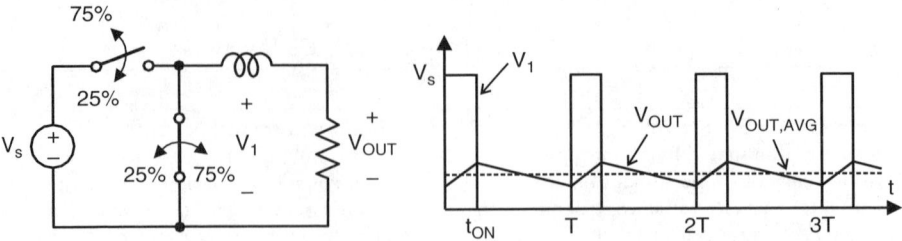

FIGURE 7.4
DC-DC converter.

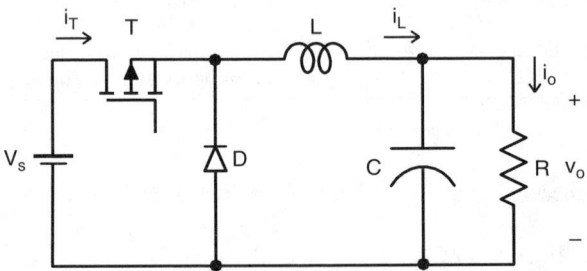

FIGURE 7.5
Buck converter.

This secondary current path is referred to as the freewheeling path. The second switch must be off when the primary switch is on and vice versa, as shown in Figure 7.5. Intuitively, a bigger filter inductor stores more energy at a given current level than a smaller one. For a given switching frequency, therefore, more inductance provides more filtering. Although the main switch in a buck converter must be fully controllable, the secondary switch does not have to be. As the secondary switch functions as the logical complement of the controllable switch, the most economical implementation of the auxiliary switch is a diode, as shown in Figure 7.5. Such a diode is called a freewheeling diode. When the main switch is conducting, current flows from the source through the inductor to the load. The diode points upwards, so there is no danger of shorting the source. When the controlled switch turns off, the inductor $L di/dt$ voltage forces the diode to conduct the current.

Some switching power supplies are based on the same idea. However, transformers are more frequently used, both for the help from their turn ratios and for electrical isolation. These DC-DC converters actually convert DC to AC, which is placed across the transformer. The secondary of the transformer is then rectified, converting AC back to the desired DC. The details are not included here. By rearranging the elements in the buck converter, a designer can create a boost converter as shown in Figure 7.6. One must add the additional capacitor to support the load while the active switch is on. When the active switch is on, the source is directly across the inductor,

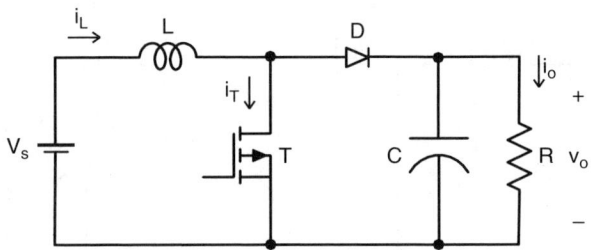

FIGURE 7.6
Boost converter.

causing the inductor current to ramp up. When the active switch is off, the diode feeds the inductor current into the load and the capacitor. If the converter is in steady state, the inductor current must go back down during this mode; thus the voltage must be higher than the source voltage. As usual, the ripple on the output voltage is determined by the switching frequency and the size of the filter elements, especially the capacitor in this case.

Buck converters and boost converters can be considered the building blocks for the structure of inverters. The buck-boost converter shown in Figure 7.7 can either generate a higher voltage at the output for a duty cycle greater than 50% or a voltage lower than the input for a duty cycle less than 50%. Figure 7.8, Figure 7.9, and Figure 7.10, respectively, show a Cuk converter,

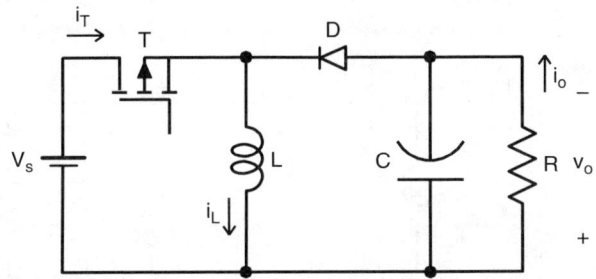

FIGURE 7.7
Buck-boost converter.

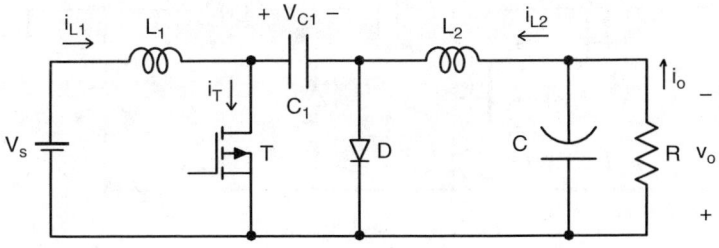

FIGURE 7.8
Cuk converter.

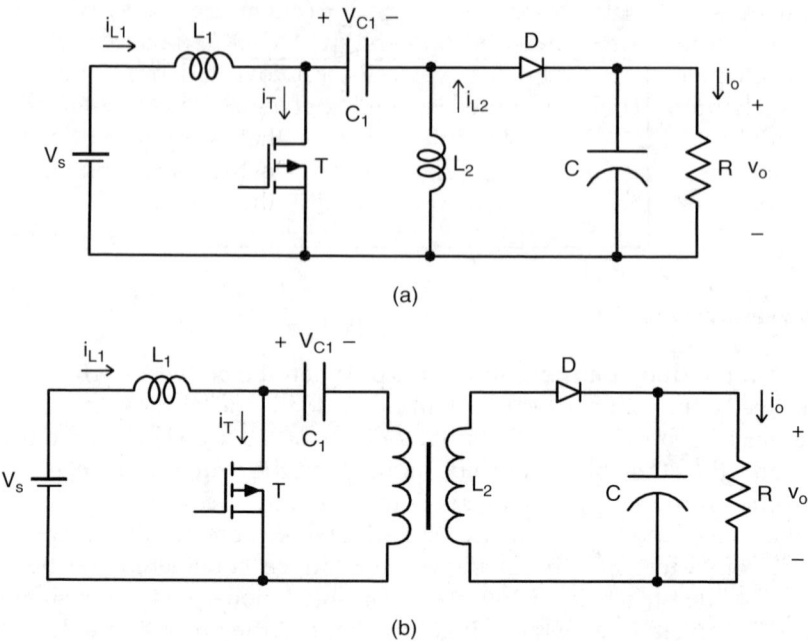

FIGURE 7.9
Sepic converter: (a) non-isolated topology, and (b) isolated topology.

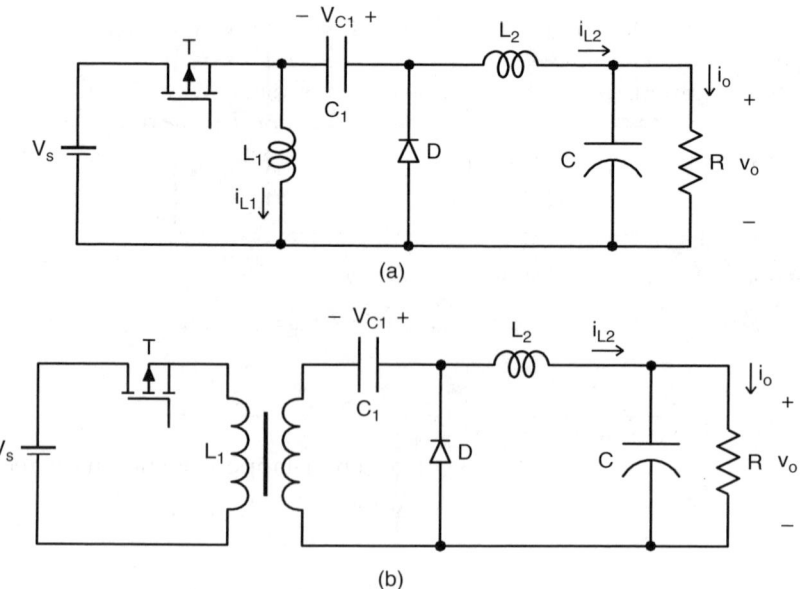

FIGURE 7.10
Zeta converter: (a) non-isolated topology, and (b) isolated topology.

a Sepic converter and a Zeta converter, all of them are used for situations where the input current must be smooth. The DC-DC converters discussed in this chapter are relevant for fuel cells, photovoltaic applications, and battery chargers. There are other isolated topologies of converters — forward, flyback, half-bridge and full-bridge — that are very common for switching mode power supplies applications. Detailed information on the operation of DC-DC converters can be found in the references.[1-3]

7.5 AC to DC conversion

For AC-DC conversion, the following parameters are required to design the circuit.

- Evaluation of AC input limitations
- Average output voltage (V_{dc})
- Output ripple voltage (V_{ripple})
- Efficiency (η)
- Circuit load, output power
- Regulation to input voltage variation
- Regulation to output load variation
- Power flow direction if required, AC to DC and DC to AC

Half-wave rectifiers are not used for medium and high power due to the severe constraints imposed on the devices and the amount of harmonics generated. In addition, due to the low power factor in half-wave rectifiers, a bulky and inefficient transformer limits their application to the range of less than 100 watts. The following single-phase full-wave rectifiers are used in practice.

7.5.1 Single-Phase Full-Wave Rectifiers, Uncontrolled and Controlled Types

Single-phase rectifiers can be constructed with a transformer with one simple secondary or with a center-tapped secondary.

7.5.1.1 Center-Tapped Single-Phase Rectifier

Figure 7.11 shows the topology of a center-tapped single-phase rectifier circuit. The center tap splits the voltage into 180°-out-of-phase voltages (v_x and v_y) with respect to the neutral terminal N. Therefore, each diode conducts for each half cycle of the input voltage $v_s = V_m \sin(\omega t)$. The circuit operation can be evaluated with thyristors (SCRs) where α is the firing angle for the SCR. For ordinary diodes it is sufficient to make $\alpha = 0$ in the description below. The average output voltage is given by

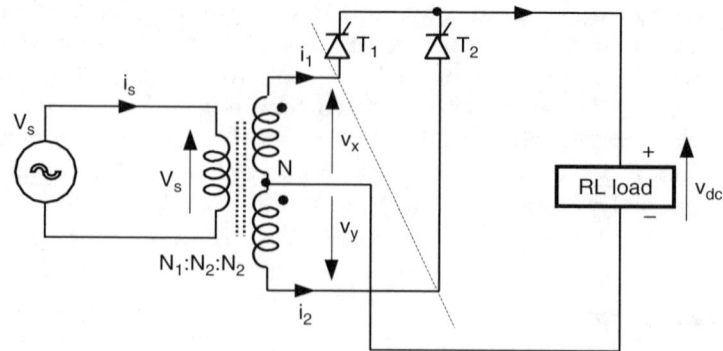

FIGURE 7.11
Center-tapped secondary single-phase rectifier.

$$V_{dc} = 2 \times \frac{1}{2\pi} \int_{\alpha}^{\pi+\alpha} V_m \sin\theta d\theta = \frac{2V_m}{\pi} \cos\alpha \qquad (7.1)$$

This equation is valid for very high reactance ωL, a condition that imposes continuous conduction mode at the output. Under these conditions each device, T1 or T2, will continue to conduct current until the next one takes over, that is, a natural commutation occurs. Figure 7.12 shows the main waveforms. The average voltage can be controlled through a time delay from the voltage zero crossing by the application of a gate pulse i_g for $\alpha = 0°$; the average voltage (V_{dc}) is maximum, and for $\alpha = 90°$ the average voltage is minimum. The ripple increases with the decrease of the average voltage. As the output current becomes flatter for highly inductive loads the thyristor currents become rectangular. The displacement factor on the input side is approximately given by $\cos\alpha$. The distortion factor can be roughly bounded by a square wave current.

7.5.1.2   Diode-Bridge Single-Phase Rectifier

Figure 7.13 shows a diode-bridge single-phase rectifier. It is an uncontrollable converter since the diodes turn on as soon as the impressed voltage becomes positive. With a pure resistive load the current waveform takes the same shape as the rectified semicycles. For a light inductive load where the current is continuous the following solution for the load current can be used.

$$L\frac{di_L}{dt} + Ri_L = \sqrt{2}V_s \sin\omega t \qquad 0° \le \omega t \le 180° \qquad (7.2)$$

and the solution for this differential equation is

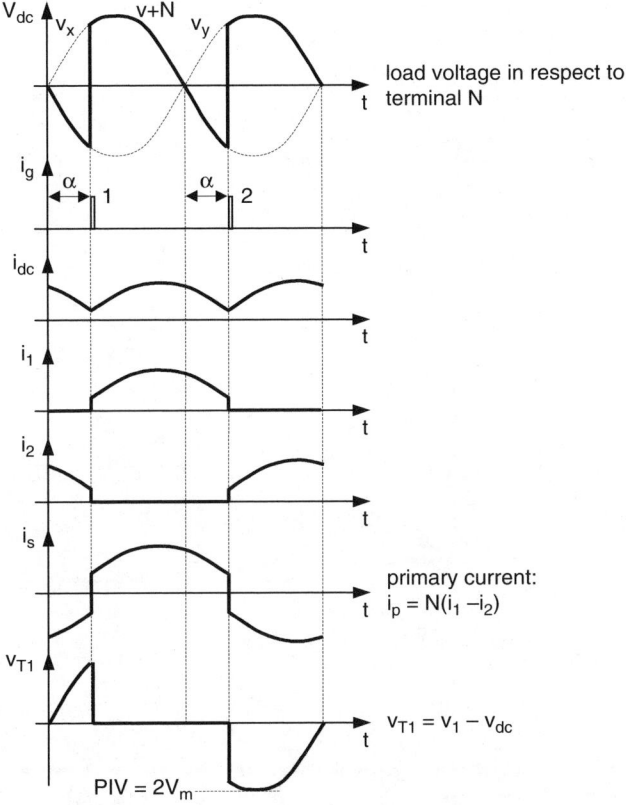

FIGURE 7.12
Main waveforms for center-tapped secondary single-phase rectifier.

$$i_L = \frac{\sqrt{2}V_s}{Z}\sin(\omega t - \theta) + A_1 e^{-Rt/L} \tag{7.3}$$

where $Z = \sqrt{R^2 + (\omega L)^2}$ and $\theta = \tan^{-1}(\omega L / R)$.

Figure 7.14 shows the output voltage and input current waveforms for a light inductive load.

7.5.1.3 Full-Controlled Bridge

Figure 7.15 shows a full-controlled single-phase converter. The thyristors T1 and T2 must be triggered at the same time — for example by using a pulse transformer with double secondary for the positive half-cycle and trigger T3 and T4 for the negative half-cycle. Assuming an inductive load, T1 and T2 keep conducting even after $\omega t = \pi$ with negative voltage; devices T3 and T4

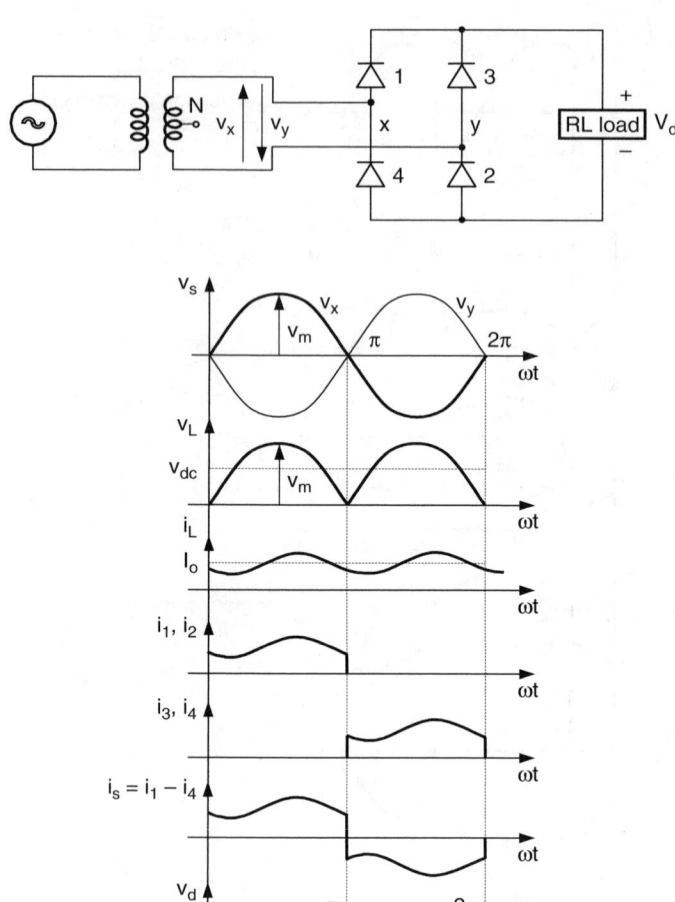

FIGURE 7.13
Circuit topology and waveforms for diode-bridge single-phase rectifier.

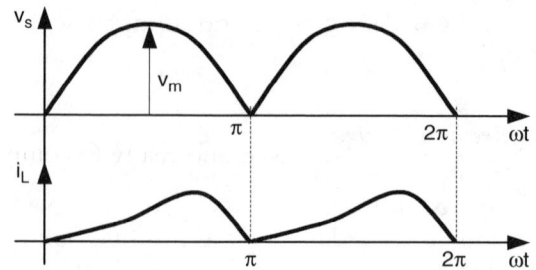

FIGURE 7.14
Output voltage and input current waveforms for light inductive load.

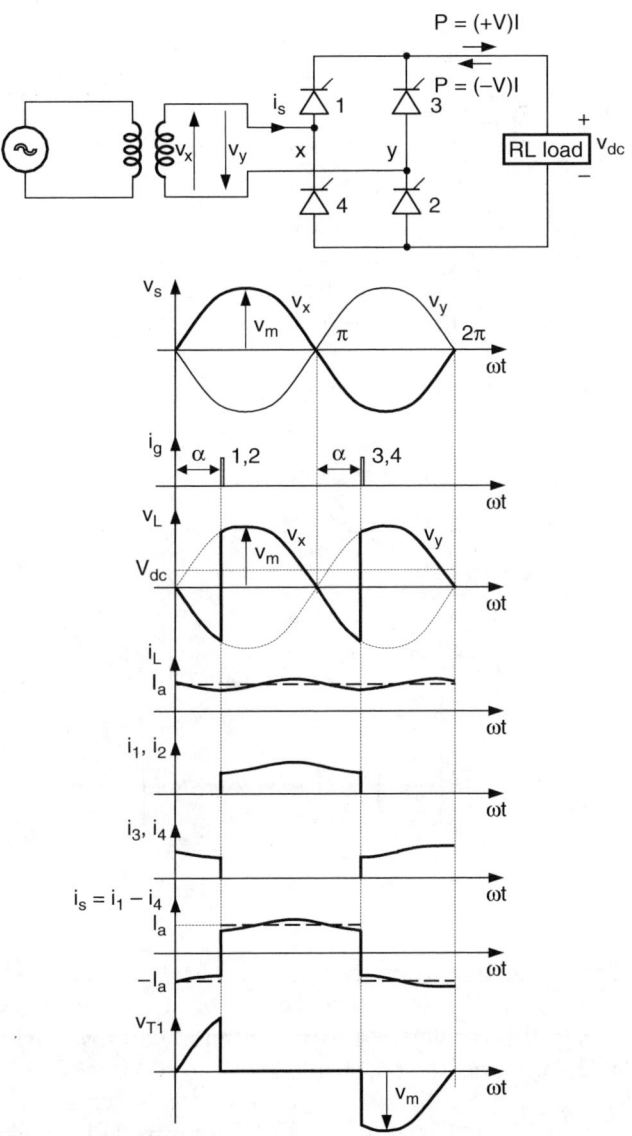

FIGURE 7.15
Full controlled single-phase converter.

during this half-cycle are forward biased and ready to commute the current when triggered at angle α. This type of single-phase converter is typical for power up to 15 kW. However, the harmonic pollution on the utility side and consequent neutral loading in three phase and a very low power factor inhibits its use for higher power. The average load voltage is the same as the full wave topology, considering two voltage diode drops of the series devices. The following equation applies for highly inductive loads.

$$V_{dc} = \frac{2}{2\pi} \int_{\alpha}^{\pi+\alpha} V_m \sin\omega t d\omega t = \frac{2V_m}{2\pi}[-\cos\omega t]_{\alpha}^{\pi+\alpha}$$

$$0 \le \alpha \le \pi \qquad (7.4)$$

$$= \frac{2V_m}{\pi}\cos\alpha = V_{do}\cos\alpha$$

V_{dc} can be varied from $-2\,V_m/\pi$ to $+2\,V_m/\pi$. Therefore, the main advantage of this full-controlled converter topology is the possibility of power regeneration to the primary side. If the thyristors are fired to apply a negative voltage for a constant positive current at the load, the power is regenerated to the primary side.

The averaged normalized output voltage is:

$$V_n = \frac{V_{dc}}{V_{do}} = \cos\alpha \qquad (7.5)$$

The rms output voltage is

$$V_{rms} = \left[\frac{2}{2\pi}\int_{\alpha}^{\pi+\alpha} V_m^2 \sin^2\omega t d\omega t\right]^{1/2}$$

$$= \left[\frac{V_m^2}{2\pi}\int_{\alpha}^{\pi+\alpha} (1-\cos 2\omega t)d\omega t\right]^{1/2} \qquad (7.6)$$

$$= \frac{V_m}{\sqrt{2}} = V_s$$

The total harmonic distortion is $THD = \sqrt{(I_{rms}/I_{rms1})^2 - 1} = 48,34\%$ with a displacement factor DF $= \cos\phi \cong \cos\alpha$. The power factor is $PF = I_{rms1}/I_{rms}\cos\alpha = 2\sqrt{2}/\pi\cos\alpha$. In this converter the fundamental value of the input current is always 90.03% of I_a and the harmonic factor is constant at 48.34%.

A simplification of the full-controlled converter is achieved by the half-converter portrayed in Figure 7.16. Only two controlled devices are used; the bottom devices are diodes, and there is a freewheeling diode across the load. The freewheeling diode increases the average voltage across the load and improves the power factor on the input side. Of course, the characteristics of bidirectionality in the previous converter are lost because the half-converter only impresses positive voltages across the highly inductive load.

Three-phase rectifiers are commonly used in high-power applications; there are more rectifying pulses resulting in more effective filtering than single-phase rectifiers; in addition the transformer utilization factor for three-phase rectifiers is improved. There are several topologies for three-phase rectifiers.

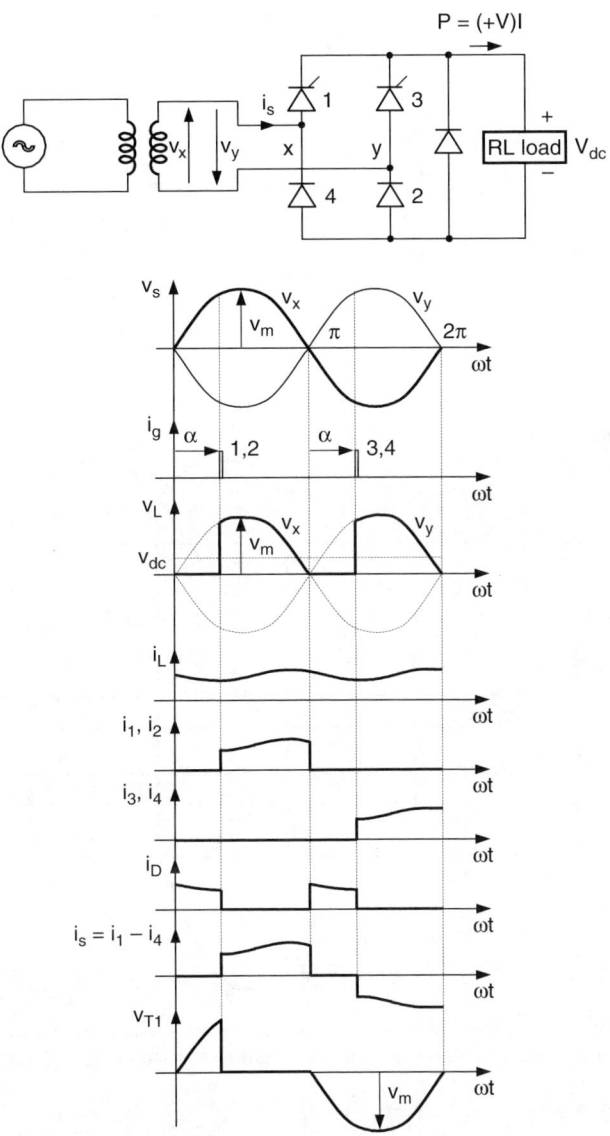

FIGURE 7.16
Semicontrolled single-phase converter.

7.5.1.4 *Half-Wave Three-Phase Bridge*

The half-wave three-phase bridge is depicted in Figure 7.17. It is a basic circuit used for understanding most polyphase circuits. Balanced three-phase voltages are assumed to be available. Every period of time that a phase voltage has the largest instantaneous voltage the correspondent diode is on as indicated by the period "a" in Figure 7.18. For a half-wave three-phase bridge there are three pulses at the rectified voltage across the load.

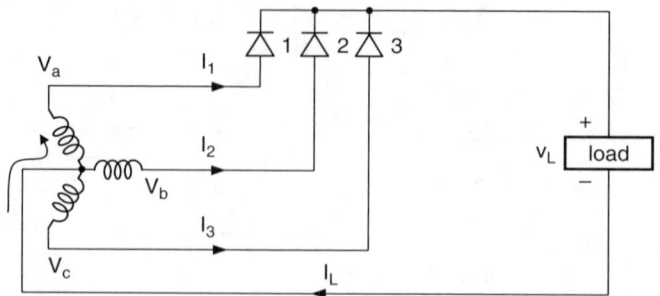

FIGURE 7.17
Half-wave three-phase bridge.

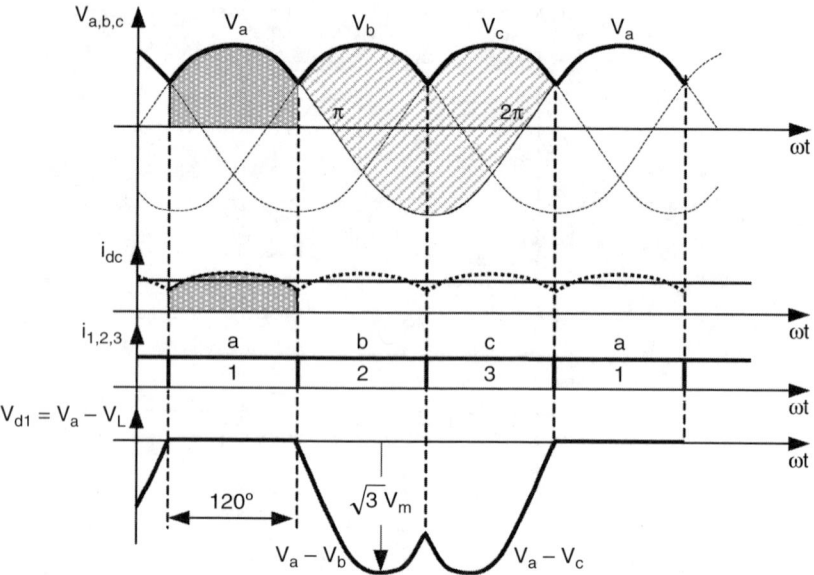

FIGURE 7.18
Waveforms for voltages and currents in a half-wave three-phase bridge.

The average value (neglecting the diode voltage drops) is

$$V_{dc} = \frac{1}{2\pi/3} \int_{\frac{\pi}{6}}^{\frac{5\pi}{6}} V_m \sin\theta d\theta = \frac{3\sqrt{3}}{2\pi} V_m \tag{7.7}$$

where V_m is the peak value of the phase voltage $= \sqrt{2}V_{rms}$.

The dashed line indicating the current i_{dc} represents the current through a resistive load. For an RL load this waveform would be smoothed. The circuit indicated in Figure 7.19 is the version of the controlled three-phase half-wave converter. Figure 7.20 shows the typical waveforms for a controlled three-phase half-wave converter.

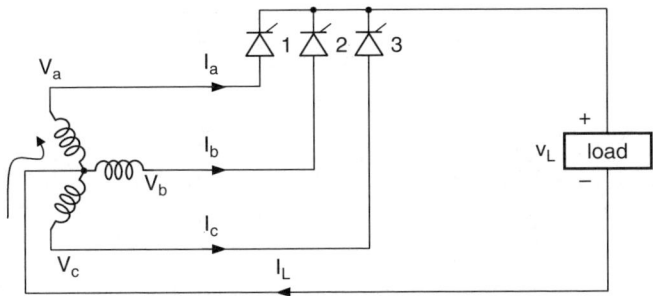

FIGURE 7.19
Controlled three-phase half-wave converter.

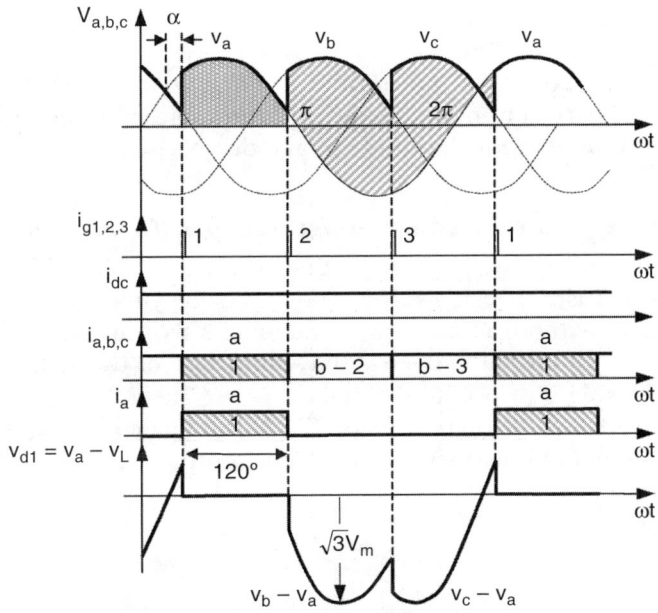

FIGURE 7.20
Waveforms for controlled three-phase half-wave converter.

Unlike single-phase rectifiers where the phase angle α is defined by the zero-crossing for three-phase rectifiers, the phase angle α is defined by the crossing of phase voltages, that is, where the diodes would naturally commute. The average voltage is calculated by

$$V_{dc} = \frac{1}{2\pi/3} \int_{\frac{\pi}{6}+\alpha}^{\frac{5\pi}{6}+\alpha} V_m \sin\theta d\theta = \frac{V_{do}}{2}\cos\alpha \qquad (7.8)$$

where $V_{do} = 3\sqrt{3}/\pi V_m$

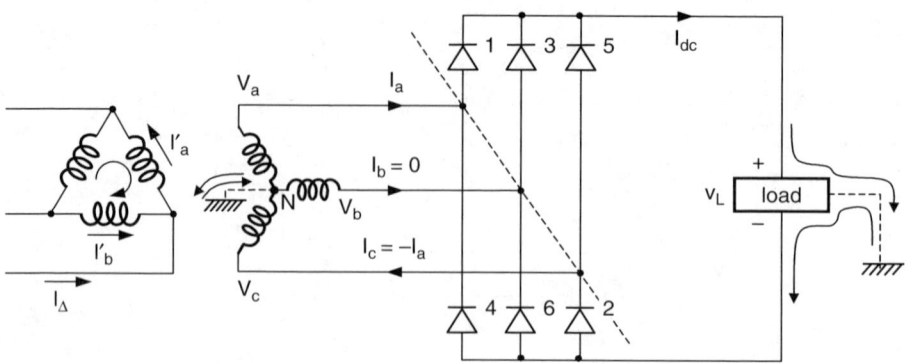

FIGURE 7.21
Graetz rectifier.

The primary side of the transformer is often δ-connected, imposing an alternating primary current and avoiding saturation. In addition, triple harmonics (3p) are kept inside the windings and do not circulate through the line.

7.5.1.5 Three-Phase Full-Wave Bridge Rectifier (Graetz Bridge)

The three-phase full-wave bridge rectifier shown in Figure 7.21 is in very wide use in industrial systems; it is also called the Graetz converter. It can operate with or without a transformer and gives a six-pulse-ripple waveform at the output. The output voltage is the subtraction of two half-wave three-phase voltages with respect to the neutral N.

The waveforms are shown in Figure 7.22. The line current i_Δ is a six-step waveform resembling a sinewave with an improved power factor on the input side. The average voltage is given by

$$V_{dc} = 2 \frac{3\sqrt{3}}{2\pi} V_{m(phase)} = \frac{3}{\pi} V_{m(line)} \tag{7.9}$$

7.5.1.6 Three-Phase Controlled Full-Wave Bridge Rectifier

The waveforms of the three-phase controlled full-wave bridge rectifier are similar to those of the Graetz bridge shifted by α. Figure 7.23 shows this. The sequence of thyristors is indicated in Figure 7.24. The average voltage is computed by

$$V_{dc} = \frac{3}{\pi} V_{m(line)} \cos \alpha = \frac{3\sqrt{2}}{\pi} V_{(line,RMS)} \cos \alpha$$

$$= \frac{3\sqrt{6}}{\pi} V_{rms(phase)} \cos \alpha = V_{do} \cos \alpha \tag{7.10}$$

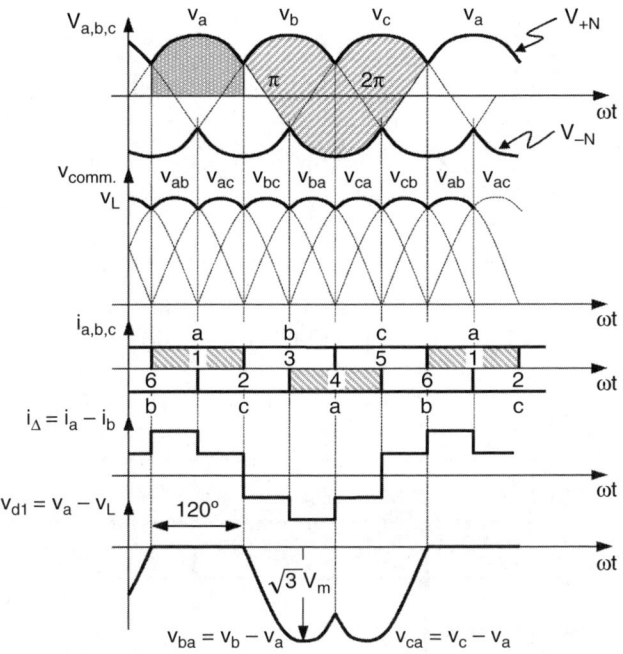

FIGURE 7.22
Main waveforms for Graetz rectifier.

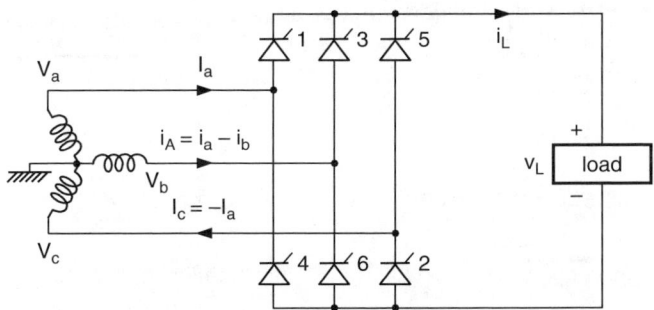

FIGURE 7.23
Three-phase controlled full-wave bridge rectifier.

The rms fundamental component of the line current is given by

$$I_{\ell 1} = \frac{\sqrt{6}}{\pi} I_{dc} \tag{7.11}$$

7.5.1.7 The Effect of Series Inductance at the Input AC Side

In practical converters it is necessary to include the effect of the AC-side inductance L_s, as depicted in Figure 7.25. Now for a given delay angle α, the

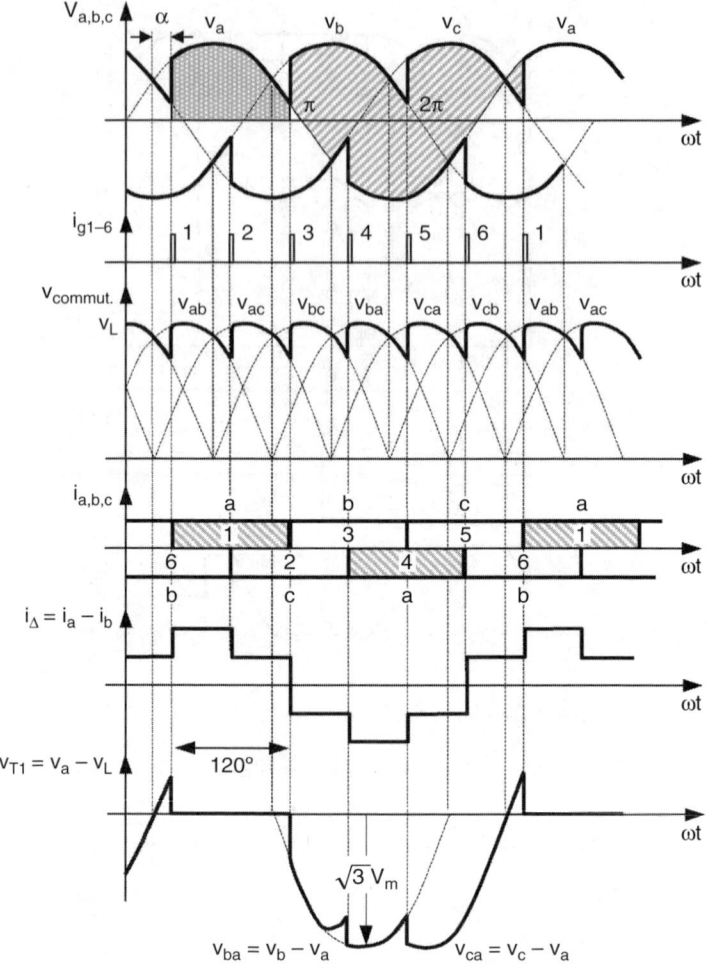

FIGURE 7.24
Waveforms for three-phase controlled full-wave bridge rectifier.

current commutation takes a finite commutation interval μ. Figure 7.26 shows the commutation process for the situation where thyristors 5 and 6 have been conducting previously, and at $\omega t = \alpha$ the current begins to commuted from thyristor 5 to 1.

The average voltage is reduced by this series inductance:

$$V_{dc} = \frac{3\sqrt{2}}{\pi} V_{(line,RMS)} \cos \alpha - \frac{3\omega L_s}{\pi} I_d$$

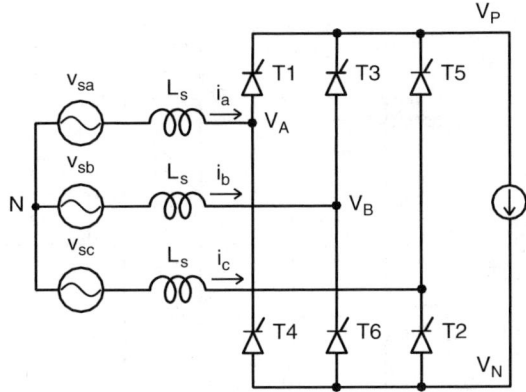

FIGURE 7.25
Graetz converter with AC-side input inductance.

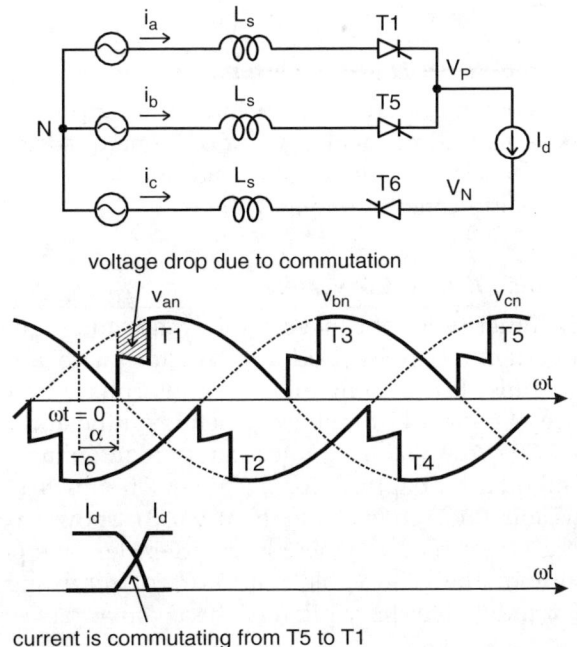

FIGURE 7.26
Commutation effect due to AC-side input inductance.

Knowing α and I_d the interval μ can be calculated by

$$\cos(\alpha + \mu) = \cos(\alpha) - \frac{2\omega L_s}{\sqrt{2}V_{(line,RMS)}} I_d$$

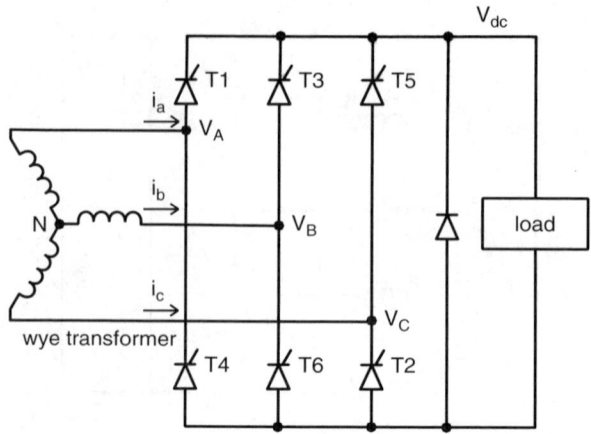

FIGURE 7.27
Three-phase semiconverter.

7.5.1.8   The Three-Phase Half-Converter

Three-phase half-converters are used in industrial applications up to the 120 kW level, where unidirectional power flow (one-quadrant) is required. A freewheeling diode is used across the load and the bottom devices are ordinary diodes, as indicated by Figure 7.27.

7.5.1.9   High-Pulse Bridge Converters

There are several other types of converters: 12-pulse and higher-pulse-number bridge converters, six-pulse bridge converters with a star-connected transformers, six-pulse bridge converters with interphase transformers, and so on. The choice of converter topology depends on the application. Figure 7.28 and Figure 7.29 show a series connection for high voltage and a parallel connection for high-current applications. Figure 7.30 shows a typical HVDC power transmission line scheme and Figure 7.31 shows a fault-tolerant scheme for energy transmission where in emergencies, one pole of the line can operate without the other pole with current returning through the ground path. For a detailed description of these converters see Kimbark.[4]

7.6   DC to AC Conversion

Conversion from DC to AC is performed by inverters. A general description of the operation of inverter topologies is valid for any power electronic device; the differences will be in hardware implementation, switching frequency, gate drivers, and, of course, power ratings. Inverters require electronic

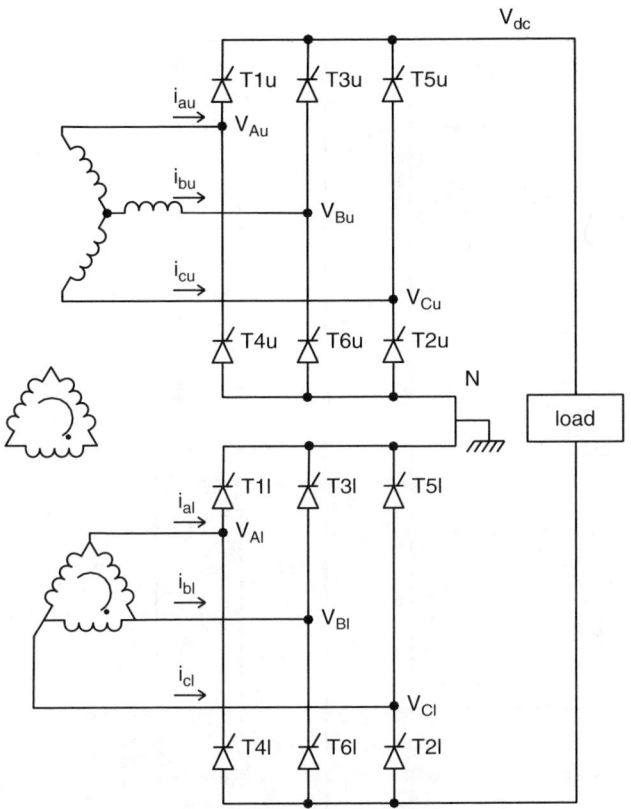

FIGURE 7.28
Series connection for high-voltage applications.

control of the pulses applied to their gates. Such control needs a prescribed pulse train that will command the frequency and voltage of the output voltage. There are several pulse-width modulation techniques and space vector techniques to synthesize the output voltage.[3]

7.6.1 Single-Phase H-Bridge Inverter

The schematic diagram of a single-phase H-bridge inverter is shown in Figure 7.32. It is called an H bridge because of the two vertical legs connecting the horizontal load. The inverter contains four switches Q1, Q2, Q3, and Q4.

Q4 is a power device with an antiparallel diode. The load, which in this case is assumed to be a passive RL load, is connected between the two legs of the inverter. The inverter is supplied by a DC source with a voltage of V_{dc}. The switches of each leg usually have complementary values; that is, when Q1, is on, Q4 is off and vice versa. Furthermore, the switches are operated in pairs. When switches Q1and Q2 are on, Q3 and Q4 are off. Similarly, when Q3 and Q4 are on, Q1, and Q2 are off. However, a small

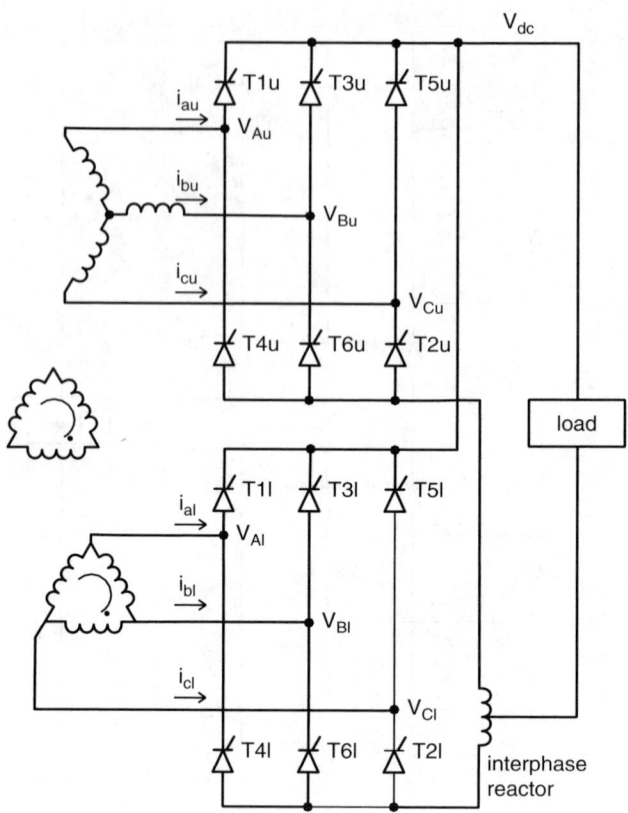

FIGURE 7.29
Parallel connection for high-current applications.

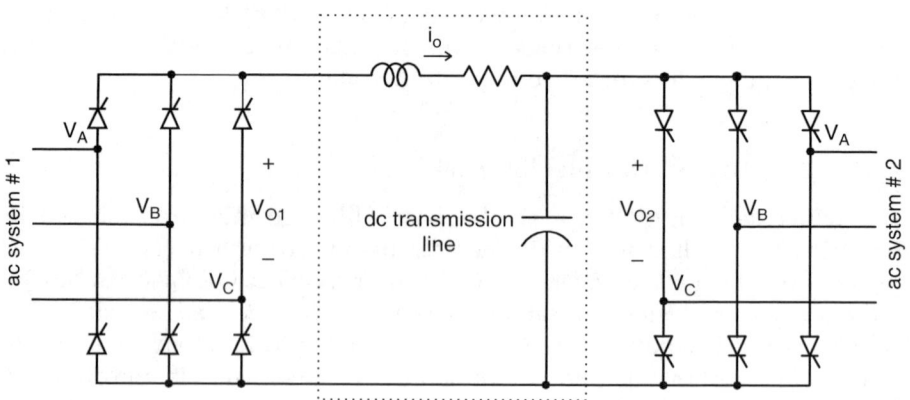

FIGURE 7.30
A six-pulse unipolar DC-transmission system.

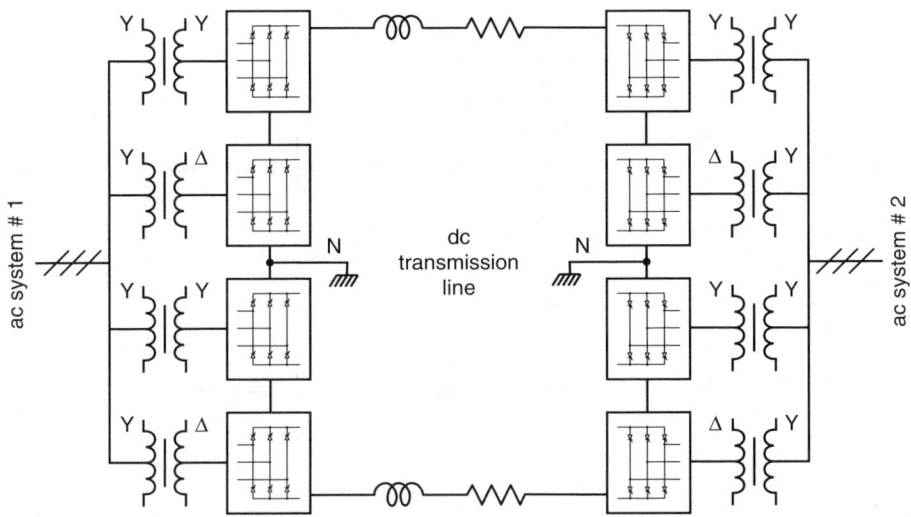

FIGURE 7.31
A 12-pulse bipolar DC-transmission system.

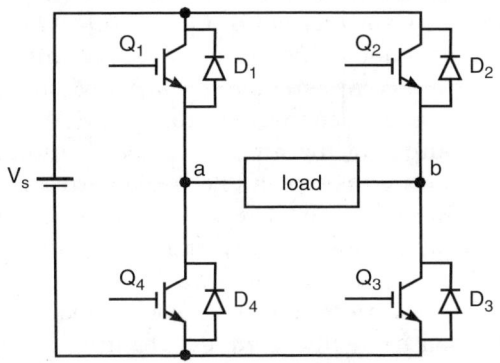

FIGURE 7.32
H-bridge inverter.

time delay is provided between the turning off of a pair of switches and the turning on of the other pair. This period, called the blanking period, is provided to prevent the DC source from being short-circuited. Consider for example the transition when Q3 and Q4 are turned off and Q1 and Q2 are turned on. During this period if Q1 gets turned on before Q4 turns off completely, then it will connect the two leads of the DC source directly. It is therefore mandatory that Q4 turn off completely before Q1 is turned on. To ensure this, a blanking period (dead time) is used. The continuity of the current during the blanking period is maintained by the antiparallel diodes. Although the blanking period is very important in practice, for evaluation of operation it is usually neglected.

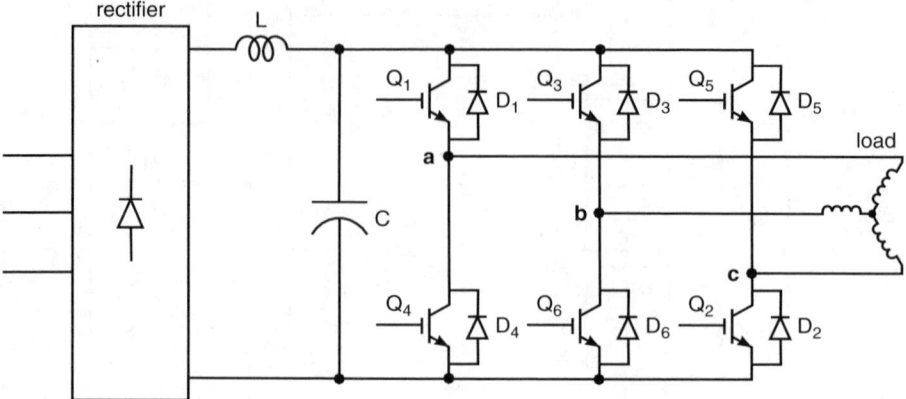

FIGURE 7.33
Three-phase voltage source inverter.

7.6.2   Three-Phase Inverter

The schematic diagram of a three-phase voltage source inverter is shown in Figure 7.33. It contains six switches Q1, Q2, Q3, Q4, Q5, and Q6; each one is made up of a power semiconductor device and an antiparallel diode. The switches of each leg are complementary: that is, when Q1 is on, Q4 is off and vice versa. The inverter is connected to a DC link and a three-phase load is connected at the output of the inverter. For a floating neutral point the phase currents will add up to zero and no triple harmonic current will flow in the load. Each phase leg must be driven with dead-time embedded signals to avoid the shoot-through fault. The inverter can be controlled by hysteresis pulse-width modulation with sinusoidal tracking or by space-vector modulation. Double PWM converters connected back-to-back have been used for AC-AC conversion with an intermediate DC-link voltage that must be boosted to a higher voltage level capable of imposing linear operation to both converters. Double PWM converters (shown in Figure 7.34) have been used for bidirectional power controllers and in applications dominated by electronic transformers, that is, AC-AC transformation without magnetic coupling.

7.6.3   Multi-Step Inverter

Multi-step inverters have been around for several years for GTO applications. They are constructed using many six-step inverters and a magnetic circuit to provide phase shift between the inverters. For example, by providing a phase shift of 30° between two inverters with a transformer connection, it is possible to generate a 12-step waveform with a spectrum that contains

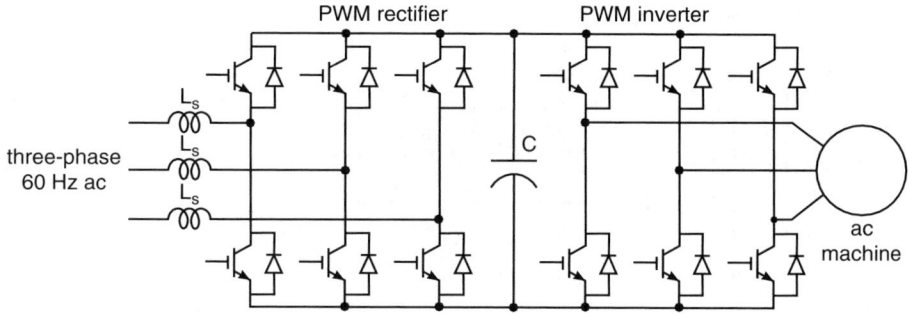

FIGURE 7.34
Double-PWM converter.

eleventh and higher harmonics only. By combining two of such inverters a 24-step inverter is achieved. With a series–parallel operation, the devices require matching, and some amount of voltage or current derating of the devices is essential. A possible solution for a higher power rating is a parallel connection of three-phase inverters through center-tapped reactors at the output. For large power applications, it is desirable that the inverter output wave be multi-stepped (approaching a sine wave) because the filter size can be reduced on both the DC and AC, sides. The lower order harmonics of a six-stepped wave (the fifth and seventh) can be neutralized by synthesizing a 12-stepped waveform as depicted in Figure 7.35. Multi-step inverters are used for bulk power transmission systems.

7.6.4 The Multi-Level Inverter

The inverter configuration shown in Figure 7.36 is basically that of a two-level inverter as the output voltage can take only two values $+V_{dc}/2$ and $-V_{dc}/2$. Multilevel inverters take more levels at the output through a special configuration of diodes and split voltages in the DC link. A very straight-forward and typical configuration is depicted in the three-level phase leg shown in Figure 7.37, capable of generating $+V_{dc}/2$, 0 and $-V_{dc}/2$. It contains four switches (S_{ua}, S_{ub}, $S_{\ell a}$ and $S_{\ell b}$), each consisting of a power device plus a freewheeling diode and two diodes to impose the zero level (D_1 and D_2). A three-phase inverter is made of three similar structures as in Figure 7.37. Table 7.3 shows the output voltage of a three-level inverter with respect to the switch command.

Multi-level inverters are constructed with capacitors splitting the DC-link voltage and with corresponding command signals for the devices that synthesize the required output voltages. Multi-level inverters demonstrated a potential for harmonic cancellation, but they have not been widely accepted due to their high-voltage wiring complexity and cost.

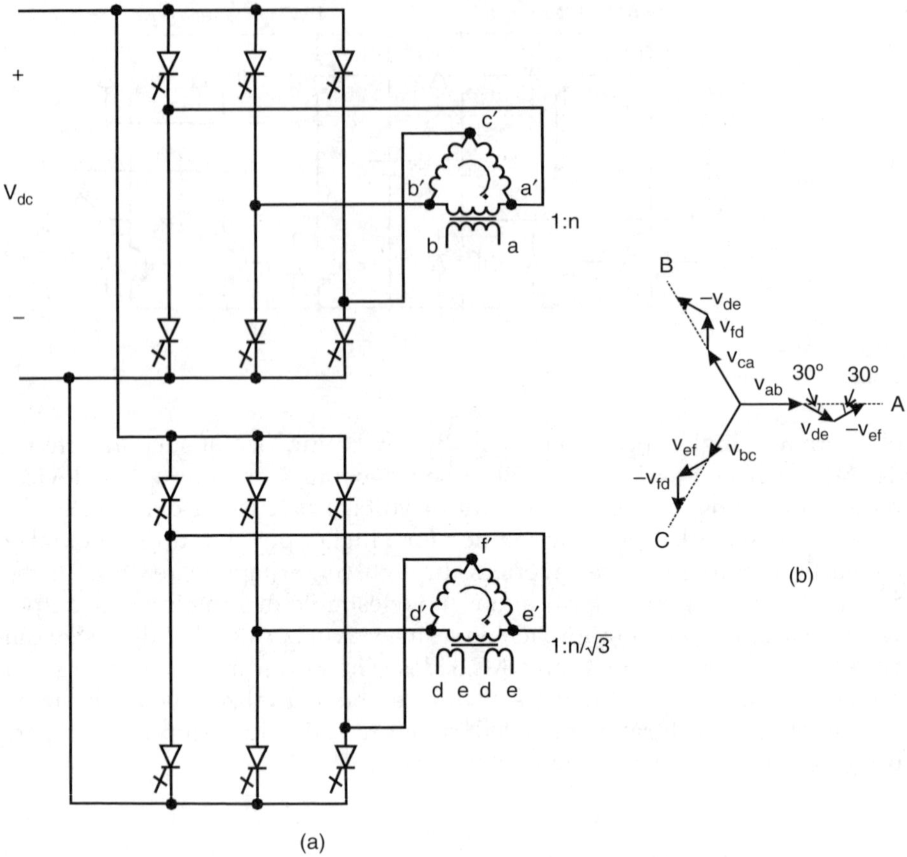

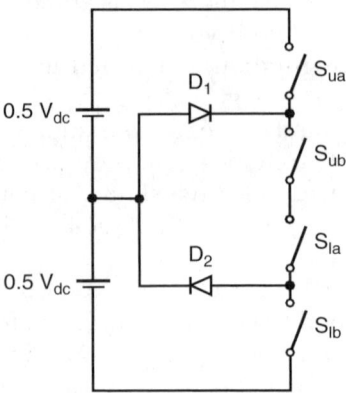

FIGURE 7.35
Example of a 12-step inverter: (a) circuit topology, and (b) synthesis of voltages.

FIGURE 7.36
Phase-leg circuit for a three-level inverter.

TABLE 7.3

Synthesis of Three-Level Voltage

Switch Command				Phase Voltage
S_{ua}	S_{ub}	$S_{\ell a}$	$S_{\ell b}$	V_{AN}
OFF	OFF	ON	ON	$-V_{dc}/2$
OFF	ON	ON	OFF	0
ON	ON	OFF	OFF	$+V_{dc}/2$

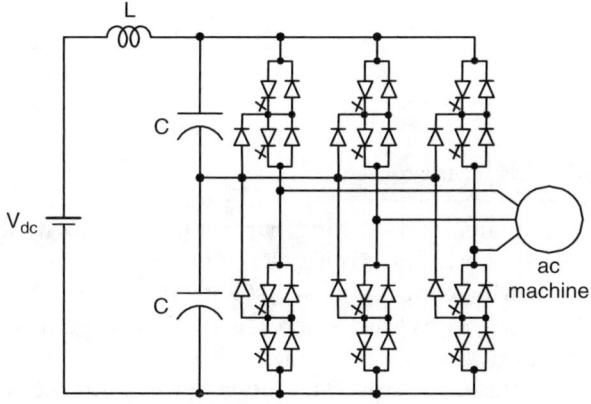

FIGURE 7.37
Three-level three-phase inverter.

7.7   AC to AC Conversion

Direct AC-AC conversion was a strong solution using a thyristor for phase-controlled regulation and a phase-controlled cycloconverter during the 1960s. With the fall of power semiconductor prices, DC-link cascaded converters became a reality during the 1990s. Recently, the high frequency matrix converter for direct conversion has been receiving worldwide attention. The matrix converter offers an all-silicon solution for AC-AC power conversion, removing all need for the reactive energy storage components used in conventional inverter-based converters like capacitors and inductors. An n-phase-to-m-phase matrix converter consists of an array of n*m bidirectional switches as shown in Figure 7.38. The power circuit is arranged in such a fashion that any of the output lines of the converter can be connected to any of the input lines. Thus the voltage at any input terminal can be made to appear at any output terminal or terminals while the current in any phase of load can be drawn from any input phase or phases. A line filter is included

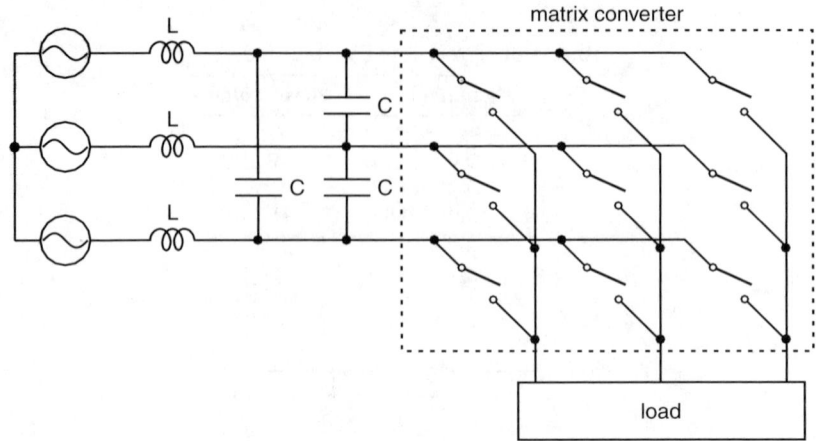

FIGURE 7.38
Three-phase to three-phase matrix converter.

to circulate high-frequency switching harmonics. The switches are modulated to generate the desired output-voltage waveform.

The matrix converter concept was first introduced in 1976; it was then considered to be a cycloconverter where the devices were fully controllable. Hence, the matrix converter is sometimes called a forced commutated cyclo-converter. A lot of research interest followed the publication of the Venturini's first matrix converter paper in 1980, which put the matrix converter control algorithm on a strong mathematical foundation. The advantages of this converter are given below.

- No DC-link capacitor or inductor. Therefore, noise, EMI, size, and weight are greatly reduced.
- Less maintenance, more durable.
- Inherently bidirectional, so it can regenerate energy back to the utility.
- High efficiency as the number of devices connected in series is less.
- Four quadrants of operation.
- Depending on modulation technique, sinusoidal input/output waveforms.
- Controllable input displacement factor independent of output load current.
- No electrolytic capacitors, hence can be used in high temperature surroundings. An ideal topology to utilize future technologies such as high temperature silicon carbide (SiC) devices.

A matrix converter requires a bidirectional switch capable of blocking voltage and conducting current in both directions. Unfortunately there are

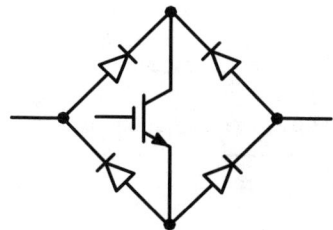

FIGURE 7.39
Diode bridge bidirectional switch.

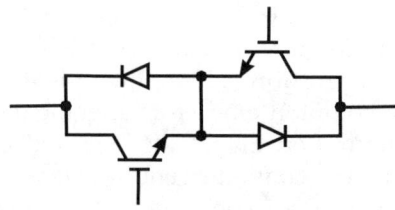

FIGURE 7.40
Common emitter back-to-back switch.

no such devices currently available to fulfill the need. So a combination of available discrete devices are generally used to get the bidirectional switches. There are some common arrangements for getting these types of switches; each has advantages and disadvantages.

7.7.1 Diode-Bridge Arrangement

The diode-bridge arrangement consists of an IGBT at the center of a single-phase diode bridge shown in Figure 7.39. The main advantage of this configuration is only one gate driver per commutation cell. The direction of current through the switch cannot be monitored, which is a disadvantage, as many highly reliable commutation methods described in the literature cannot be used.

7.7.2 Common-Emitter Antiparallel IGBT Diode Pair

The common-emitter antiparallel IGBT diode-pair arrangement is shown in Figure 7.40. It consists of two diodes and two IGBTs connected in antiparallel. Here it is possible to control the direction of the current in each of the devices independently. Each of the switch cells needs an isolated power supply for the gate drives.

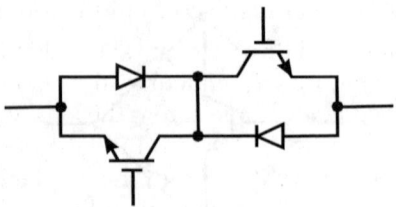

FIGURE 7.41
Common collector back-to-back switch.

7.7.3   Common-Collector Antiparallel IGBT Diode Pair

The common-collector antiparallel IGBT diode pair, shown in Figure 7.41, is similar to the previous structure, but here the IGBTs are arranged in the common-collector configuration. Here fewer isolated gate drives are required than with the common emitter arrangement. But practical implementation of this configuration makes the common emitter configuration the preferred one. The matrix converter requires very intensive digital signal computation to enable the correct connection of the power devices and very specialized modulation techniques that are beyond the scope of this book. It is expected that future systems will pervasively incorporate matrix converters for controlling AC machines and AC systems.

7.8   Problems

To do the problems below, you will need to consult power electronics books available in your library. This research will complement what you have learned in this chapter.

7.1 Research manufacturers' datasheets and prepare a table comparing the devices SCR, GTO, IGCT, TRIAC, Power-MOSFET, IGBT, and MCT with respect to the following parameters: maximum voltage and current ratings, symmetric or asymmetric voltage blocking, voltage or current gating, junction temperature range, conduction drop, drop sensitivity with temperature, safe operating area (SOA), switching frequency, turn-off gain, reapplied dv/dt, turn-on time, turn-off time, leakage current, snubber/protection needs, and main application.

7.2 Discuss methods of cooling power converters. What do you understand about thermal resistance of a heat sink?

7.3 Induction generators need DC-AC, AC-DC and AC-AC converters. Describe one application for each conversion.

7.4 Where are DC-DC converters used in induction generator systems?

7.5 Assume $L_s = 0$ and an HIL (high inductive load) for the half-converter indicated in Figure 7.27. Calculate the relationship of the delay angle α with the average voltage across the load V_{dc}. Draw the output voltage waveform across the load and identify the devices that conduct during various intervals. Obtain the displacement factor (DPF) and the power factor (PF) in the input-line current and compare results with a full-bridge converter operating at $V_{dc} = 3\sqrt{2}/2p$ $V_{(line,RMS)}$.

7.6 Research pulse-width modulation technology and write a summary report.

7.7 Describe the features and characteristics of multi-level converters and multi-step converters.

7.8 Matrix converters require specialized modulation techniques. Conduct research and write a summary report.

7.9 A full-bridge three-phase rectifier supplies energy to a DC load of 300 V and 60 A from a three-phase bridge of 440 V through a ΔY transformer. Select a diode and specify the transformer for a voltage drop across each diode of 0.7 V and continuous current.

7.10 In a conventional thyristor 12-pulse converter, estimate the internal drop of voltage, regulation, and power factor for the following data set.

Source	Load
$V = 220$ V	$V_{dc} = 48$ V
$f = 60$ Hz	$\alpha = 20°$
$L = 0.1$ mH	$R_{load} = 20\ \Omega$
$R = 0.1\ \Omega$	$L_{load} = 0.2$ mH

7.11 A single-phase diode bridge is supplied by an AC source of 257 V, 60 Hz, which in turn supplies a DC load of $I_{dc} = 60$ A. Estimate the voltage drops due to (a) source with an inductance of 0.02 mH, (b) each diode with forward voltage drop of $\Delta V_f = 0.6 + 0.0015(I_{dc})$, and (c) source and wiring resistances of $R_r = 0.0015\ \Omega$. Draw an equivalent circuit to represent such rectifier.

References

1. Mohan, N., Undeland, T.M., and Robbins, W.P., *Power Electronics: Converters, Applications, and Design*, 3rd edition, John Wiley & Sons, New York, 2002.
2. Erickson, R.W. and Maksimovic, D., *Fundamentals of Power Electronics*, Kluwer Academic Publishers, Dordrecht, the Netherlands, 2001.
3. Rashid, M. H., *Power Electronics: Circuits, Devices, and Applications*, Prentice Hall, New York, 2003.

4. Kimbark, E.W., *Direct Current Transmission, Volume 1*, Wiley Interscience, New York, 1971.

5. Bose, B.K., *Modern Power Electronics and AC Drives*, Prentice Hall, New York, 2001.

6. Ghosh, A. and Ledwich, G., *Power Quality Enhancement Using Custom Power Devices*, Kluwer Academic Publishers, Dordrecht, the Netherlands, 2002.

7. Chattopadhyay, A., AC-AC converters, in *Power Electronics Handbook*, Rashid, M.H., Ed., 2001, pp. 327–331.

8. Youm, J.H. and Kwon, B. H., Switching technique for current-controlled AC-to-AC converters, *IEEE Trans. On Industrial Electronics*, 46(2), 309–318, April 1999.

8

Scalar Control for Induction Generators

8.1 Scope of This Chapter

Scalar control of induction motors and generators means control of the magnitude of voltage and frequency so as to achieve suitable torque and speed with an impressed slip. Scalar control can be easily understood based on the fundamental principles of induction-machine steady-state modeling. A power electronic system is used either for a series connection of inverters and converters between the induction generator and the grid or for a parallel-path system capable of providing reactive power for isolated operation. This chapter describes such principles, laying out a foundation for understanding the more complex vector controlled systems.

8.2 Scalar Control Background

Scalar control disregards the coupling effect on the generator; that is, the voltage will be set to control the flux and the frequency in order to control the torque.[1] However, flux and torque are also functions of frequency and voltage, respectively. Scalar control is different from vector control in that both magnitude and phase alignment of the vector variables are controlled. Scalar control drives give somewhat inferior performance, but they are easy to implement and are very popular for pumping and industrial applications. The importance of scalar control has diminished recently because of the superior performance of the vector-controlled drives and the introduction of high-performance inverters. High-performance inverters offer prices competitive to scalar-control-based inverters (volts/Hertz inverters), although they are cheap and widely available.

The principles of scalar control will be presented in this chapter along with a discussion of some simple enhancements that may be retrofit onto existing commercial volts/hertz drives.

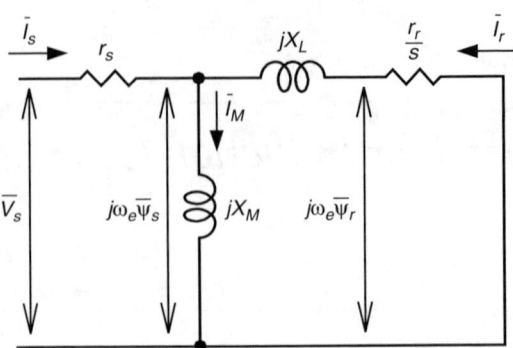

FIGURE 8.1
Γ-stator equivalent model.

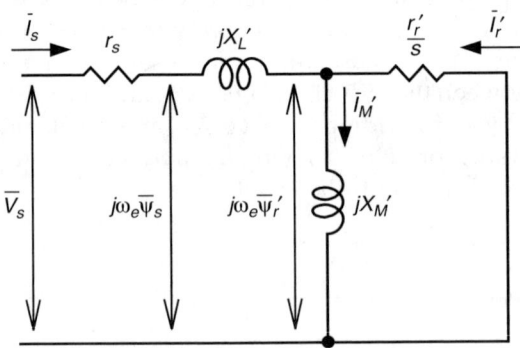

FIGURE 8.2
Γ-rotor equivalent model.

The main constraint on the use of a scalar control method for induction motors and generators is related to the transient response.[2] If shaft torque and speed are bandwidth-limited and torque varies slowly, tracking required speed variations (within hundreds of milliseconds up to the order of almost a second), scalar control may work appropriately. Hydropower and wind power applications have slower mechanical dynamics than that. Therefore, it seems that scalar control is still a good approach for renewable energy applications.

As a background to understanding the scalar control method, a steady-state two-inductance per-phase equivalent circuit for an induction motor can be used.[3] There are two models: the Γ-stator equivalent model, presented in Figure 8.1, and the Γ-rotor equivalent model, presented in Figure 8.2. The Γ-stator equivalent model is related to the ordinary per-phase equivalent model by the transformation coefficient $\gamma = X_s/X_m$, as shown in Table 8.1. The Γ-rotor equivalent model is related to the ordinary per-phase equivalent model by the transformation coefficient $\rho = X_m/X_r$, as in Table 8.2.

TABLE 8.1

Γ-Stator Equivalent Parameters

Rotor resistance referred to stator	$R_{r,\gamma} = \gamma^2 R_r$
Magnetizing reactance	$X_{m,\gamma} = \gamma X_m = X_s$
Total leakage reactance	$X_{l,\gamma} = \gamma X_{ls} + \gamma^2 X_{lr}$
Rotor current referred to stator	$I_{r,\gamma} = \dfrac{I_r}{\gamma}$
Rotor flux	$\psi_{r,\gamma} = \gamma \psi_r$
Angular frequency of rotor currents (rotor frequency)	$\omega_r = s\omega_e$
Rotor angular frequency related to the slip frequency	$\omega_r = \dfrac{P}{2}\omega_{sl}$

TABLE 8.2

Γ-Rotor Equivalent Parameters

Rotor resistance referred to stator	$R_{r,\rho} = \rho^2 R_r$
Magnetizing reactance	$X_{m,\rho} = \rho X_m$
Total leakage reactance	$X_{l,\rho} = X_{ls} + \rho X_{lr}$
Rotor current referred to stator	$I_{r,\rho} = \dfrac{I_r}{\rho}$
Rotor flux	$\psi_{r,\rho} = \rho \psi_r$
Angular frequency of rotor currents (rotor frequency)	$\omega_r = s\omega_e$

In the Γ-stator equivalent model the stator flux (ψ_s) is considered to be approximated by the air-gap flux (ψ_m), while the stator resistance (r_s) effects are usually neglected. Considering that $R_{r,\gamma}/s = R_{r,\gamma}\,\omega_e/\omega_r$, the rotor current in the Γ-stator equivalent model can be expressed as

$$\left| I_{r,\gamma} \right| = \frac{\psi_m}{R_{r,\gamma}} \frac{\omega_r}{\sqrt{\left(\tau_r \omega_r\right)^2 + 1}} \tag{8.1}$$

where $\tau_r = L_{l,\gamma}/R_{r,\gamma}$, $L_{l,\gamma}$ is the total leakage inductance in this model; that is, $L_{l,\gamma} = X_{l,\gamma}/\omega$.

Neglecting the resistive losses in the rotor ($3R_{r,\gamma}|I_{r,\gamma}|^2$), the electrical power transferred from the mechanical shaft to the rotor circuit is approximated by

$$P_{elec} = 3R_{r,\gamma}\frac{\omega}{\omega_r}\left| I_{r,\gamma} \right|^2 \approx P_{mech} \tag{8.2}$$

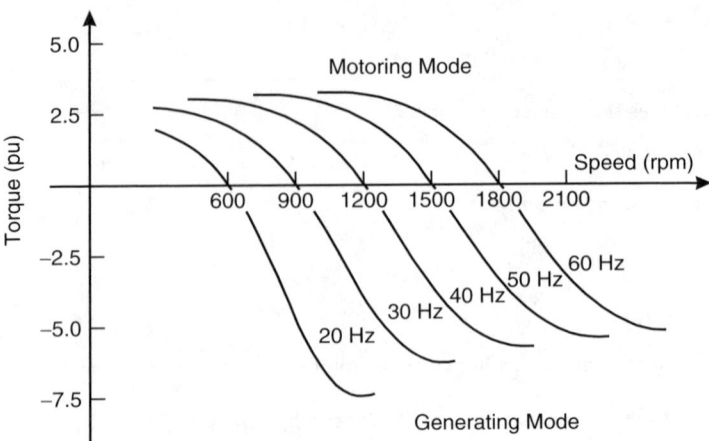

FIGURE 8.3
Torque/speed performance for constant V/Hz.

The generator torque shaft can be calculated as the ratio of P_{mech} to the mechanical angular speed. For scalar voltage control, the angular shaft speed is leading the stator angular speed by $\omega_m = 2/P\,\omega_e\,(1 + s)$ and the developed electrical torque on the shaft is

$$T_e = 3\frac{P}{2}\frac{\psi_m^2}{R_{r,\gamma}}\frac{\omega_r}{\left(\tau_r\omega_r\right)^2 + 1} \tag{8.3}$$

For $\omega_r = 1/\tau_r$ the critical slip is $s_{critical} = 1/\tau_r\omega_e$ and the maximum torque developed is given by

$$T_{M,max} = 3\frac{P}{2}\frac{\psi_m^2}{\left(L_{1,\gamma}\right)}.$$

The machine will operate above (generator) and below (motor) the synchronous frequency (ω_s) as in Figure 8.3 with its operating slip well below the critical slip. Therefore, the denominator of Equation 8.3 is close to unity and the torque is virtually proportional to the rotor angular frequency for constant flux. Equation 8.4 is derived from Equation 8.3 and transformed back to the standard machine T model, considering that flux is $\psi_m = V_s/\omega_e$.

$$T_e = 3\frac{P}{2}\frac{1}{r_r}\psi_m^2\omega_{sl} \tag{8.4}$$

Equation 8.4 indicates that for a fixed V/Hz ratio one characteristic curve is locked, and the torque crossover operating point changes in accordance with the impressed stator frequency as shown in Figure 8.3. Therefore, depending on the shaft speed, one point along the torque-speed curve will be developed. For negative slip the machine will develop negative torque, and the operation will be as a generator. If variable speed is required, the V/Hz ratio must be maintained to keep rated air-gap flux. The firmness of air-gap flux results in good utilization of the motor iron, a high torque per stator ampere, and rated torque is possible for a wide speed range. It is important to note that in the previous discussion we neglected stator resistance and the saturation effect, both playing a very important role for a realistic quantitative evaluation. For motoring operation, the stator resistance drop is opposed to the impressed voltage, and a boost scheme is required for low-frequency operation. For generating mode, the stator voltage is reversed, resulting in an increased back-emf with increased air-gap flux. In that case the torque initially increases with speed reduction, but saturation of the magnetizing inductance offsets such an increase.

8.3 Scalar Control Schemes

The fundamental objective behind the scalar method is to provide a controlled slip operation. The scheme depicted in Figure 8.4 with a grid-connected induction generator is a possible one. The static frequency converter in the figure can be a cycloconverter, a matrix converter (bidirectional in nature), or an inverter/inverter connection with a DC-link interfacing the 60 Hz grid with the generator stator. The simplified scheme shown in Figure 8.4 requires a programmed slip, which will be dependent on machine parameter variation, temperature variation, and mechanical losses. Therefore, a closed-loop control will improve the drive performance. Since the slip frequency (ω_{sl}) is proportional to torque, an outer speed control loop will generate a signal proportional to the required slip — for example, through a PI regulator.

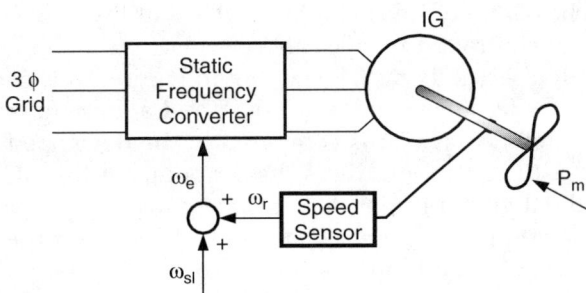

FIGURE 8.4
Principle of slip control for induction generators.

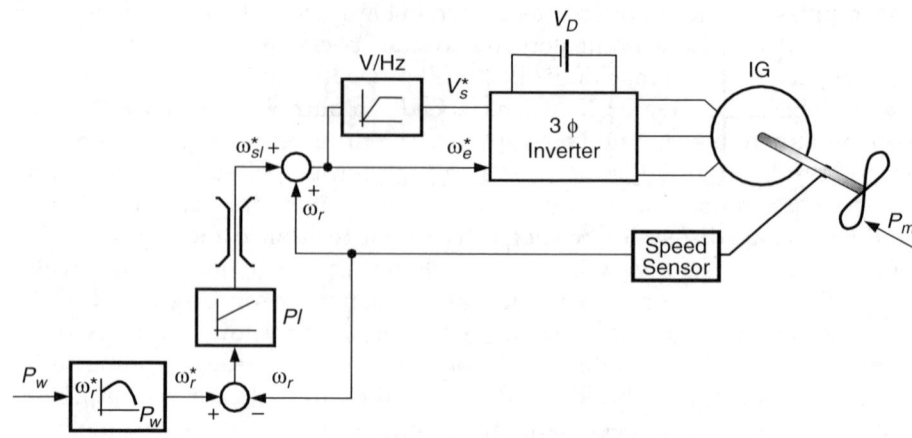

FIGURE 8.5
Closed-loop speed control of IG with volts/Hz.

The system portrayed in Figure 8.5 has an outer-loop speed control and a PI regulator that generates slip, which is added to the shaft speed to generate the stator frequency. The machine terminal voltage is also programmed through a look-up table. The converter receives both inputs ω_e^* and V_s^*, which command a three-phase sinusoidal generator by pulse-width modulation in the inverter. The scheme considers that external torque is applied on the generator shaft and the shaft speed reference ω_r^* is computed in order to seek a shaft operating speed that keeps the slip signal negative ω_{sl}^*, maintaining the machine in generating mode. The three-phase inverter receives a negative phase sequence command, and the power delivered across the battery is indicated in the block diagram. Since most systems are grid connected, a DC-link is required to interface the machine converter to a grid inverter and pump energy back to the AC side, as indicated in Figure 8.6.

A current-fed link system for grid connection with V/Hz is depicted in Figure 8.6. The DC link current allows easy bidirectional flow of power. Although the DC link current is unidirectional, a power reversal is achieved by a change in polarity of the mean DC link voltage and symmetrical voltage blocking switches are required. The three-phase utility side is managed by a thyristor-based controlled rectifier and the machine-side inverter can use a transistor with a series diode. The system in Figure 8.6 is commanded by the machine stator frequency reference (ω_e^*), and a look-up table for V/Hz sets a voltage reference, which is compared to the developed machine terminal voltage. A PI control produces the set point for the DC-link current, and firing for the thyristor bridge controls the power exchange (P_o) with the grid. The stator frequency reference ω_e^* can be varied in order to optimize the power tracking of the induction generator or may be programmed in accordance with the input power availability at the generator shaft (P_M).

An enhanced voltage-link double-PWM converter is depicted in Figure 8.7. The DC-link capacitor voltage is kept constant by the converter connected

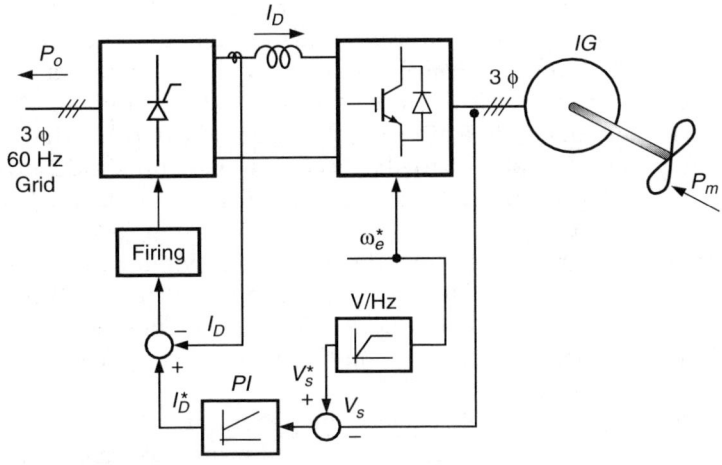

FIGURE 8.6
Current-fed link system for grid connection with V/Hz.

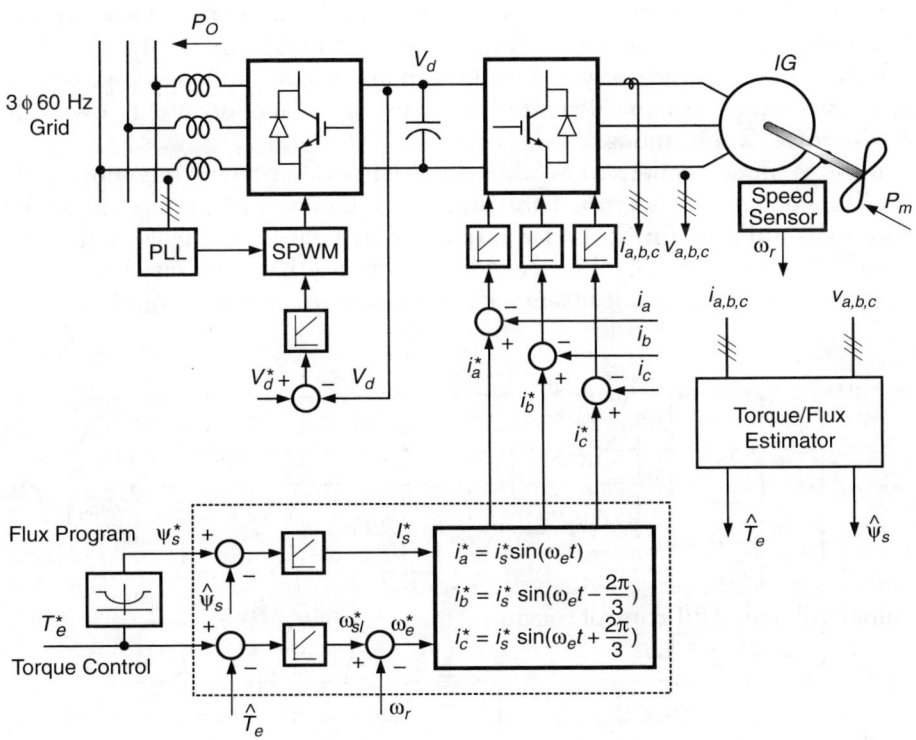

FIGURE 8.7
Voltage-fed link system for grid connection with current control for generator side.

to the utility grid. The series inductances that connect the system to the grid keep the converter control within safe limits by inserting a current-source-like feature. When power is transferred from the induction generator, the DC link voltage will increase slightly and the feedback control of the grid-side converter will generate a sinusoidal pulse-width modulation (SPWM) in order to pump this power to the grid. The generator-side inverter is current-controlled with either PI controllers on the stationary frame (with SPWM) or with hysteresis-band controllers. Three-phase reference currents are generated through a programmed sine wave generator that receives the stator electrical angular speed reference (ω_e^*) and the current peak amplitude ($\hat{I}_s^*$) supplied by corresponding torque and flux loops. Those loops are closed with estimation of actual torque flux in the generators by feeding back the generator current and voltage.

In order to optimize the efficiency of the generator, a look-up table reads the torque command and programs the optimum flux reference for the system. A start-up sequence initially boosts the DC-link voltage to a higher voltage than the peak value of the grid, in order for the pulse-width modulation to work properly. The overall system is very robust due to the current control in the inverters, the DC-link capacitor (with lower losses and faster response than DC-link inductors), the on-line estimation of generator torque and flux. The system may be implemented with the last generation of microcontrollers since internal computations are not so mathematically intensive. All these scalar-based control schemes can also incorporate speed governor systems on the mechanical shaft in order to control the incoming power for hydropower applications.

If the induction generator is primarily used in stand-alone operation, reactive power must be supplied for proper excitation, and the use of power electronics allows for several approaches such as variable impedance,[4,5] series and shunt regulators. Recently, p–q theory has been used for these applications.[6] The overall scheme is based on the diagram of Figure 8.8. The

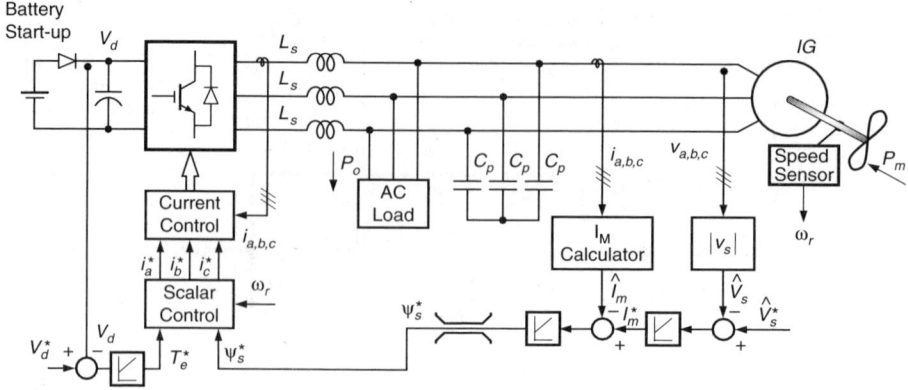

FIGURE 8.8
Static VAR compensator scalar-based control with voltage regulation.

capacitor bank is a bulk uncontrolled source of reactive current. The static VAR compensator is an inverter providing a controlled source of reactive current. The inverter uses a scalar control approach, as previously discussed; the dashed subsystem of Figure 8.7 constitutes the block called "scalar control" in Figure 8.8.

Machine torque and flux input will be used in this application to regulate both DC-link voltage at the capacitance C_p and generator voltage supplied for the AC load. These regulators have to reject the disturbances produced by load and speed variations. A DC-link voltage regulator has been implemented to achieve a high-enough DC-link voltage for proper current-controlled inverter operation. The regulator input is the difference between the DC link voltage reference and the measured value.

At the generator-side terminal current and voltages are measured to calculate the magnetizing current needed for the generator, and the instantaneous peak voltage is compared to a stator voltage reference (V_s^*), which generates a set-point for flux through the feedback loops on the inverter side. This system requires a charged battery for start-up. That can be recharged with an auxiliary circuit (not shown) after the system is operational. Since stator voltage is kept constant, the frequency can be stabilized at about 60 Hz for the AC load, but some slight frequency variation is still observed, and the range of turbine shaft variation should be within the critical slip in order to avoid instability. Therefore, AC loads should not be too sensitive for frequency variation in this stand-alone application.

8.4 Problems

8.1 This chapter showed that there is a critical slip value where machine torque peaks. Discuss why scalar control is in danger of instability due to a particular characteristic and what measures should be taken to avoid unstable operation.

8.2 The current-fed converter scheme explained in Figure 8.6 has a thyristor-based controlled rectifier interfacing with the electrical grid. Describe why this scheme does not operate at unity power factor on the grid side and how the scheme can be improved in this respect.

8.3 Figure 8.3 shows an interesting feature that generating torque tends to increase with decreasing stator frequency. Describe why this trend is desirable for renewable energy conversion systems and what limits the actual performance of induction machines operating at very low speed.

8.4 The torque equation (8.4) shows that for a fixed air-gap flux machine torque is directly proportional to slip frequency. Variable rotor resistance due to temperature suggests that a derating factor could be applied for correction. Discuss the real implications of parameter variation in the scalar control method.

8.5 Simulate one of the schemes with system simulator software.

References

1. Boys, J.T. and Walton, S.J., Scalar control: an alternative AC drive philosophy, *IEEE Proc. B, Electric Power Applications*, 135(3), 151–158, May 1988.
2. Murphy, J.M.D. and Turnbull, F.G., *Power Electronic Control of AC Motors*, Oxford University Press, New York, 1988.
3. Trzynadlowski, A.M., *Control of Induction Motors*, Academic Press, San Diego, 2000.
4. Brennen, M.B. and Abbondati, A., Static exciters for induction generators, *IEEE Trans. on Industry Applications*, Vol. IA–13, 422–428, Sept./Oct. 1977.
5. Bonert, R. and Hoops, G., Stand alone induction generator with terminal impedance controller and no turbine controls, *IEEE Trans. Energy Conversion*, 5(1), 28–31, 1990.
6. Leidhold, R., Garcia, G., and Valla, M.I., Induction generator controller based on the instantaneous reactive power theory, *IEEE Transactions on Energy Conversion*, 17(3), 368–373, Sept. 2002.

9

Vector Control for Induction Generators

9.1 Scope of This Chapter

Vector control is the foundation of modern high-performance drives. It is also known as decoupling, orthogonal, or transvector control. Vector control techniques can be classified according to their method of finding the instantaneous position of the machine flux, which can be either (1) the indirect or feedforward method or (2) the direct or feedback method. This chapter will discuss field orientation — the calculation of stator current components for decoupling of torque and flux — while laying out the principles for rotor- and stator-flux-oriented control approaches for induction generator systems. A discussion of parameter sensitivity, detuning, sensorless systems, PWM generation, and signal processing issues can be found in the available technical literature.

9.2 Vector Control for Induction Generators

The induction machine has several advantages: size, weight, rotor inertia, maximum speed capability, efficiency, and cost. However, induction machines are difficult to control because being polyphase electromagnetic systems they have a nonlinear and highly interactive multivariable control structure. The technique of field-oriented or vector control permits an algebraic transformation that converts the dynamic structure of the AC machine into that of a separately excited decoupled control structure with independent control of flux and torque.

The principles of field orientation originated in West Germany in the work of Hasse[1] and Blaschke[2] at the technical universities of Darmstadt and Braunschweig and in the laboratories of Siemens AG. A variety of other derived implementation methods are now available, but two major classes are very important and will be discussed in this chapter: direct control and indirect control. Both approaches calculate the on-line instantaneous magnitude and position of either the stator or the rotor flux vector by imposing alignment of the machine current vector so as to obtain decoupled control. Therefore, a very broad and acceptable classification of vector control is based on the procedure used to determine the flux vector.

Although Hasse proposed electromagnetic flux sensors, currently direct vector control is based on the estimation through state observers of the flux position. The technique proposed by Blaschke is still very popular and depends on feedforward calculation of the flux position with a crucial dependence on the rotor electrical time constant, which may vary with temperature, frequency, and saturation, as will be discussed in this chapter. The principles of vector control can be employed for control of instantaneous active and reactive power. Therefore, it is probably fair to say that the modern p – q instantaneous power theory has also evolved from the vector control theory.[3] If a totally symmetrical and three-phase balanced induction machine is used, a d – q transformation is used to decouple time-variant parameters, helping to lay out the principles of vector control.

9.3   Axis Transformation

The following discussion draws on voltage transformation, but other variables like flux and current are also similarly transformed. Figure 9.1 shows a physical distribution of the three stationary axes a^s, b^s and c^s, 120° apart from each other, symbolizing concentrated stator windings of an induction machine. Cartesian axes are also portrayed in Figure 9.1 where q^s is a horizontal axis aligned with phase a^s, and a vertical axis rotated by –90° is indicated in the figure by d^s. Three-phase voltages varying in time along the axes a^s, b^s and c^s can be algebraically transformed onto two-phase voltages varying in time along the axes d^s and q^s.

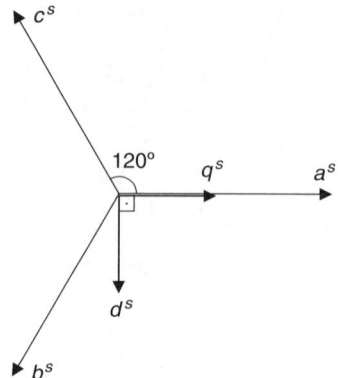

FIGURE 9.1
Physical distribution of three stationary axes and a quadrature stationary axis.

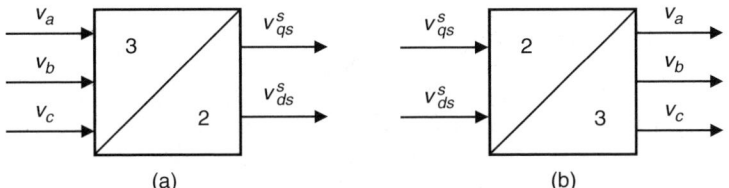

FIGURE 9.2
Simplified block diagrams for stationary frame conversion: (a) three-phase to two-phase conversion and (b) two-phase to three-phase conversion.

This transformation is indicated by Equation 9.1. The inverse of such a transformation can be taken by assuming that the zero sequence v_{os}^s is equal to zero for a balanced system and Equation 9.2 is valid for the inverse transformation. Equation 9.1 takes a two-phase quadrature system and transforms it to three-phase, while Equation 9.2 takes a three-phase system and converts it to a two-phase quadrature system. Figure 9.2 shows a simplified block diagrams representing transformations of this kind commonly used for vector-control block diagrams.

$$
\begin{bmatrix} v_{as} \\ v_{bs} \\ v_{cs} \end{bmatrix} =
\begin{bmatrix} 1 & 0 & 1 \\ -\dfrac{1}{2} & -\dfrac{\sqrt{3}}{2} & 1 \\ -\dfrac{1}{2} & \dfrac{\sqrt{3}}{2} & 1 \end{bmatrix}
\begin{bmatrix} v_{qs}^s \\ v_{ds}^s \\ v_{os}^s \end{bmatrix}
\tag{9.1}
$$

$$
\begin{bmatrix} v_{qs}^s \\ v_{ds}^s \\ v_{0s}^s \end{bmatrix} =
\begin{bmatrix} 1 & 0 & 0 \\ 0 & -\dfrac{\sqrt{3}}{3} & \dfrac{\sqrt{3}}{3} \\ \dfrac{1}{2} & \dfrac{1}{2} & \dfrac{1}{2} \end{bmatrix}
\begin{bmatrix} v_{as} \\ v_{bs} \\ v_{cs} \end{bmatrix}
\tag{9.2}
$$

In addition to the stationary $d^s - q^s$ frame, an induction machine has two other frames depicted in Figure 9.3: one aligned with one internal flux (stator, air-gap, or rotor) denoted by $d^e - q^e$ synchronous rotating reference frame and the other aligned with a hypothetical shaft rotating at the electrical speed dependent on the number of poles, denoted by $d^r q^r$ rotor frame. Machine instantaneous variables (voltages, currents, fluxes) can be transformed from the stationary frame $d^s q^s$ to the synchronous rotating reference frame $d^e - q^e$ through Equation 9.3. For convenience of notation the variables referred to the synchronous reference frame do not take the index e, that is, the voltage along the axis q^e is v_{qs} instead of v_{qs}^e. Equation 9.3 takes a two-phase quadrature system along the stationary frame and transforms it onto a two-phase

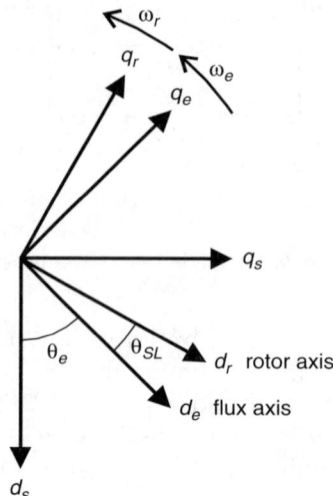

FIGURE 9.3
Stationary, synchronous, and rotor frames.

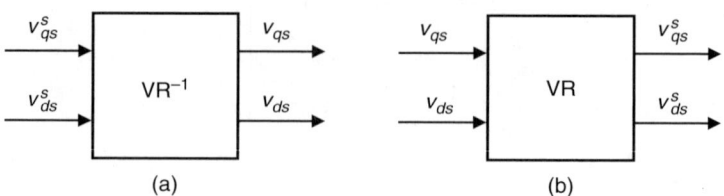

(a)   (b)

FIGURE 9.4
Block diagrams for vector-rotation conversion, (a) from stationary to synchronous frame and (b) from synchronous to stationary frame.

reference frame, while Equation 9.4 takes a two-phase reference frame and converts it back to the stationary frame. The angle θ_e is considered the instantaneous angular position of the flux where the reference frame will be aligned. Therefore, sinusoidal time variations along the stationary frame $d^s - q^s$ will appear as constant values onto the synchronous rotating reference frame $d^e - q^e$. Those operations are also called vector rotation (VR) and inverse vector rotation (VR^{-1}). Figure 9.4 shows simplified block diagrams representing transformations of this kind usually used for vector control block diagrams.

$$\begin{bmatrix} v_{qs} \\ v_{ds} \end{bmatrix} = \begin{bmatrix} \cos(\theta_e) & -\sin(\theta_e) \\ \sin(\theta_e) & \cos(\theta_e) \end{bmatrix} \begin{bmatrix} v_{qs}^s \\ v_{ds}^s \end{bmatrix} \tag{9.3}$$

$$\begin{bmatrix} v_{qs}^s \\ v_{ds}^s \end{bmatrix} = \begin{bmatrix} \cos(\theta_e) & \sin(\theta_e) \\ -\sin(\theta_e) & \cos(\theta_e) \end{bmatrix} \begin{bmatrix} v_{qs} \\ v_{ds} \end{bmatrix} \tag{9.4}$$

In an induction machine, the mutual inductances between stator and rotor windings vary with rotor position. The transformation from three-phase equations for stator and rotor to either reference frame eliminates this time variance. Thus a set of two equivalent circuits along the d-axis and q-axis can represent the dynamics of the three-phase machine. Figure 9.5 shows the d – q stationary model, that is, onto $d^s – q^s$ stationary frame. In this model, the variables are transformed between $a – b – c$ and $d^s – q^s$ through Equations 9.1 and 9.2, which are valid for both voltages and currents. In this equivalent d–q circuit, the steady-state response of voltages and currents leads to sinusoidal and cosinusoidal waveforms, which represent the machine three-phase 120° phase-shifted variables. Figure 9.6 depicts the d–q synchronous rotating model, that is, onto $d^e – q^e$ reference frame. In this model, the variables are transformed between $a – b – c$ and $d^e – q^e$ through Equations 9.1 to 9.4.

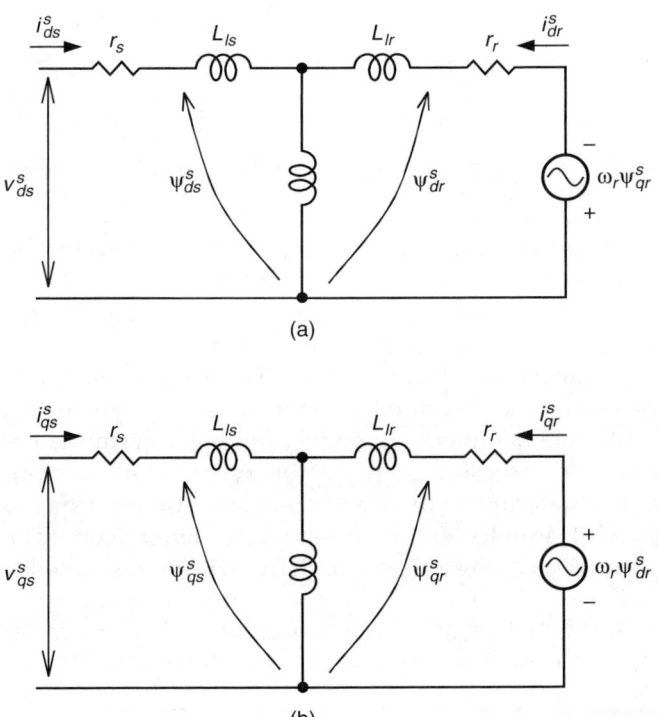

FIGURE 9.5
Induction machine model in stationary frame, (a) direct axis equivalent circuit and (b) quadrature axis equivalent circuit.

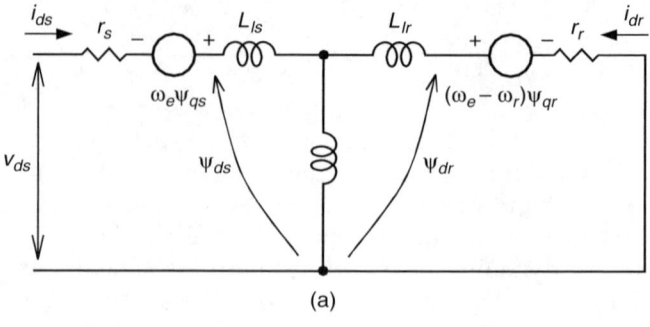

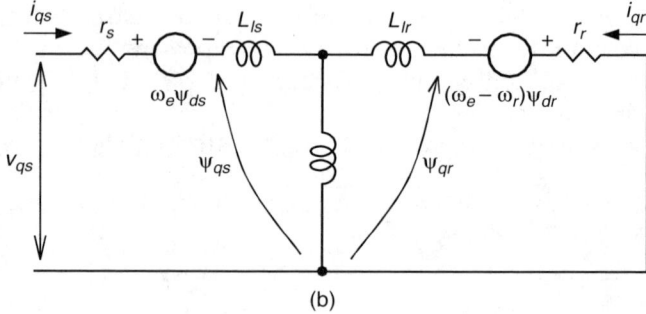

FIGURE 9.6
Induction machine model in synchronous rotating frame, (a) direct axis equivalent circuit and (b) quadrature axis equivalent circuit.

The instantaneous electrical angular speed ω_e is required for the additional terms indicated in Figure 9.6. The vector rotation (VR) and inverse-vector-rotation (VR^{-1}) transformations are taken after integration of the given angular speed ω_e signal. In this equivalent d – q circuit the steady-state response of voltages and currents leads to constant (DC) signals, which represent the machine three-phase 120°-shifted variables. Such a synchronous rotating reference frame equivalent circuit model runs faster in simulators due to the DC nature of the impressed signals. The synchronous rotating reference frame is easily converted to the stationary frame by imposing $\omega_e = 0$ in the equivalent circuit. In addition, such equivalent circuit concepts are fundamentals for the vector control approach, as will be discussed later in this chapter.

9.4   Space Vector Notation

A three-phase stator voltage excitation impressed on three-phase geometrically distributed windings results in three magnetomotive forces (mmf),

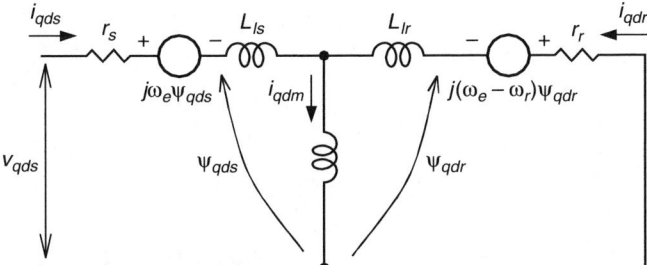

FIGURE 9.7
Space vector equivalent transient model of an induction machine, synchronous rotating reference frame equivalent.

which are added to the result in a traveling wave, that is, a magnetic field across the air gap rotating at synchronous speed. Such a spatial conception is implicit in traditional electrical machinery analysis. If we apply the same concept to voltages and currents, the resulting space vectors will be virtual rather than physical, but conveniently express a d – q model through complex variables. For example, the stator current space vector can be computed by Equation 9.5, where

$$\alpha = e^{j\frac{2\pi}{3}}.$$

$$\underline{i}_{qds} = \frac{2}{3}\left(i_a + \alpha i_b + \alpha^2 i_c\right) \tag{9.5}$$

The notation used in Equation 9.5 shows an underscore on the variable indicating a complex number. The index s indicates a stator quantity, and indexes d and q in the subscript show the merging of direct and quadrature representation in just one variable, the coefficient 2/3 scales the result in relation to the peak values of the three-phase variable. The transformation from the space vector back to the three-phase original currents is performed through matrix equation 9.6.

The relations of the space vector with the synchronous rotating reference frame depend upon the axis orientation of the d – q frame. In this book, when the q-axis is at horizontal, the d-axis is at –90°. Therefore, the current space vector can be expanded as $\underline{i}_{qds} = i_{qs} - ji_{ds}$ where j represents the $\pi/2$ rotational operator. Real and imaginary operators, as indicated by Equations 9.7 and 9.8, can be used to indicate this mapping between the reference frame and the space vector notation. The same space vector approach can be used to handle voltages and fluxes in either stator or rotor circuits. Figure 9.7 indicates the space vector complex algebraic equivalent model of a squirrel cage machine, it is a very compact representation that embeds the transient nature.

$$\begin{bmatrix} i_a & i_b & i_c \end{bmatrix}^T = \mathrm{Re}\begin{bmatrix} 1 & \alpha & \alpha^2 \end{bmatrix}^T \underline{i}_{qds} \tag{9.6}$$

$$i_{qs} = \mathrm{Re}[\underline{i}_{qds}] \tag{9.7}$$

$$i_{ds} = \mathrm{Im}[\underline{i}_{qds}] \tag{9.8}$$

Space vector Equations 9.9 and 9.10 represent the transient response of an induction machine.[3] Each equation is expanded in two axis equations. Torque is indicated in Equation 9.11 with an expanded real time equation in Equation 9.12. The machine also needs an electromechanical equation as indicated by Equation 9.13.

$$\underline{v}_{qds} = r_s \underline{i}_{qds} + j\omega_e \underline{\psi}_{qds} + \frac{d}{dt}\underline{\psi}_{qds} \tag{9.9}$$

$$0 = r_r \underline{i}_{qdr} + j(\omega_e - \omega_r)\underline{\psi}_{qdr} + \frac{d}{dt}\underline{\psi}_{qdr} \tag{9.10}$$

$$T_e = \frac{3}{2}\left(\frac{P}{2}\right)\mathrm{Re}\left(j\underline{\psi}_{qds} \cdot \overline{\underline{i}_{qds}}\right) = \frac{3}{2}\left(\frac{P}{2}\right)\mathrm{Re}\left(j\underline{\psi}_{qdr} \cdot \overline{\underline{i}_{qdr}}\right) \tag{9.11}$$

$$T_e = \frac{3}{2}\left(\frac{P}{2}\right)\left(\psi_{ds}i_{qs} - \psi_{qs}i_{ds}\right) = \frac{3}{2}\left(\frac{P}{2}\right)\left(\psi_{dr}i_{qr} - \psi_{qr}i_{dr}\right) \tag{9.12}$$

$$\frac{d\omega_r}{dt} = \frac{1}{J}\left(T_e - T_L\right) \tag{9.13}$$

9.5 Field-Oriented Control

The objective of field-oriented or vector control is to establish and maintain an angular relationship between the stator current space vector and one internal field vector, usually rotor flux or stator flux as suggested by Equation 9.12. This angular relationship may be achieved by regulating the slip of the machine to a particular value that will fix the orientation through feedforward on-line calculations. In this approach, often called indirect vector control (IVC), rotor flux is aligned with the d-axis. IVC is very popular for industrial drives: it is inherently four-quadrant work until zero speed and is suitable for speed-control loop but very dependent on machine parameters. Under flux-aligned conditions along the d-axis of the synchronous rotating reference frame, the stator current is decoupled into two components; i_{qs} is proportional to machine electrical torque whereas i_{ds} is proportional to the machine flux, similarly to a DC machine control. Therefore, field orientation implies that on-line transformation of machine variables take into account a hypothetical rotating frame equivalent to a DC machine. By real-time transformation back to the three-phase stationary frame, impressed stator currents will impose a DC-like transient response.

Field orientation may also be achieved by on-line estimation of the field vector position, and usually the stator-flux will be aligned with the d-axis. Under this approach, called direct vector control (DVC), the control is less sensitive to parameters. It is dependent only on the stator resistance, which is easier to correct. It is very robust due to real-time tracking of machine parameter variation with temperature and core saturation, but it typically does not work at zero shaft speed. For induction generator applications, DVC seems to be a more effective approach because zero speed operation is not required. There are other ways to control an induction machine based on direct torque control (DTC), as developed by ABB based on the theory developed by Depenbrock,[7,8] where two different loops, corresponding to the magnitudes of the stator flux and torque, directly control the switching of the converter. Although DTC boasts a lot of successful applications in drives, its use for induction generators is not yet fully proven and will not be discussed in this book.

9.5.1 Indirect Vector Control

In indirect vector control (IVC), the unit vector, $\sin(\theta_e)$, and $\cos(\theta_e)$, used for vector rotation of the terminal variables is calculated in a feedforward manner through the use of the internal signals ω_r (rotor speed) and i_{qs}^* (quadrature stator current component proportional to torque). The control principles required to implement indirect vector control in the rotor flux oriented reference frame can be derived from the rotor voltage equation (Equation 9.10) and the torque equation (Equation 9.12). The magnetizing rotor current is defined as[3]

$$\begin{cases} 0 = R_r i_{dr} + \dfrac{d}{dt} \psi_{dr} - \left(\omega_e - \omega_r \right) \psi_{qr} \\[2mm] 0 = R_r i_{qr} + \dfrac{d}{dt} \psi_{qr} + \left(\omega_e - \omega_r \right) \psi_{dr} \end{cases} \tag{9.14}$$

$$\begin{cases} i_{dr} = \dfrac{1}{L_r} \psi_{dr} - \dfrac{L_m}{L_r} i_{ds} \\[2mm] i_{qr} = \dfrac{1}{L_r} \psi_{qr} - \dfrac{L_m}{L_r} i_{qs} \end{cases} \tag{9.15}$$

Substituting Equation 9.15 into Equation 9.14, where $\omega_{sl} = \omega_e - \omega_r$,

$$\begin{cases} \dfrac{d}{dt}\psi_{dr} + \dfrac{R_r}{L_r}\psi_{dr} - \dfrac{L_m}{L_r}R_r i_{ds} - \omega_{sl}\psi_{qr} = 0 \\ \dfrac{d}{dt}\psi_{qr} + \dfrac{R_r}{L_r}\psi_{qr} - \dfrac{L_m}{L_r}R_r i_{qs} + \omega_{sl}\psi_{dr} = 0 \end{cases} \tag{9.16}$$

By imposing alignment of rotor flux with the d-axis of the synchronous rotating reference frame, one can consider that flux projection on the q-axis is zero, and so is its derivative, $\psi_{qr} = d/dt\,\psi_{qr} = 0$. Therefore, the vector will be completely along the d-axis, $\psi_{qdr} = \psi_{dr}$. Substituting such requirements into Equation 9.16, Equations 9.17 and 9.18 (below) establish the control laws to implement in the synchronous rotating reference frame. The torque equation (Equation 9.19) will need only ψ_{dr} and i_{qs}. By observing such equations, it is possible to see that by maintaining i_{ds} constant the rotor flux will be kept constant and by controlling i_{qs} the machine torque will respond proportionally. Therefore, a current controlled converter, as depicted in Figure 9.8, will cause the induction machine to have a fast transient response under decoupled commands, like a DC machine. The q-axis stator reference i_{qs}^* generates the reference slip frequency ω_{sl} (Equation 9.18), which is added to the instantaneous rotor speed ω_r in order to generate the required angular frequency ω_e, which after integration the flux angle θ_e generates the sine and cosine waveforms for vector rotation.

$$\frac{L_r}{R_r}\frac{d}{dt}\psi_{dr} + \psi_{dr} = L_m i_{ds} \tag{9.17}$$

$$\omega_{sl} = \frac{L_m}{\psi_{dr}}\frac{R_r}{L_r}i_{qs} \tag{9.18}$$

$$T_e = \frac{3}{2}\left(\frac{P}{2}\right)\frac{L_m}{L_r}i_{qs}\psi_{dr} \tag{9.19}$$

Figure 9.8 shows an IVC-based system with i_{ds}^* and i_{qs}^* command. The power circuit consists of a front-end rectifier connected through a DC link to a three-phase inverter. For generation purposes, the grid-connected rectifier must be operational in four quadrants, delivering incoming machine power to the utility side. By maintaining i_{ds}^* constant, the machine flux will be constant and a positive i_{qs}^* will drive the machine in a motoring mode, whereas for $i_{qs}^* < 0$, the slip frequency ω_{sl} is negative, and the corresponding flux angle θ_e will be such that sine/cosine waveforms will phase-reverse the vector rotation automatically for generating mode.

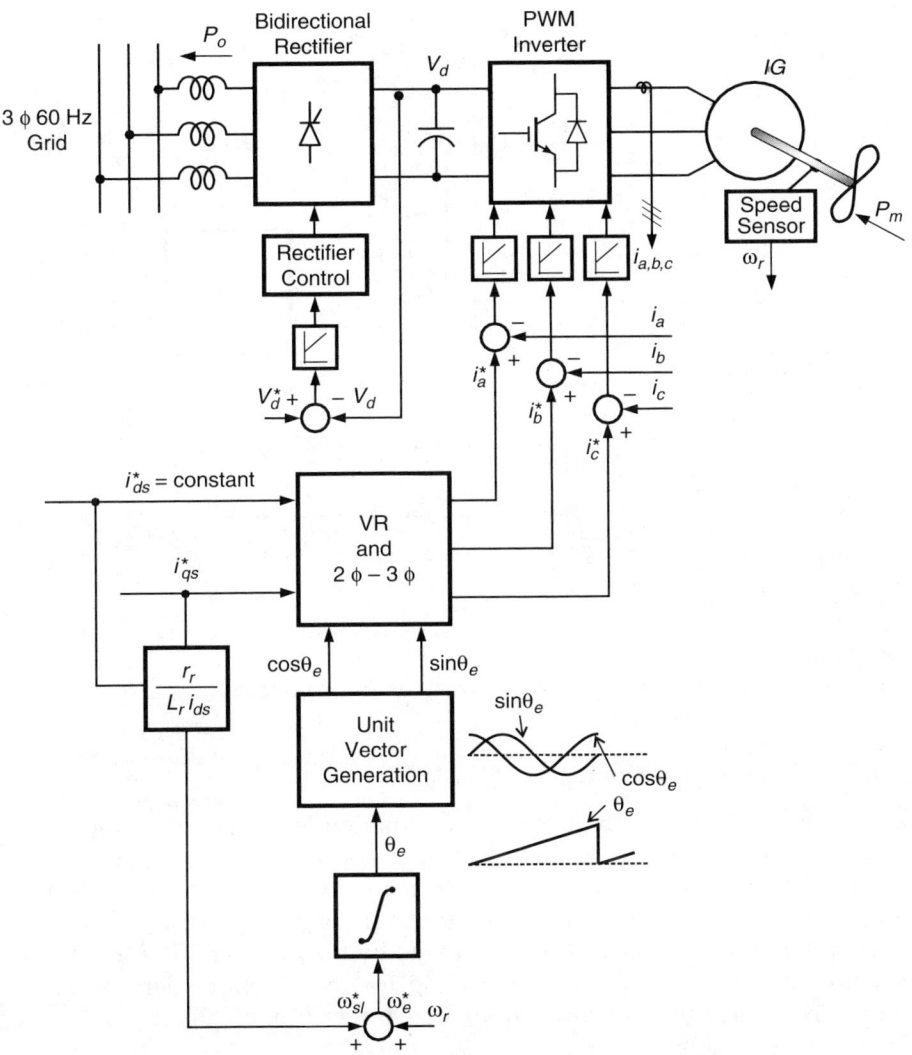

FIGURE 9.8
Basic IVC structure with decoupled stator current input command.

The block diagram indicates a PWM current control. There are several possibilities when implementing modulators; the reader is referred to Holmes and Lipo.[6] Synchronous current control with feedforward machine counter-EMF is preferable for an optimized operation, as indicated in Figure 9.9, where the reference voltages v_a^*, v_b^*, and v_c^* will drive a voltage PWM controller. Speed and flux outer loops can also be incorporated as outer loops on the basic structure. IVC is a very popular and powerful approach to control induction motors and generators. However, exact machine parameters are needed for improved performance.

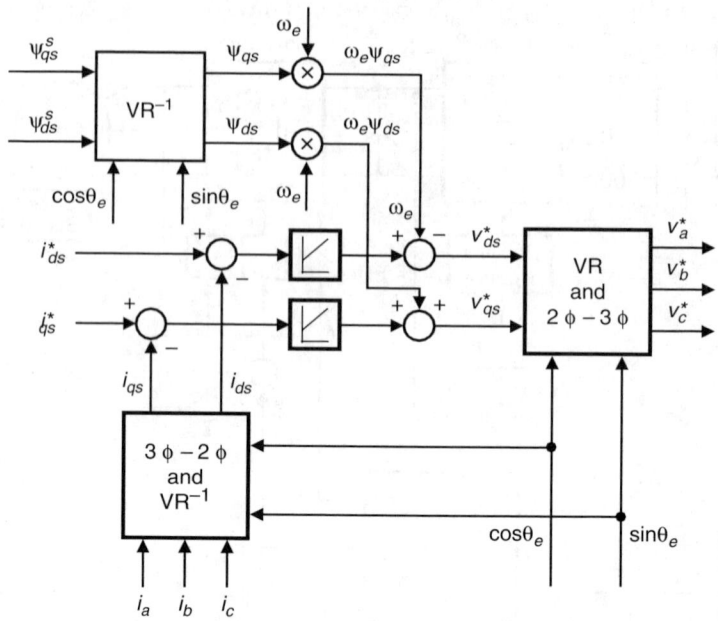

FIGURE 9.9
Synchronous current control for IVC system with flux and speed loops.

9.5.2   Direct Vector Control

Direct vector control with stator flux orientation is preferred for induction generator systems for several reasons: it is not necessary to use the speed signal in the fabrication of the unit-vector signals, and the estimation is robust against parameter variations for operating speed greater than 5% of base speed. The unit vectors are calculated by the integration of the d- and q-axis voltages minus the required drop for the correspondent reference frame. The stator flux reference frame has larger flux voltage for extended low-speed operation.

There is a coupling that can be easily compensated for with a feedforward network; the decoupler is in the forward path of the system. Detuning parameters will not affect the steady-state response, and the stator resistance variation can be taken into account with thermocouples installed in the machine slots.

Stator flux orientation also improves transient response with detuned leakage inductance, and there is a limited pull-out torque for a given flux, which improves the stability.

The equations for stator flux orientation are derived from the Blaschke equations, where the d − q rotor flux vector (ψ_{dr}, ψ_{qr}) is algebraically replaced by stator flux (ψ_{ds}, ψ_{qs}). Equations 9.20 and 9.21 are the Blaschke rotor equations in the reference frame; the relationship between stator flux and rotor flux is given by Equations 9.22 and 9.23, where $\sigma = 1 - L_m^2/L_s L_r$ is the total leakage factor.

$$(1 + sT_r)\psi_{dr} = L_m i_{ds} + \omega_{sl} T_r \psi_{qr} \tag{9.20}$$

$$(1 + sT_r)\psi_{qr} = L_m i_{qs} - \omega_{sl} T_r \psi_{dr} \tag{9.21}$$

$$\psi_{qr} = \frac{L_r}{L_m}\psi_{qs} - \frac{\sigma L_s L_r}{L_m} i_{qs} \tag{9.22}$$

$$\psi_{dr} = \frac{L_r}{L_m}\psi_{ds} - \frac{\sigma L_s L_r}{L_m} i_{ds} \tag{9.23}$$

After substitution and algebraic manipulations, the Blaschke equations in terms of stator currents become

$$(1 + sT_r)\psi_{ds} = L_s i_{ds}(1 + s\sigma T_r) + \omega_{sl} T_r (\psi_{qs} - \sigma L_s i_{qs}) \tag{9.24}$$

$$(1 + sT_r)\psi_{qs} = L_s i_{qs}(1 + s\sigma T_r) - \omega_{sl} T_r (\psi_{ds} - \sigma L_s i_{ds}) \tag{9.25}$$

In order to attain vector control, the q component of the stator flux must be set to zero, and Equations 9.26 and 9.27 follow.

$$i_{qs} L_s (1 + s\sigma T_r) = \omega_{sl} T_r (\psi_{ds} - \sigma L_s i_{ds}) \tag{9.26}$$

$$(1 + sT_r)\psi_{ds} = L_s i_{ds}(1 + s\sigma T_r) - \omega_{sl} T_r \sigma L_s i_{qs} \tag{9.27}$$

Equations 9.26 and 9.27 indicate coupling between the torque component current i_{qs} and the stator flux ψ_{ds}. The injection of a feedforward decoupling signal i_{dq} in the flux loop eliminates these effects as is indicated in Equation 9.28. The transient nature of i_{dq} is not included in Equation 9.28 since it was found that it does not improve the overall response much.[10] The i_{dq} decoupling signal is given by Equation 9.29 and the angular slip frequency ω_{sl} is given in Equation 9.30.

$$(1 + sT_r)L_s i_{dq} - \omega_{sl} T_r \sigma L_s i_{qs} = 0 \tag{9.28}$$

$$i_{dq} = \frac{\sigma L_s i_{qs}^2}{\psi_{ds} - \sigma L_s i_{ds}} \tag{9.29}$$

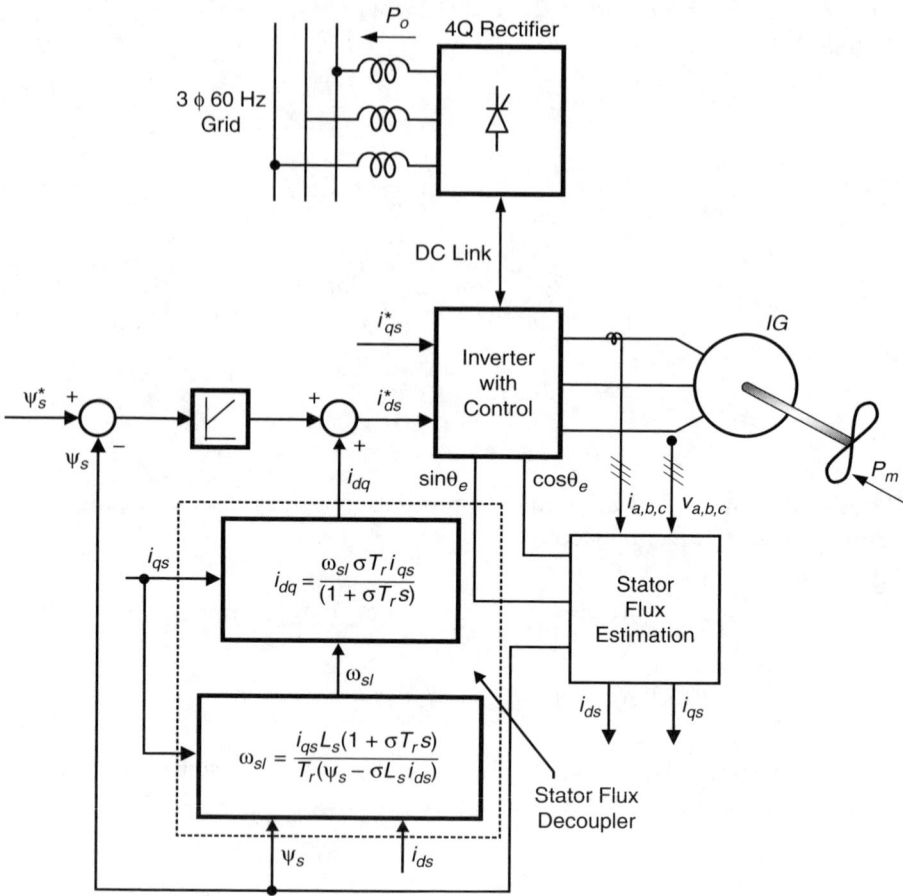

FIGURE 9.10
DVC based induction generator control with torque and flux command.

$$\omega_{sl} = \frac{i_{qs}L_s\left(1+\sigma T_r s\right)}{T_r\left(\psi_{ds}-\sigma L_s i_{ds}\right)} \tag{9.30}$$

The control block diagram in stator-flux oriented DVC mode is shown in Figure 9.10. The system uses open-loop torque control with a stator-flux control loop. The stator flux remains constant in the constant torque region but can also be programmed in the field weakening region. In addition to Equation 9.29 and 9.30. the key estimation equations can be summarized as follows.[10]

$$\psi_{ds}^{s} = \int\left(v_{ds}^{s}-i_{ds}^{s}R_s\right) \tag{9.31}$$

$$\psi_{qs}^{s} = \int \left(v_{qs}^{s} - i_{qs}^{s} R_{s} \right) \tag{9.32}$$

$$\psi_{s} = \sqrt{ \left(\psi_{ds}^{s} \right)^{2} + \left(\psi_{qs}^{s} \right)^{2} } \tag{9.33}$$

$$\cos \theta_{e} = \frac{\psi_{ds}^{s}}{\psi_{s}} \tag{9.34}$$

$$\sin \theta_{e} = \frac{\psi_{qs}^{s}}{\psi_{s}} \tag{9.35}$$

$$\omega_{e} = \frac{ \left(v_{qs}^{s} - i_{qs}^{s} R_{s} \right) \psi_{ds}^{s} - \left(v_{ds}^{s} - i_{ds}^{s} R_{s} \right) \psi_{qs}^{s} }{ \psi_{s}^{2} } \tag{9.36}$$

The control block diagram of Figure 9.10 shows the principles for controlling the induction generator. The most important parameter to be corrected is the stator resistance, which can be accomplished with a machine thermal model and measurement of terminal voltage and current. Power electronic systems can be either connected in series with the machine, as is discussed in Chapter 10, where a fuzzy-based control optimizes a double-pwm converter system, or they can be used in parallel as static VAR compensation. Figure 9.11 shows the scheme for induction generators. An economic evaluation for each solution should be conducted.

For an induction generator operating in stand-alone mode, a procedure that regulates the output voltage is required, as shown in the control system indicated in Figure 9.11. The DC-link voltage across the capacitor is kept constant and the machine impressed terminal frequency will vary with variable speed. For motoring applications flux is typically kept constant and only for operating speeds above rated value is a flux-weakening command required. However, induction generators require constant generated voltage. Since the frequency of the generated voltage and current are dependent on the rotor speed, the product of the rotor speed and the flux linkage should remain constant so that the terminal voltage will remain constant, where the minimum rotor speed will correspond to the maximum saturated flux linkage. A look-up table is used to program the right flux level in Figure 9.11. The system starts up with a battery connected to the inverter. Then, as the DC-link voltage is regulated to a higher value across the capacitor, the battery will be turned off by the diode and a DC load can be applied across the capacitor. The machine terminals are also capable of supplying power to an auxiliary load at variable frequency.

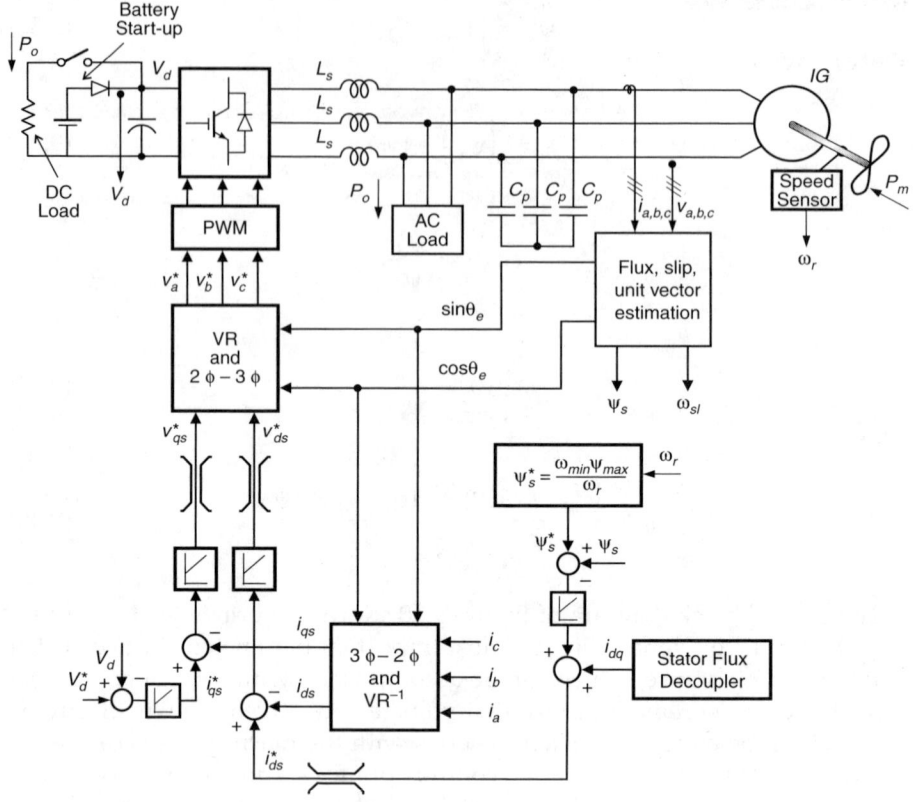

FIGURE 9.11
Stand-alone DVC stator flux based induction generator control.

9.6   Problems

9.1 Summarize the features of indirect vector control and direct vector control as applied to induction generator systems.

9.2 Do research in the literature on parameter sensitivity for vector-control-based systems and write a summary report.

9.3 Do research in the literature on sensorless vector-control-based systems and write a summary report.

9.4 Do research in the literature on space vector modulation as applied to vector-control-based systems and write a summary report.

9.5 Simulate one of the schemes using system simulation software.

References

1. Hasse, K., About the Dynamics of Adjustable-Speed Drives with Converter-Fed Squirrel-Cage Induction Motor, Ph.D. dissertation in German, Darmstadt Technische Hochschule, Darmstadt, Germany, 1969.
2. Blaschke, F., The principle of field-orientation as applied to the new transvektor closed-loop control system for rotating-field machines, *Siemens Review*, Vol. 34, 217–220, 1972.
3. Kazmierkowski, M.P., Krishnan, R., and Blaabjerg, F., *Control in Power Electronics: Selected Problems*, Academic Press, San Diego, 2002.
4. Murphy, J.M.D. and Turnbull, F.G., *Power Electronic Control of AC Motors*, Oxford University Press, New York, 1988.
5. Trzynadlowski, A.M., *Control of Induction Motors*, Academic Press, San Diego, 2000.
6. Holmes, D.G and Lipo, T.A., *Pulse Width Modulation for Power Converters: Principles and Practice*, John Wiley & Sons, 2003.
7. Novotny, D.W. and Lipo, T.A., *Vector Control and Dynamics of AC Drives*, Oxford University Press, New York, 1996.
8. Bose, B.K., *Modern Power Electronics and AC Drives*, Prentice Hall, New York, 2001.
9. Krishnan, R., *Electric Motor Drives: Modeling, Analysis, and Control*, Prentice Hall, New York, 2001.
10. Xu, X., De Doncker, R.W., and Novotny, D.W., A stator flux oriented induction machine drive, *IEEE/PESC Conf. Rec.*, 870–876, 1988.
11. Seyoum, D., Rahman, F., and Grantham, C., Terminal voltage control of a wind turbine driven isolated induction generator using stator oriented field control, *IEEE APEC Conf. Rec.*, Miami Beach, Florida, Vol. 2, 846–852, 2003.
12. Seyoum, D., The Dynamic Analysis and Control of a Self-Excited Induction Generator Driven by a Wind Turbine, Ph.D. dissertation, School of Electrical Engineering and Telecommunications, The University of New South Wales, Australia, 2003.

10

Optimized Control for Induction Generators

10.1 Scope of This Chapter

Efficient electric generation and viable and effective load control for alternative or renewable energy sources usually requires heavy investment and is usually accompanied by an ambiguous decision-making analysis that takes into consideration the intermittence of input power and demand. Therefore, in order to amortize the installation costs in the shortest possible time, optimization of electric power transfer, even for the sake of relatively small incremental gains, is of paramount importance. This chapter presents practical approaches based on hill climbing control and fuzzy logic control for peak power tracking control of induction-generator-based renewable energy systems.

10.2 Why Optimize Induction-Generator-Based Renewable Energy Systems?

General approaches to optimization utilize linear, nonlinear, dynamic, and stochastic techniques, while in control theory, the calculus of variations and optimal control theory are important. Operations research is the specific branch of mathematics concerned with techniques for finding optimal solutions to decision-making problems. However, these mathematical methods are, to a large degree, theoretically oriented and thus very distant from the backgrounds of the decision makers and management personnel in industry.

Recently, practical applications of heuristic techniques like hill climbing and fuzzy logic have been able to bridge this gap, thereby creating the possibility of applying practical optimization to industrial applications. One field that has proven to be attractive for optimization is the generation and control of electric power from alternative and renewable energy resources like wind, hydro and solar energy.[1,2] Installation costs are very high, while the availability of alternative power is by its nature intermittent, which tends to constrain efficiency. It is therefore of vital importance to optimize the efficiency of electric power transfer, even for the sake of relatively small incremental gains, in order to amortize the installation costs within the shortest possible time. The following characteristics of induction-generator-based control motivate the use of heuristic optimization:

- Parameter variation that can be compensated for by design judgment
- Processes that can be modeled linguistically but not mathematically
- A need to improve efficiency as a matter of operator judgment
- Dependence on operator skills and attention
- Process parameters affect one another
- Effects cannot be attained by separate PID control
- A fuzzy controller can be used as an advisor to the human operator
- Data intensive modeling

There are many mathematical algorithms to solve optimization problems for a local minimum based on gradient projection requiring the evaluation of gradient information with regard to a certain objective function. An uncomplicated but powerful technique known as hill climbing is based on defining an algorithmic function and choosing actions so as to achieve a next state with an evaluation closer to that of the goal. Fuzzy and neurofuzzy techniques have by now become efficient tools for modeling and control applications because of their potential benefits in optimizing cost effectiveness.

10.3 Optimization Principles: Optimize Benefit or Minimize Effort

Examples of state evaluation functions and hill climbing are plentiful in everyday problems.[3] When someone plans a trip on a map the roads that go towards the right direction are observed. Then, the ones that reduce the distance at the fastest rate are selected. Other considerations may come into the decision, such as scenery, facilities along the way, and weather conditions. In electrical engineering problems the effort required or the benefit desired in any practical situation can be expressed as a function of certain decision variables, the *objective* or *cost function*, and optimization is defined as the process of finding the conditions that generate a maximum (or minimum) value of such a function.

Figure 10.1 shows a typical function of this kind. The *y*-axis can represent any physical variable like power, force, velocity, voltage, current, or resistance, while the *x*-axis might indicate the amplitude of the physical variable

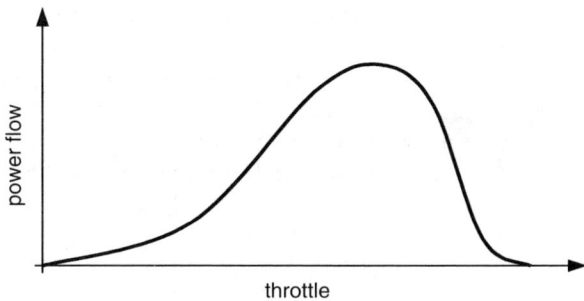

FIGURE 10.1
Objective function for power flow optimization.

to be maximized — for example, a throttle controlling a water-pump system (*x*-axis). As the throttle opens, the water flow increases but the pressure decreases, and the power (*y*-axis) initially goes up, reaching a maximum and decreasing as the throttle continues opening. Although finding the maximum of such a simple function seems to be trivial, it is not; the maximum can change under some parameter variations: temperature, density, ageing, part replacement, impedance, nonlinearities like dead-band and time delays, and cross dependence of input and output variables. The typical way of dealing with so many interdependencies would be by utilizing sensors to improve parameter robustness, analytical and experimental preparation of look-up tables, and mathematical feedforward de-coupling. Such approaches increase costs and development time due to the extra sensors, more laboratory design phases, and the necessity of precise mathematical models.

A metarule is a rule that describes how other rules should be used and establishes how to build the knowledge framework. For example, there is an optimum pump speed for the corresponding maximum of power flow where minimum input pump effort maximizes output power. The heuristic way of finding the maximum might be based on the following metarule. "If the last change in the input variable *x* has caused the output variable *y* to increase, keep moving the input variable in the same direction; if it has caused the output variable to drop, move it in the opposite direction."

10.4 Application of Hill Climbing Control (HCC) for Induction Generators

Figure 10.2 shows the systems involved in an induction generator installation. There are several controls related to such systems, classified as mechanical, electromechanical; electroelectronic, and electronic. Mechanical controls use centrifugal weights to close or open the inflow of water for the hydro turbines or to exert pitch control of the blades in wind turbines. Among these are the

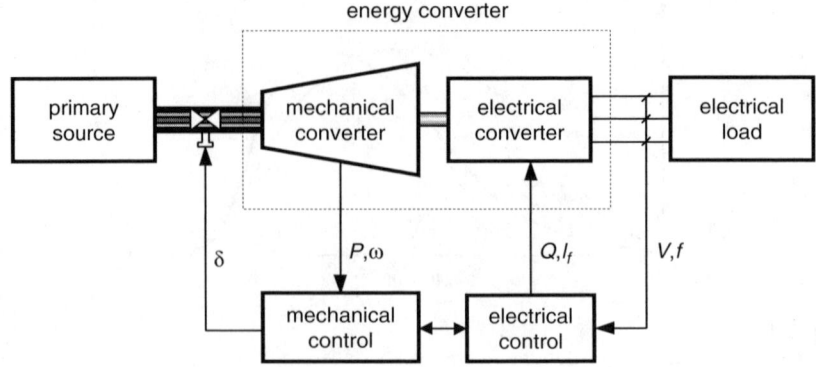

FIGURE 10.2
Transformation of primary energy into electrical energy.

flywheels used to store transient energy, anemometers to determine the wind intensity, and wind tails and wind vanes for the direction of the winds. They act to control and limit speed. They are slow and rough, yet robust, reliable and efficient.

Electromechanical controls are more accurate, relatively lighter, and faster since they have less volume and weight than mechanical controls. They use governors, actuators, servomechanisms, electric motors, solenoids, and relays to control the primary energy. Some modern self-excited induction generators also use internal compensatory windings to keep the output voltage constant.

Electroelectronic controls use electronic sensors and power converters to regularize energy and power. They can be quite accurate in adjusting voltage, speed, and frequency.

Electronic controls usually exert their action on the load, which is mechanically coupled to the turbine, and so the load also controls them.

The output characteristics of nearly all induction generators resemble each other enough to allow us to describe the mathematical relationships among mechanical power and phase voltage, rotor current, and machine parameters as discussed in Chapter 5. Equation 5.10 is repeated here as

$$P_{mec} = 3I_2^2 \left[\sqrt{\left(\frac{V_{ph}}{I_2} \right)^2 - (X_1 + X_2)^2} - (R_1 + R_2) \right] \tag{10.1}$$

Equation 10.1 is plotted in Figure 10.3 for an induction generator of 20 kW, 220/380 V, 60/35 A, 60 Hz, and four poles. Notice that the power characteristic of the induction generator has a maximum point that in a hydro or wind power plant will be the maximum power to be extracted from nature, which may be different from maximum efficiency. For this, there is no need to know

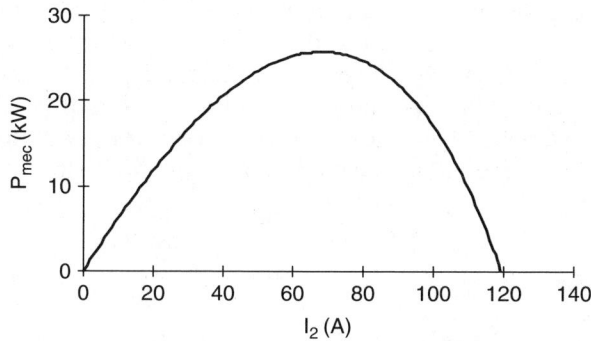

FIGURE 10.3
Mechanical power characteristic of an induction generator with respect to the load.

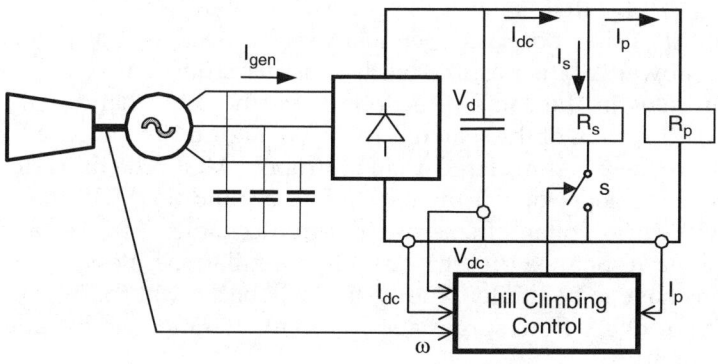

FIGURE 10.4
Induction generator with electronic control by the load.

the model of the generator or the turbine, the behavior of the wind, the deterioration of the material due to ageing, or other factors difficult to measure. The right value of the load current is, then, set up in such a way as to maximize the power extracted from nature. The speed at which the operating point reaches maximum generated power (MGP) is the most important consideration because if the primary energy is not used at the instant it is available from nature, it will be gone.

This speed can be obtained by a step modulation of the load current increment, also called electronic control by the load,[3] depicted in Figure 10.4. The load excess is dissipated by transferring the generator's energy to the public network, either for battery charge, back pumping of water, irrigation, hydrogen production, heating, or freezing. Dissipation can also occur by simply burning out the excess of energy in bare resistors to the wind or to the flow of water. Electronic control by the load for electric micropower plants is the purpose of using hill climbing control (HCC). The HCC algorithm will adapt the load current to the available primary energy. There are several methods of using the metarule described in Section 10.3 for induction generators; some of them will be described in Section 10.5. However, the most common techniques to climb to the maximum generated point are: fixed step, divided step, adaptive step and exponential step.

10.5 HCC-Based Maximum Power Search

10.5.1 Fixed Step

The most straightforward HCC method of reaching maximum power peak is the fixed step approach. The logic of this algorithm is based on the output power $P_{dc} (= V_{dc} I_{dc})$ that is measured for each new increment (or decrement) of current ΔI_{dc} being observed per increase (or decrease) of power. If the change in power ΔP_{dc} is positive, with a last variation $+\Delta I_{dci}$, the search for MGP continues in the same direction; if, on the other hand, $+\Delta I_{dci}$ causes a $-\Delta P_{dc}$, the direction of the search is reversed. Figure 10.5 shows a fixed step search where there is some large oscillation about MGP. The fixed step search has the inconvenience of causing oscillations around the MGP, that, if maintained, will cause voltage flickering, electromechanical stresses and, when there is some resonance present, sustained oscillations. Besides that, if the oscillations have to be limited, the system will be slow at the beginning and during the recovery from surges and transient alterations in the amounts of primary energy.

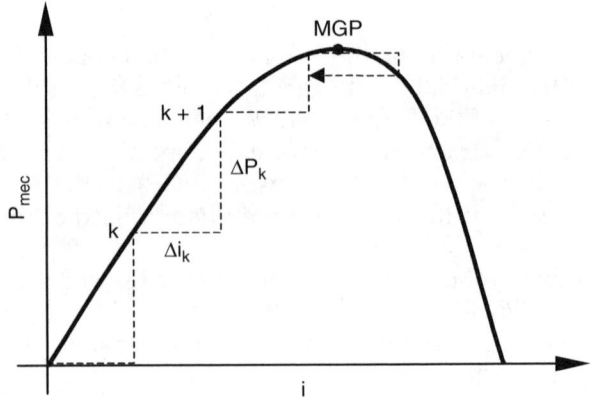

FIGURE 10.5
HCC fixed step.

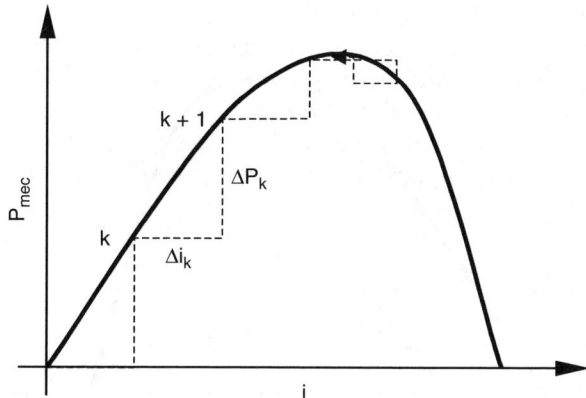

FIGURE 10.6
HCC divided step.

10.5.2 Divided Step

Figure 10.6 illustrates the divided step approach, which minimizes the oscillations around MGP. It is similar to the fixed step approach, but the divided step increment of load current, besides having its sign changed, is divided by two after having reached the MGP. There is a tendency to dissipate the oscillations around that point after the step is applied a certain number of times. The value of the load current is determined by

$$i_{k+1} = i_k + \Delta i_{k+1} \qquad (10.2)$$

and every time that MGP is reached, the step changes by

$$\Delta i_{MGP} = \Delta i_k / 2 \qquad (10.3)$$

When there are variations in the load or in the primary energy, the step resumes its original fixed value and, because of this, each recovery provokes load current oscillations, repeating the inconveniences of the fixed step approach. Another inconvenience of the fixed step and the divided step approaches is that they do not take into account machine inertia, and it is difficult to define after how many applications of the same step, the increment should be changed to a new size. That is, how many steps should be applied before the machine can fully respond to the new situation of control variable Δi? Many alternatives using fixed and divided steps have been implemented in practice with very poor results with regard to stable operation and load current step size. For a reliable operation, it still remains a question of the automatic establishment of a relationship between the inertia of the machine and the application of the current step.

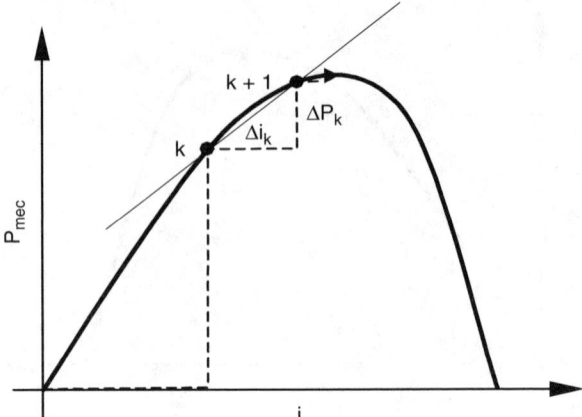

FIGURE 10.7
HCC adaptive step.

10.5.3   Adaptive Step

The adaptive step approach is based on an acceleration factor obtained from the tangent to the *PxI* curve operating point of the generator. This tangent passes through points k and $k+1$ in Figure 10.7. The accelerating factor at the point $k+1$ can be defined as:

$$k_{ak+1} = \frac{\Delta P_{k+1}}{\Delta i_{k+1}} \tag{10.4}$$

and the defined instantaneous current for it is: $i_{k+1} = i_k + k_{ak+1}\Delta i$.

As the operation point approaches the MGP, tangent slope (load voltage) tends to zero, the current increment also tends to zero this way, practically eliminating power oscillations, though sometimes there is a relatively slow oscillation due to the derivative nature of this algorithm. During transient states or under abrupt variations of load or primary power, the algorithm generates an acceleration factor proportional to the tangent slope. Quickly, a new point of stable operation is reached. To accelerate the increment of the generator current with the adaptive step, it generates a large V_c which decreases as the MGP comes closer. The result is a softer and more stable operation of the load current increase. Two inputs must be carefully selected: the size of the initial step and the rule for the step variation.

10.5.4   Exponential Step

Another possible solution to this problem is the exponential step approach as depicted in Figure 10.8. The exponential step Δi using an exponential

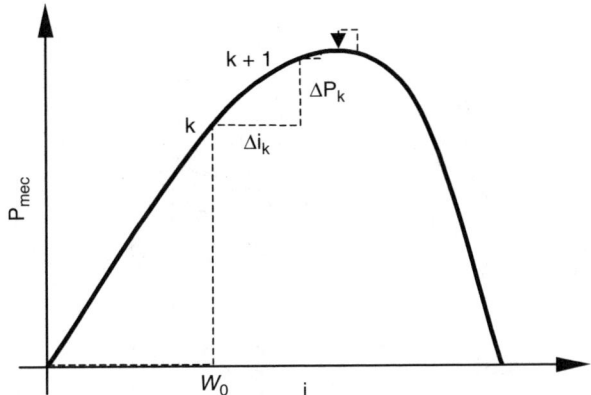

FIGURE 10.8
HCC exponential step.

incremental approximation gap ξ_i to the maximum generating point (MGP) given by

$$f_i = 1 - \xi_i = 1 - e^{-ki} \text{ for } i = 1, 2, 3, \ldots, n \quad (10.5)$$

The first iteration ($i = 1$) will give the initial current increment as

$$f_1 = 1 - e^{-k} = W_0 \quad (10.6)$$

from which it is possible to obtain

$$k = -\ln(1 - W_o) \quad (10.7)$$

A generic step of current i is given by

$$\Delta i = f_i - f_{i-1} = \left(1 - e^{-ki}\right) - \left[1 - e^{-k(i-1)}\right] = e^{-ki}\left(e^k - 1\right) \quad (10.8)$$

From the definition of ξ_i, for $i = n$ it is possible to determine that

$$n = -\frac{1}{k} * \ln(\xi_n) \quad (10.9)$$

The MGP exponential approach leads to a very fast transient response and a soft approach to the MGP. For implementation of this method it is enough to determine the initial increment W_0 and the approaching gap ξ that determines the iteration number of incremental variations n and W_0, which determine the constant k. In the implementation of this method, two commitments

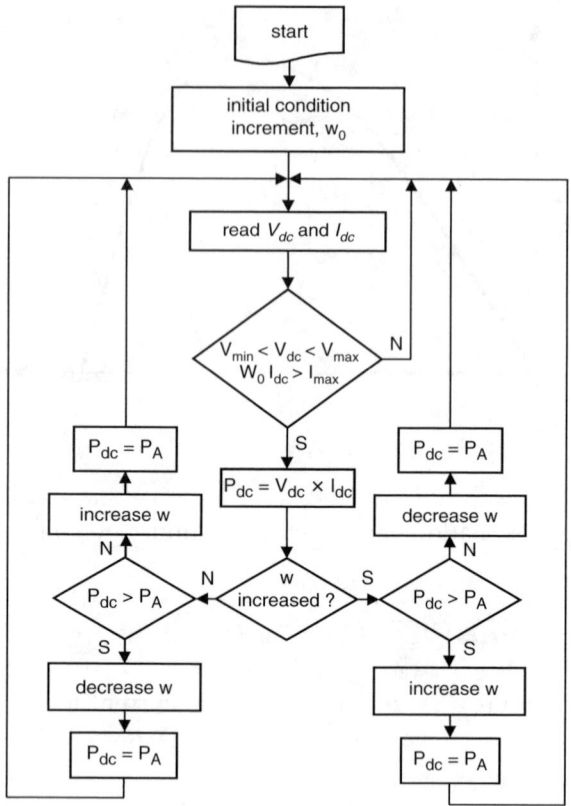

FIGURE 10.9
HCC flowchart.

also exist: (1) the transient state is related to the initial increment and W_0; and (2) the operating state is related to the approaching gap, ξ.

The flowchart in Figure 10.9 illustrates the routine sequence used to execute the incremental steps ΔV_c in the control program, which was implemented in LabView®. Small modifications in this program can be made for the fixed, divided, adaptive and exponential steps as discussed below. In the laboratory tests, the size of each current step was established as a gradual application of a secondary load, R_2, in steps of 2% or so, according to the schematic diagram of Figure 10.4. The primary resistance, R_1, had a fixed value of 90 Ω while the variation of the secondary resistance went from 45 Ω until open circuit.

As a result of the application of fixed step, the output power of the generator went according to the curve of Figure 10.10. As can be observed, the gradual application of this step of current went from 425 W, increasing quickly during 12 seconds, approximately, to stabilize in an oscillatory way around 475 W.

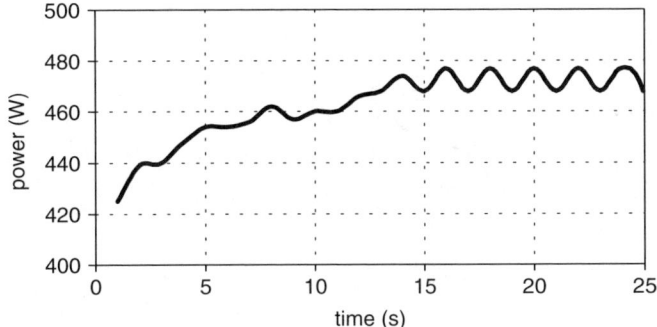

FIGURE 10.10
HCC with fixed step.

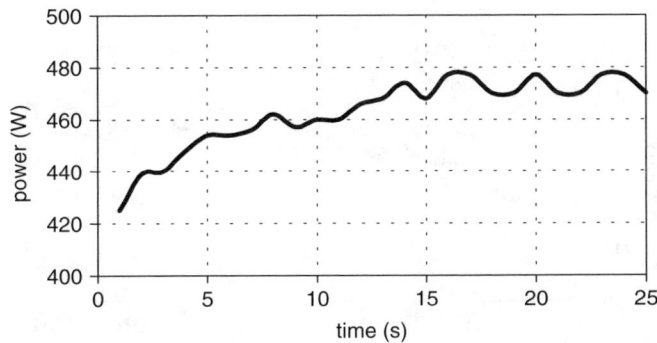

FIGURE 10.11
HCC with divided step.

Similarly to what was done with the fixed step, the application of the divided step resulted in the curve shown in Figure 10.11. Here a similarity was observed in the settling time to reach the set power, only with a difference in the oscillations around the steady state power, 475 W. The power oscillations, in this case, had a much wider period at similar conditions.

With the adaptive step, it was observed in Figure 10.12 that the initial step was much larger than the subsequent ones, tending to zero at the same time as it approached the steady state values and that the oscillation period around this point was much larger.

Finally, the exponential step, in the oscillatory curve of the output power shown in Figure 10.13, presented smaller oscillations along all the evolution of power to the steady state. In spite of this, at the beginning of the power variation, it took a form similar to the other types of step. The oscillations around the MGP were due to several factors, among which can be mentioned: (1) the coupling between the turbine and the generator (a rubber belt in these laboratory tests); (2) the inertia and power differences between the

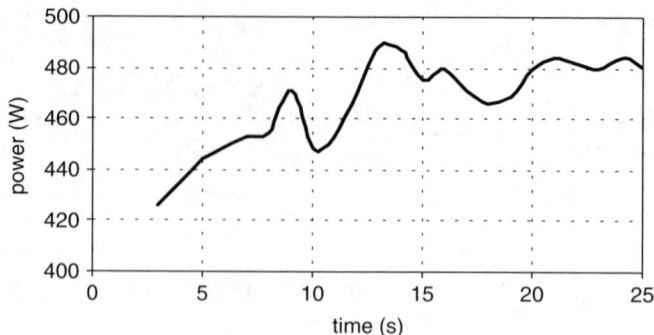

FIGURE 10.12
HCC with adaptive step.

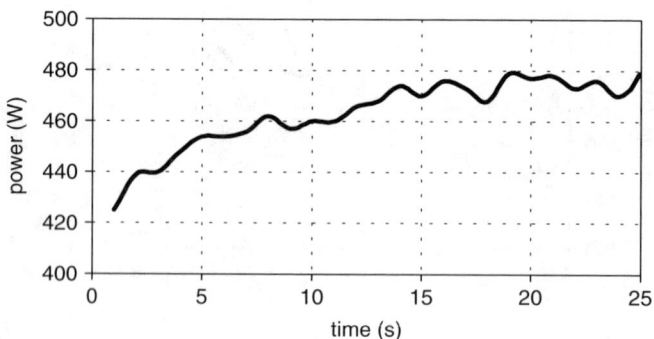

FIGURE 10.13
HCC with exponential step.

turbine and the generator with a small load (19.8 W to 7.5 kW); 3) the coincidence or near-coincidence of the period of self-oscillation of the generator-turbine set with that of the increment steps used in the control.

In order to better observe the phenomena, no measures were taken to reduce such phenomena during the tests. The major inconvenience of the fixed step and divided step approaches is that they do not take into account the inertia of the machine, and it is difficult to define after how many applications of the same step it should be altered. Besides that, this step recovers slowly in load transients or during alterations in the amounts of the primary energy or of the load. With the adaptive and exponential steps, the generator causes an initial large step of load current and then decreases as the point of maximum power gets closer. The result is a softer and stable operation.

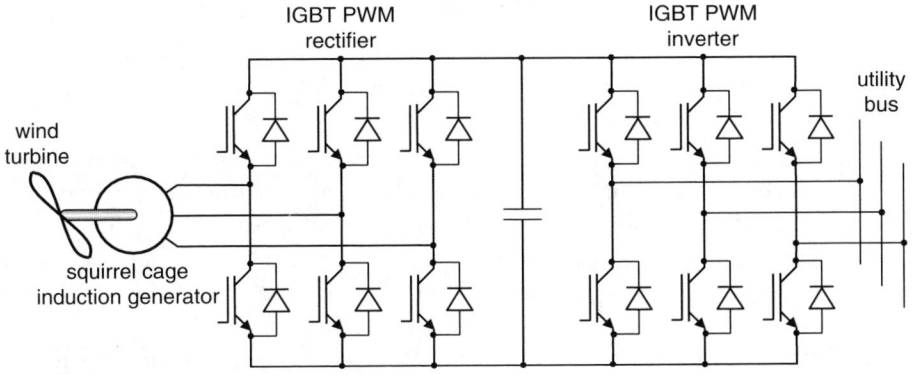

FIGURE 10.14
A double-PWM converter-fed wind energy system.

10.6 Fuzzy Logic Control (FLC)-Based Maximum Power Search

Fuzzy Logic Control (FLC) is an extension of the methods of load control discussed above. Appendix A3 reviews the concepts of fuzzy logic control required for this chapter. The squirrel-cage induction generator with a variable-speed wind turbine (VSWT) system and double-sided PWM converter with fuzzy logic control is used to maximize the power output and enhance system performance. All the control algorithms have been validated by simulation and implementation.[4-6] System performance has been evaluated in detail, and an experimental study with a 3.5 kW laboratory drive system was constructed to evaluate performance.

The voltage-fed converter scheme used in this system is shown in Figure 10.14. This work was particularly applied to a vertical-axis wind turbine, but horizontal turbines can be coupled to the shaft as well. A PWM-insulated gate-bipolar transistor (IGBT) rectifies the variable-frequency variable-voltage power from the generator. The rectifier also supplies the excitation needs for the induction generator. The inverter topology is identical to that of the rectifier, and it supplies the generated power at 60 Hz to the utility grid. Salient advantages of the double PWM converter system include:

- The line-side power factor is unity with no harmonic current injection (satisfies IEEE 519).

- The cage-type induction generator is rugged, reliable, economical, and universally popular.

- Induction generator current is sinusoidal.

- There is no harmonic copper loss.
- The rectifier can generate programmable excitation for the machine.
- Continuous power generation from zero to highest turbine speed is possible.
- Power can flow in either direction, permitting the generator to run as a motor for start-up (required for vertical turbine). Similarly, regenerative braking can quickly stop the turbine.
- Autonomous operation of the system is possible with a start-up capacitor charging the battery.

The wind turbine is characterized by the power coefficient (C_p), which is defined as the ratio of actual power delivered to the free stream power flowing through a similar but uninterrupted area. The tip speed ratio (TSR or λ) is the ratio of turbine speed at the tip of a blade to the free stream wind speed. The power coefficient is not constant but varies with the wind speed, rotational speed of the turbine, and turbine blade parameters such as angle of attack and pitch angle. C_p is defined as a nonlinear function of the tip-speed-ratio as

$$C_p(\lambda) = f(\lambda) \tag{10.10}$$

where $\lambda = r_w \omega_w / V_w$

The weighted torque by the power coefficient at the shaft is

$$T_m = \frac{1}{\omega_m} \left[C_p(\lambda) P_m \right] \tag{10.11}$$

The turbine is coupled to the induction generator (IG) through a step-up gear box (1:η_{GEAR}) so that the IG runs at a higher rotational speed, despite the low speed ω_w of the wind turbine. Therefore, the aerodynamic torque of a vertical turbine is given by the equation

$$T_m = C_p(\lambda) \cdot \left[0.5 \frac{\lambda \rho_o R_w^3}{\eta_{gear}} \right] \cdot V_w^2 \tag{10.12}$$

where

C_p = power coefficient
λ = tip speed ratio (TSR) ($R_w \omega_w / V_w$)
ρ_o = air density
R_w = turbine radius
η_{GEAR} = speed-up gear ratio
V_w = wind speed
ω_w = turbine angular speed

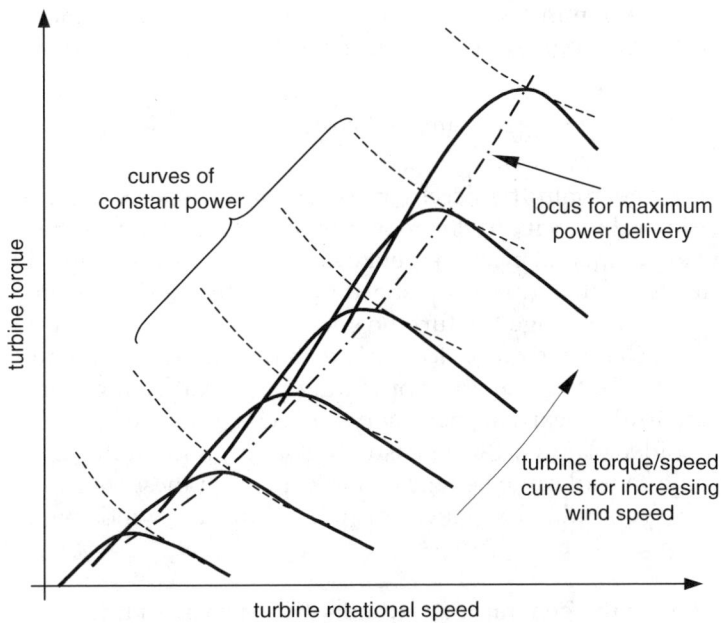

curves of
constant power

locus for maximum
power delivery

turbine torque

turbine torque/speed
curves for increasing
wind speed

turbine rotational speed

FIGURE 10.15
Family of turbine torque characteristics for variable wind velocities.

Figure 10.15 shows a set of constant power lines (dashed) superimposed on the family of curves indicating the region of maximum power delivery for each wind speed. This means that, for a particular wind speed, the turbine speed (or the TSR) is to be varied to get the maximum output power, and this point deviates from the maximum torque point, as indicated. Since the torque–speed characteristics of the wind generation system are analogous to those of a motor-blower system (except that the turbine runs in reverse direction), the torque follows the square-law characteristics and the output power follows the cube law, as indicated above. This means that, under reduced-speed light-load steady-state conditions, generator efficiency can be improved by programming its magnetic flux.

The static torque as given by Equation 10.12 is the average torque production, which must be added to the intrinsic torque pulsation. An actual turbine is quite complex to model since the moving surface of the turbine slips with respect to the wind and there are also extra components of wind in the x, y, and z directions. The blade that moves with the wind during one half revolution moves against the wind during the other half revolution, and some turbulence occurs behind the moving surface. There are various inertial modes in the system such as rotor blades, hub, gearbox, shaft, generator, and damping caused by the wind itself and the oil in the gearbox. The most important source of torque pulsation taken in account for the present model is that of the rotor blades passing by the tower. The oscillatory torque of the turbine is more dominant at the first, second, and fourth harmonics of fundamental turbine angular velocity (ω_w) and is given by the Equation 10.13.

$$T_{osc} = T_m \left[A\cos(\omega_w) + B\cos(2\omega_w) + C\cos(4\omega_w) \right] \qquad (10.13)$$

In order to maintain the necessary optimum tip speed ratio, that is, the quotient of wind velocity to the generator speed that extracts the maximum power, an experimental measurement can be conducted so as to build a look-up table for the actual turbine tip speed ratio profile, and with wind velocity measurements, to change the turbine speed set-up. This procedure requires experimental characterization for each turbine and an anemometer should be used to on-line track the best operating point. Changing from one optimum point to the next is a necessary stepwise procedure, and resonance must be avoided. The step size is usually not optimum in this approach. So, a fuzzy logic control can be suggested to cope with these uncertainties. The reasons to use an induction generator fuzzy control in wind energy systems are:

- To change the generator speed adaptively, so as to track the power point as the wind velocity changes (without wind velocity measurement)
- To reduce the generator rotor flux, boosting the induction generator efficiency when the optimum generator speed set-up is attained (in steady-state)
- To have robust speed control against turbine torque pulsation, wind gusts and vortices

Figure 10.16 shows the control block diagram of an induction generator with a double PWM converter-fed wind energy system that uses the power circuit of Figure 10.14. The induction generator and inverter output currents are sinusoidal due to the high frequency of the pulse-width modulation and current control as shown in the figure. The induction generator absorbs lagging reactive current, but the current is always maintained in phase at the line side; that is, the line power factor is unity. The rectifier uses indirect vector control in the inner current control loop, whereas the direct vector control method is used for the inverter current controller.[7,8]

Vector control permits fast transient response of the system as discussed in Chapter 9. A fuzzy-logic-based vector control is used to enhance three characteristics in this system: (1) the search for the best generator speed (FLC-1) command to track the maximum power of the wind, (2) the search for the best flux intensity in the machine (FLC-2) to optimize inverter and induction generator losses, and (3) robust control of the speed loop (FLC-3) to overcome possible shaft resonances due to wind gusts and vortex. Fuzzy controllers are described in detail in Simões et al.[4,5]

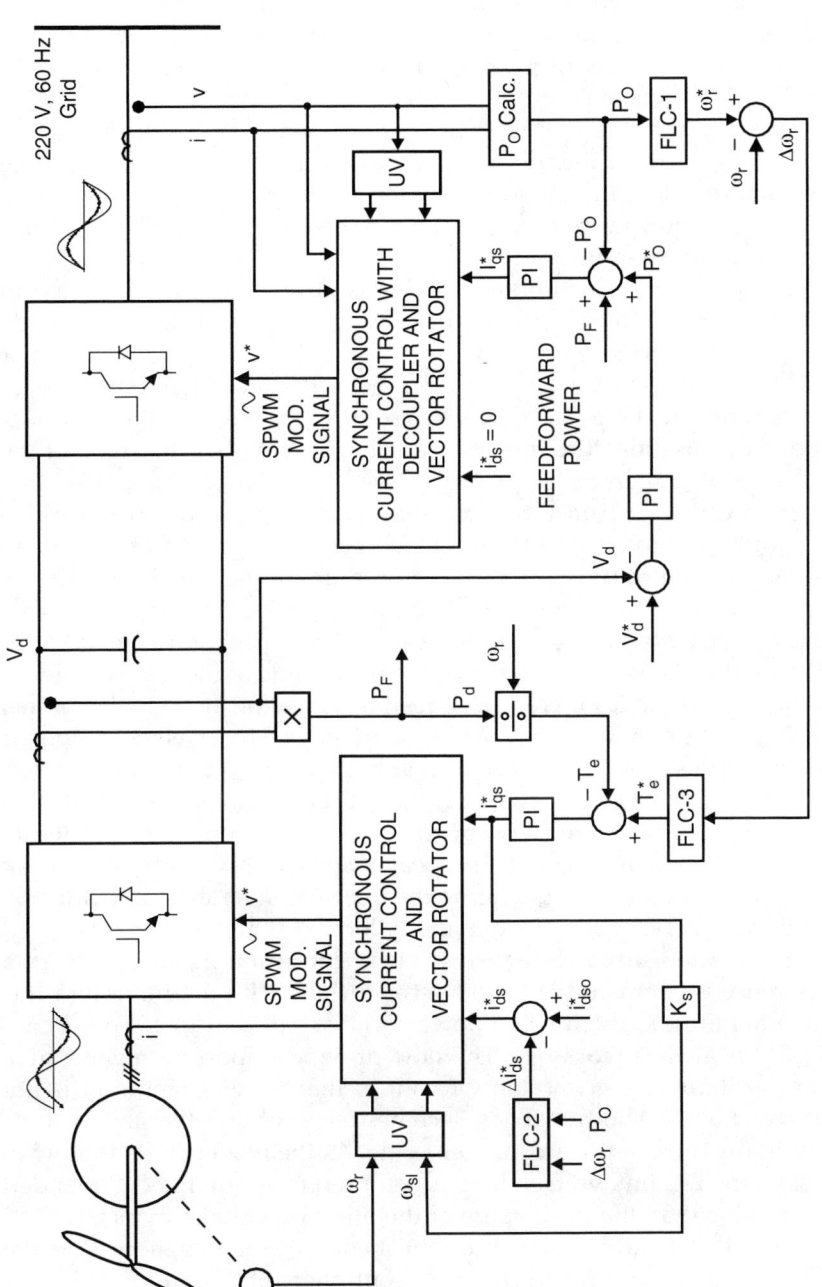

FIGURE 10.16
Fuzzy controlled double PWM converter for wind energy system.

For a particular wind velocity V_w, there is an optimum setting of the generator speed ω_r^*. The speed loop control generates the torque component of induction generator current so as to balance the developed torque with the load torque. The variable-voltage variable-frequency power from the supersynchronous induction generator is rectified and pumped to the DC link. The DC-link voltage controller regulates the line power P_o (i.e., the line active current) so that the link voltage always remains constant. A feedforward power signal from the induction generator output to the DC voltage loop prevents transient fluctuation of the link voltage.

There is a local inductance L_s connected between the line-side inverter output and the three-phase utility bus. Such inductance is very important for stable operation of the synchronous current controller, and it is selected in such a way that the maximum modulation index at which the line-side inverter operates is as close to one as is permitted by the minimum pulse or notch width capability of the device. Under these conditions the control varies over a wide range, making the control less sensitive to errors in the controller gains and compensation. The inductance value is such that the slope of the PI output is smaller than the slope of the triangular carrier of the SPWM.[9]

The system can be satisfactorily controlled for start-up and regenerative braking shutdown modes besides the usual generating mode of operation. The flowchart in Figure 10.17 shows the procedure for the initial start-up of a wind turbine. Both inverters are initially disabled during charging of the DC-link capacitor; that is, all the gate pulses are off. The capacitor is charged from the diodes on the line-side inverter with the peak value of the line voltage. To prevent damage to diodes there is a series resistance for the capacitor initial charging R_s in the DC link. Such resistance is by-passed by an electromagnetic relay when the bus voltage reaches something like 95% of the line voltage. When the capacitor has been charged, the DC-link voltage loop control is exercised, the pulse gates are enabled, and the control loops are activated to establish a smaller voltage than the bus peak value (typically, 75% of the peak value), therefore, successfully operating the line-side inverter in PWM mode.

With the DC-link bus voltage fixed, the induction generator can be excited with i_{ds}, as the rated flux is established in the induction generator. Next, a speed reference is commanded to rotate the turbine with the minimum turbine speed required to catch some power from wind (the fuzzy speed controller FLC-3 is always working). The flow of wind imposes a regenerative torque in the induction generator. Of course, the power generation is not optimum yet, but the slip frequency becomes negative and the power starts to flow from the turbine towards the line side. As the power starts to flow to the line side, the DC-link voltage loop control can be gradually commanded to a higher value than the peak value of the line-side voltage (typically, 75% higher). After the DC-link voltage is established in the new higher value, the system is ready to be controlled by fuzzy controllers FLC-1 and FLC-2.

An IGBT PWM-bridge rectifier that also supplies lagging excitation current to the induction generator rectifies the variable-frequency variable-voltage power generated by the induction generator. The DC-link power is inverted

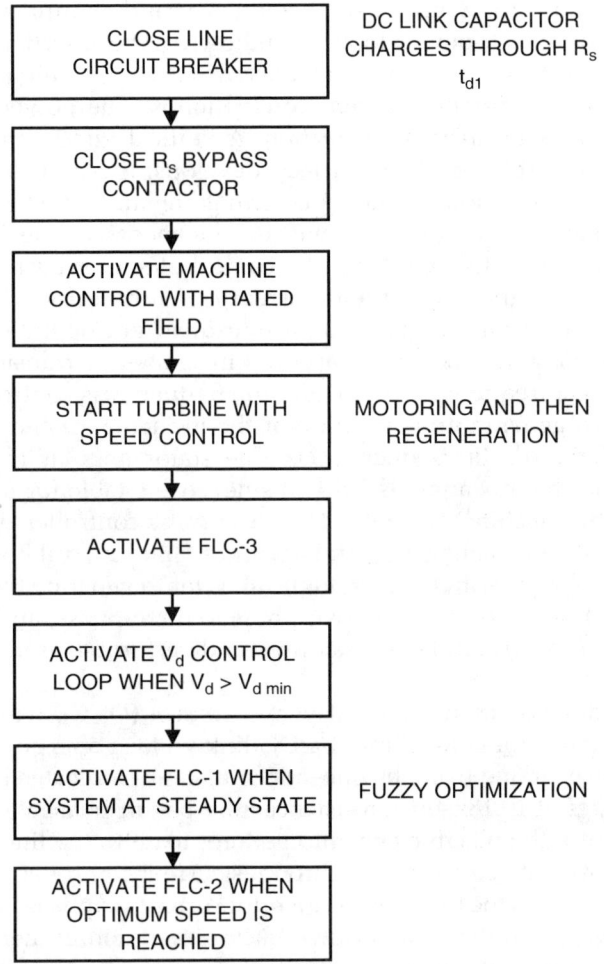

FIGURE 10.17
Initialization of control system.

to 60 Hz, 220 V AC through an IGBT PWM inverter and fed to the utility grid at unity power factor, as indicated. The line-power factor can also be programmed to lead or lag by static VAR compensation, if desired. The generator speed is controlled by indirect vector control with torque control and synchronous current control in the inner loops. The induction generator flux is controlled in open loop by control of the i_{ds} current, but in normal condition, the rotor flux is set to the rated value for fast transient response. The line-side converter is also vector-controlled using direct vector control and synchronous current control in the inner loops. The output power P_0 is controlled to control the DC-link voltage V_d. Since increase of P_0 causes decrease of V_d, the voltage loop error polarity has been inverted.

Tight regulation of V_d within a small tolerance band requires a feedforward power injection in the power loop, as indicated. The insertion of the filter inductance L_s creates some coupling effect, which is eliminated by a de-coupler in the synchronous current control loops. The power can be controlled to flow easily in either direction. A vertical wind turbine requires start-up motoring torque. As the speed develops, the induction generator goes into generating mode. Regenerative braking shuts down the machine.

For the system shown in Figure 10.16, the machine-side inverter uses indirect vector control (IVC). With the control of vector currents i_{ds} and i_{qs}, the rotor flux is aligned with the d-current i_{ds}. With vector control there is no danger of an instability problem; a four-quadrant operation, including zero speed is possible. The drive has a DC-machine-like transient response: there is no direct frequency control, and the frequency is controlled through the unit vector generated by addition of the induction generator speed ω_{rt} with a feedforward slip-frequency ω_{sl}. The stator angular frequency ω_e is integrated and the flux angle is fed to a sine/cosine table for inverse vector rotation of the machine currents. The PI current controller generates the quadrature reference voltages v_{qs}^* and v_{ds}^*, and a PWM modulator that vector-rotates such voltages with the flux angle, in order to generate the gate pulses for the IGBT. The vector-rotation, two-phase to three-phase, and three-phase to two-phase transformations are standard and performed as in the literature on vector control.[7,8]

The line-side inverter uses direct vector control (DVC), with a very fast transient response for controlling the DC-link voltage and power flow. The unit vector is generated from the line-side voltages as indicated by the phasor diagram of Figure 10.18, so the vector current I_s is in phase with the vector voltage V_s. Since the reactive current i_{ds} is kept to zero, the line-side current is inverse-vector-rotated, and a synchronous current controller with a decoupling network places the three-phase inverter currents in phase with the three-phase voltages, permitting unity power factor operation. If there is an available inverter power rating, the line-side power factor can be programmable for leading or lagging as required by the utility power system. The equations used in the line-side DVC strategy are available in the literature.[9–11]

10.6.1   Description of Fuzzy Controllers

The optimum fuzzy logic operating control of an induction generator requires three different fuzzy logic controllers (FLC). The controller here designated as FLC-1 tracks the maximum wind speed. The controller FLC-2 establishes the optimum magnetic flux for the induction generator. FLC-3 is the optimum generator's torque-speed controller.

10.6.1.1   Speed Control with the Fuzzy Logic Controller FLC-1

The product of torque and speed is turbine power, and it equals the line power (assuming a steady-state lossless system). The curves in Figure 10.19

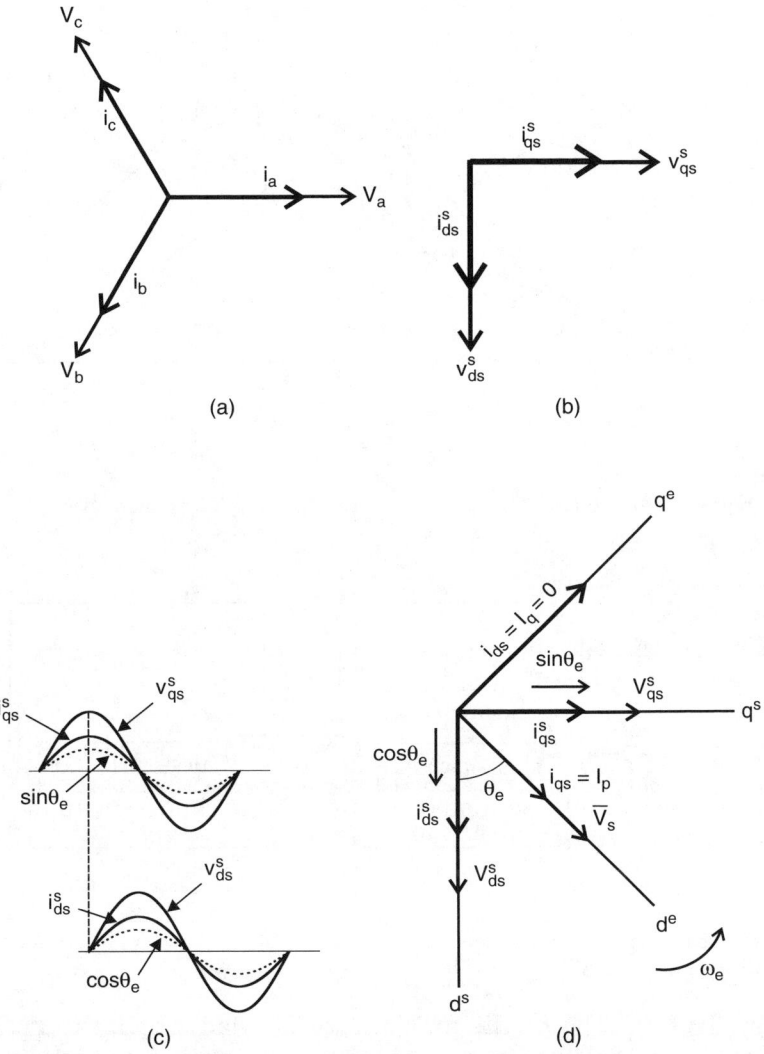

FIGURE 10.18
Phasor diagram for line-side DVC, (a) three-phase line phasors, (b) two-phase line phasors, (c) signal voltage and current waves, and (d) signals in d^s – q^s and d^e – q^e frames.

show the line power P_0 over a family of generator speed ω_r in terms of wind velocities. For a particular value of wind velocity, the function of the fuzzy controller FLC-1 is to seek the generator speed until the system settles down at the maximum output power condition. At a wind velocity of V_{w4}, the output power will be at A if the generator speed is ω_{r1}. The FLC-1 will alter the speed in steps until it reaches the speed ω_{r2} where the output power is maximized at B. If the wind velocity increases to V_{w2}, the output power will jump to D, and then FLC-1 will bring the operating point to E by searching

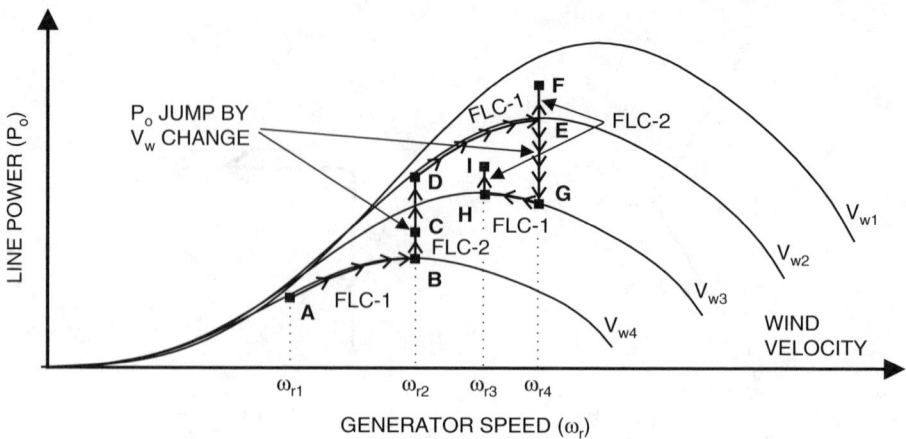

FIGURE 10.19
Fuzzy control FLC-1 and FLC-2 operation showing maximization of line power.

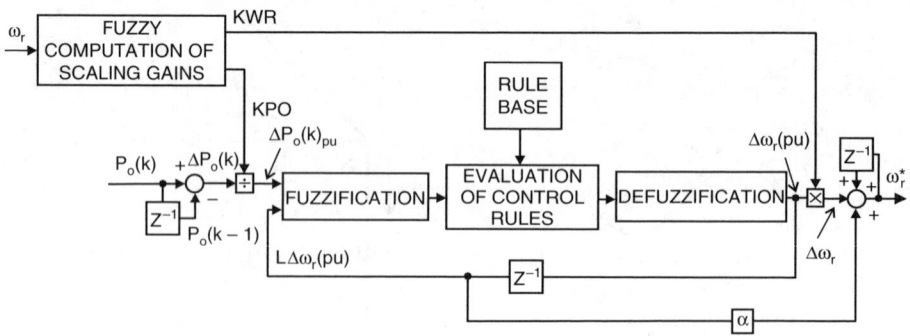

FIGURE 10.20
Block diagram of fuzzy control FLC-1.

the speed to ω_{r4}. The profile for decrease of wind velocity to V_{w3} is also indicated. Therefore, the principle of the fuzzy controller is to increment (or decrement) the speed in accordance with the corresponding increment (or decrement) of the estimated output power P_0. After reaching the optimum generator speed, which maximizes the turbine aerodynamic performance, the machine rotor flux i_{ds} is reduced from the rated value, to reduce the core loss, thereby further increasing the machine-converter system efficiency.

If ΔP_o is positive with the last positive $\Delta \omega_r$, the search is continued in the same direction. If, on the other hand, $+\Delta \omega_r$ causes $-\Delta P_o$, the direction of search is reversed. The variables ΔP_o (variation of power), $\Delta \omega_r$ (variation of speed) and $L\Delta \omega_r$ (last variation of speed) are described by membership functions given in Figure 10.20 and the inference is given in the rule in Table 10.1.

In the implementation of fuzzy control, the input variables are fuzzified in accordance with the membership functions indicated in Figure 10.21. The

TABLE 10.1

Rule Table for FLC-1

ΔP_0 \ $\Delta\omega_{r(last)}$	P	ZE	N
PVB	PVB	PVB	NVB
PBIG	PBIG	PVB	NBIG
PMED	PMED	PBIG	NMED
PSMA	PSMA	PMED	NSMA
ZE	ZE	ZE	ZE
NSMA	NSMA	NMED	PSMA
NMED	NMED	NBIG	PMED
NBIG	NBIG	NVB	PBIG
NVB	NVB	NVB	PVB

valid control rules are evaluated and combined through rule Table 10.1, and finally the output is back defuzzified to convert it to a crisp value. The output $\Delta\omega_r$ is added by some amount of $L\Delta\omega_r$ in order to avoid local minima due to wind vortex and torque ripple. The controller operates on a per-unit basis so that the response is insensitive to system variables and the algorithm is universal to any system.

The scale factors KPO and KWR as shown are functions of the generator speed, so that the control becomes somewhat insensitive to speed variation. The scale factor is generated by fuzzy computation. The speed is first evaluated into seven fuzzy sets as shown in membership functions indicated in Figure 10.22; the scale factors KPO and KWR are generated in according to the rule in Table 10.2. From this explanation, the advantages of fuzzy control are obvious. It provides adaptive step size in the search that leads to fast convergence, and the controller can accept inaccurate and noisy signals. The FLC-1 operation does not need any wind velocity information, and its real-time-based search is insensitive to system parameter variation.

10.6.1.2 Flux Intensity Control with the Fuzzy Logic Controller FLC-2

Since most of the time the generator is running at light load, the induction generator rotor flux i_{ds} can be reduced from the rated value to reduce the core loss and thereby increase the machine-converter system efficiency. The principle of on-line search-based flux programming control by a second fuzzy controller FLC-2 is explained in Figure 10.23. At a certain wind velocity V_w and at the corresponding optimum speed $\Delta\omega_r$ established by FLC-1, which operates at rated flux (ψ_R), the rotor flux (ψ_R) is reduced by decreasing the magnetizing current i_{ds}. This causes an increasing torque current i_{qs} by the speed loop for the same developed torque. As the flux is decreased, the induction generator iron loss decreases with the concurrent increase of copper loss. However, the total system (converters and machine) loss decreases, resulting in an increase of the total generated power P_o. The search is continued until the system settles down at the maximum power point A, as indicated

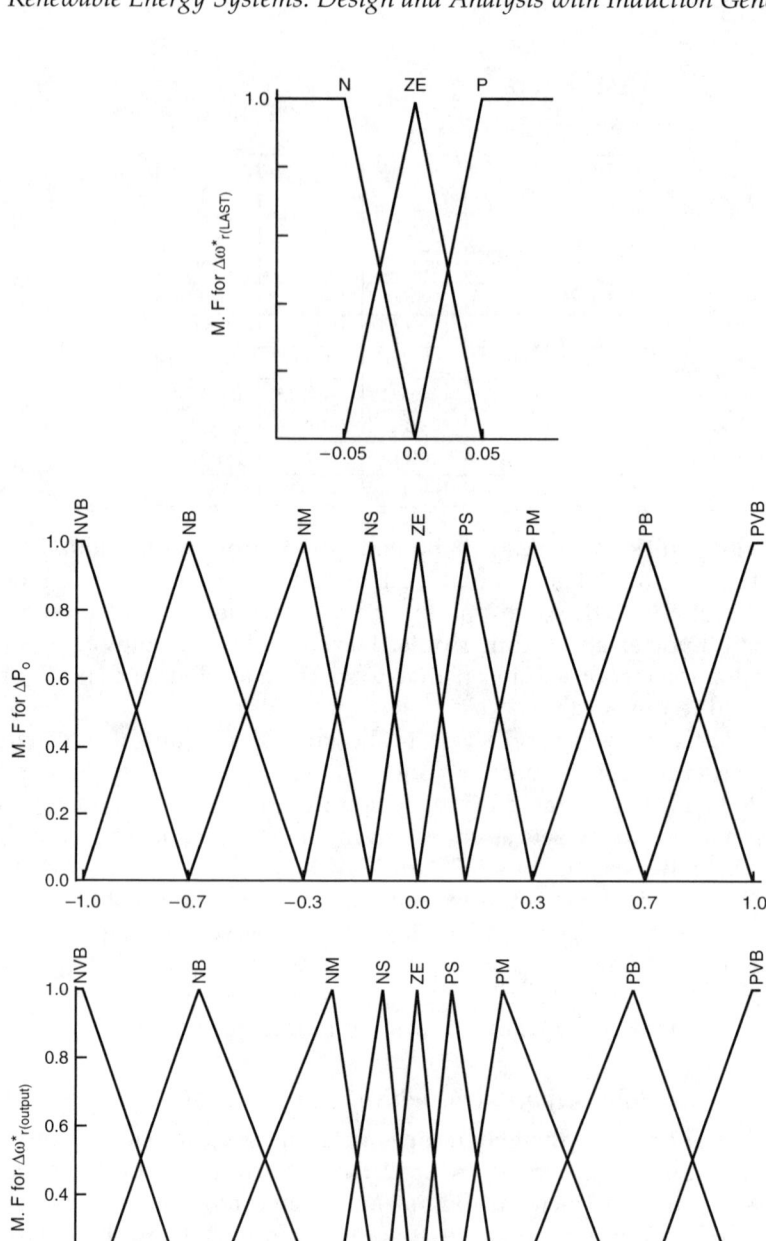

FIGURE 10.21
Membership function for FLC-1.

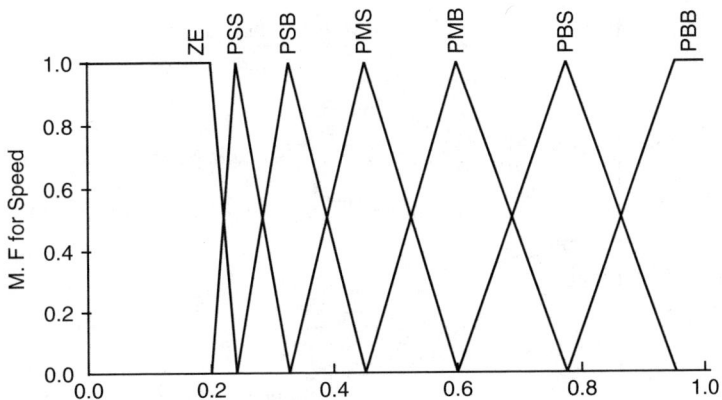

FIGURE 10.22
Membership function for computation of scaling gains.

TABLE 10.2

Rule Table for Computation of Scaling Gains

ω_r	KPO	KWR
PSS	40	25
PSB	210	40
PMS	300	40
PMB	375	50
PBS	470	50
PBB	540	60

in Figure 10.23. Any attempt to search beyond point A will force the controller to return to the maximum power point.

The principle of fuzzy logic controller FLC-2 is somewhat similar to that of FLC-1 and is explained in Figure 10.24. The system output power $P_o(k)$ is sampled and compared with the previous value to determine the increment ΔP_o. In addition, the last excitation current decrement $L\Delta i_{ds}$ is reviewed. The membership functions for variation of power ΔP_o, last variation of i_{ds}, $L\Delta i_{ds}$ and change in i_{ds}, Δi_{ds}, are given in Figure 10.25. On these bases, the decrement step of i_{ds} is generated from fuzzy rules given in rule Table 10.3 through fuzzy inference and defuzzification.

The adjustable gains KP and KIDS, which convert the actual variables to per-unit variables, are given by the respective expressions:

$$KP = a\omega_r + b \tag{10.14}$$

$$KIDS = c_1\omega_r - c_2 T + c_3 \tag{10.15}$$

where a, b, c_1, c_2 and c_3 are derived by trial and error. The effect of controller FLC-2 is to boost the power output.

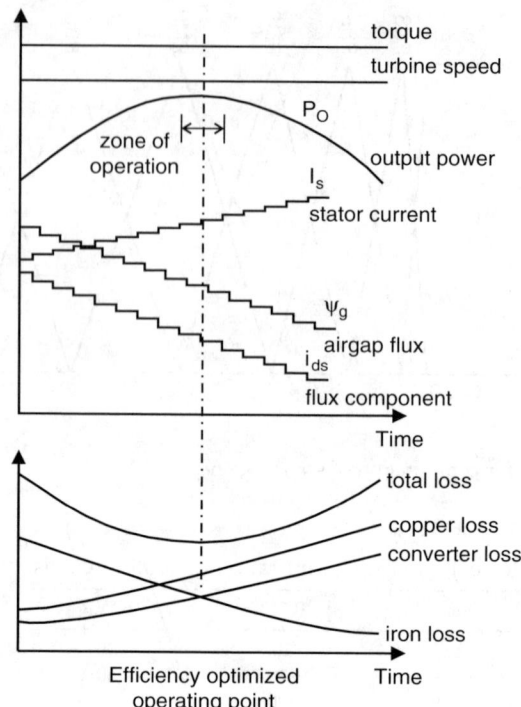

FIGURE 10.23
Fuzzy control FLC-2 operation showing optimization of flux level.

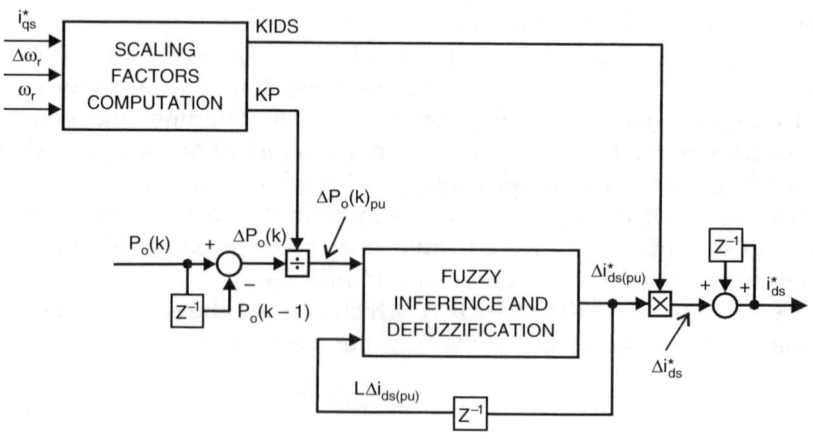

FIGURE 10.24
Block diagram of fuzzy control FLC-2.

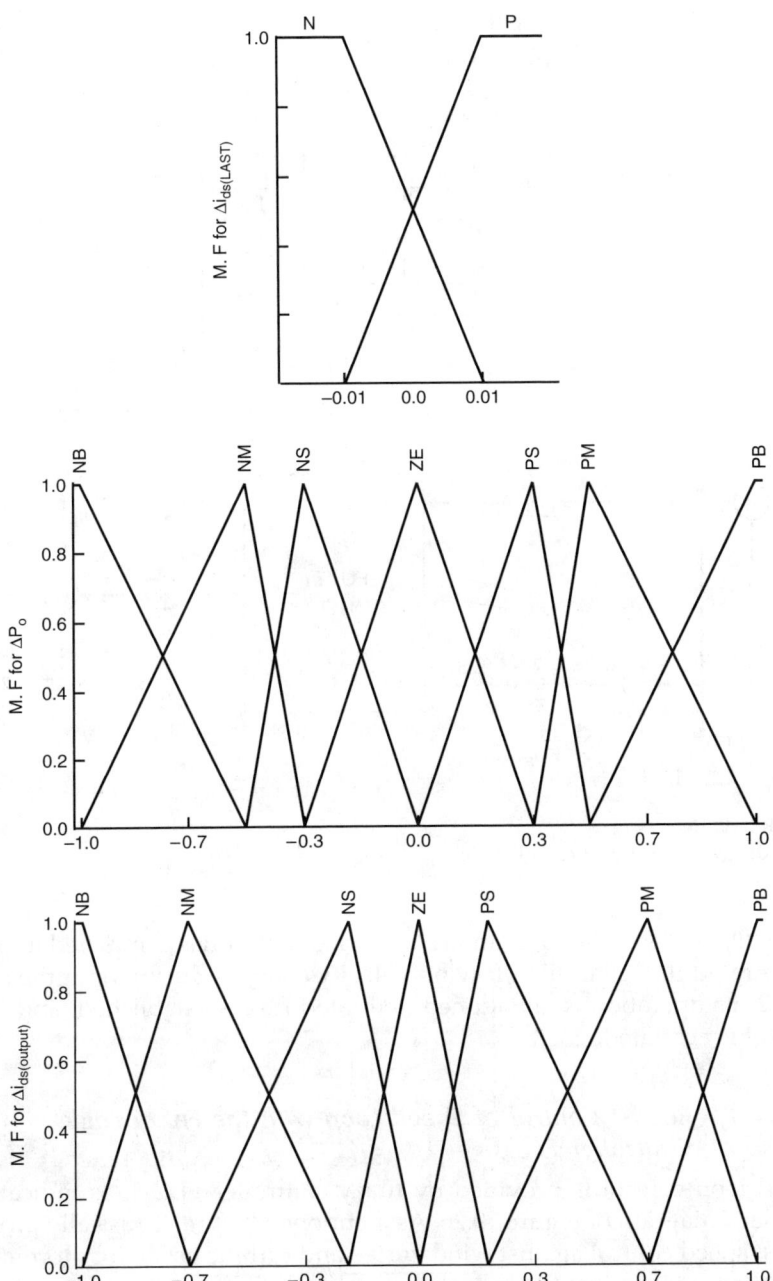

FIGURE 10.25
Membership function for FLC-2.

TABLE 10.3

Rule Table for FLC-2

ΔP_0 ＼ $\Delta i_{ds(last)}$	N	P
PB	NM	PM
PM	NS	PS
PS	NS	PS
NS	PS	NS
NM	PM	NM
NB	PB	NB

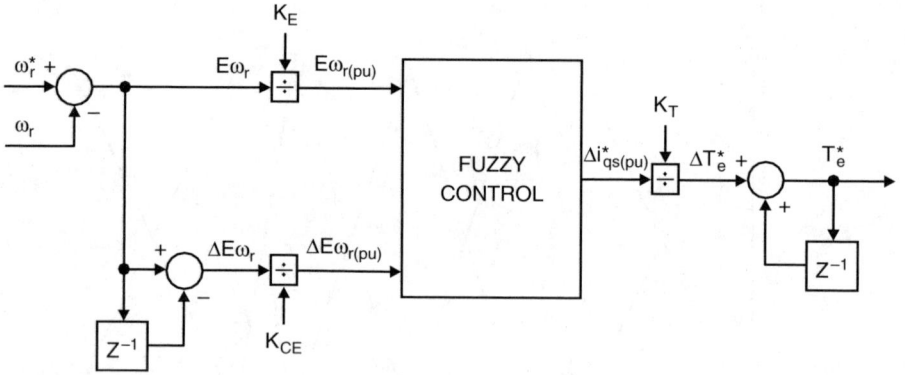

FIGURE 10.26
Block diagram of fuzzy control FLC-3.

The FLC-2 controller operation starts when FLC-1 has completed its search at the rated flux condition. If wind velocity changes during or at the end of FLC-2, its operation is abandoned, the rated flux is established, and FLC-1 control is activated.

10.6.1.3　*Robust Control of Speed Loop with the Fuzzy Logic Controller FLC-3*

Speed loop control is provided by fuzzy controller FLC-3, as indicated in the block diagram of Figure 10.26. As mentioned before, it basically provides robust speed control against wind vortex and turbine oscillatory torque. The disturbance torque on the induction generator shaft is inversely modulated with the developed torque to attenuate modulation of output power and prevent any possible mechanical resonance effect. In addition, the speed control loop provides a deadbeat-type response when an increment of speed is commanded by FLC-1. Figure 10.26 shows the proportional-integral (PI) type of fuzzy control used in the system. The speed loop error $E\omega_r$ and error change $\Delta E\omega_r$ signals are converted to per-unit signals, processed through fuzzy control in accordance to the membership functions given in Figure 10.27 and the rule in Table 10.4. The output of the fuzzy controller FLC-1 is summed to produce the generator torque reference T_e^*.

10.6.2 Experimental Evaluation of the Fuzzy Optimization Control

The induction machine used in the experimental setup for this chapter was a standard NEMA Class B type with 220 V, 3.5 hp rating. A 7.5 hp four-quadrant laboratory dynamometer was used to emulate the wind turbine (programmable shaft torque). The system parameters are given in Table 10.5.

While the speed fuzzy controller FLC-3 was always active during system operation, the controllers FLC-1 and FLC-2 operated in sequence at steady (or small turbulence) wind velocity. Besides, a start-up procedure was required for complete activation and a shutdown sequence in case of a fault. There was control coordination for sequencing and the start-up/shut-down procedures.[4-6] For the start-up, the line-side circuit breaker was closed. The DC-link capacitor charged to the peak value of the line voltage through a series resistance, which avoided the inrush charging current. After a delay of 0.5 sec, the resistance was by-passed with a relay, and the rated flux was imposed on the induction machine.

The turbine was started with speed control and as the power started to flow, the DC-link voltage rose. The DC-link voltage control was activated when the DC-link voltage was above the limit value and FLC-1 started to search the optimum speed reference ω_r^*. As ω_r^* was altered the power generation went up, until FLC-1 settled down in the steady-state condition, indicated by a small variation in $|\Delta\omega_r^*|$ with alternating polarity.

In this condition the system was transferred to FLC-2 in order to optimize the flux by decreasing i_{ds}^*. During the speed search, any deviation from the expected variation of power indicated that the system was subject to large wind variation and the system was transferred from FLC-1 to non-fuzzy operation, waiting for the transient to vanish. During the flux optimization the excitation current i_{ds}^* was decreased adaptively by FLC-2. Control was then transferred to optimum operation when the variation $|\Delta i_{ds}^*|$ was small with alternating polarity. The search was finished and the optimum operation state kept the optimum ω_r^* and i_{ds}^*.

The power and variation of power were recorded in order to see if there was any variation in the wind velocity to restart the search. Again, during the operation of FLC-2, any load transient, indicated by variation of torque, transferred the system to the non-fuzzy operation state, waiting for the transient to vanish in order to restart from FLC-1.

The system had external fault indications from the turbine and inverters, which could indicate dangerous operation during too-high wind velocity, or when wind velocity was too low and the power could not satisfactorily be generated. Inverters could indicate a tripping by short-circuit: too much

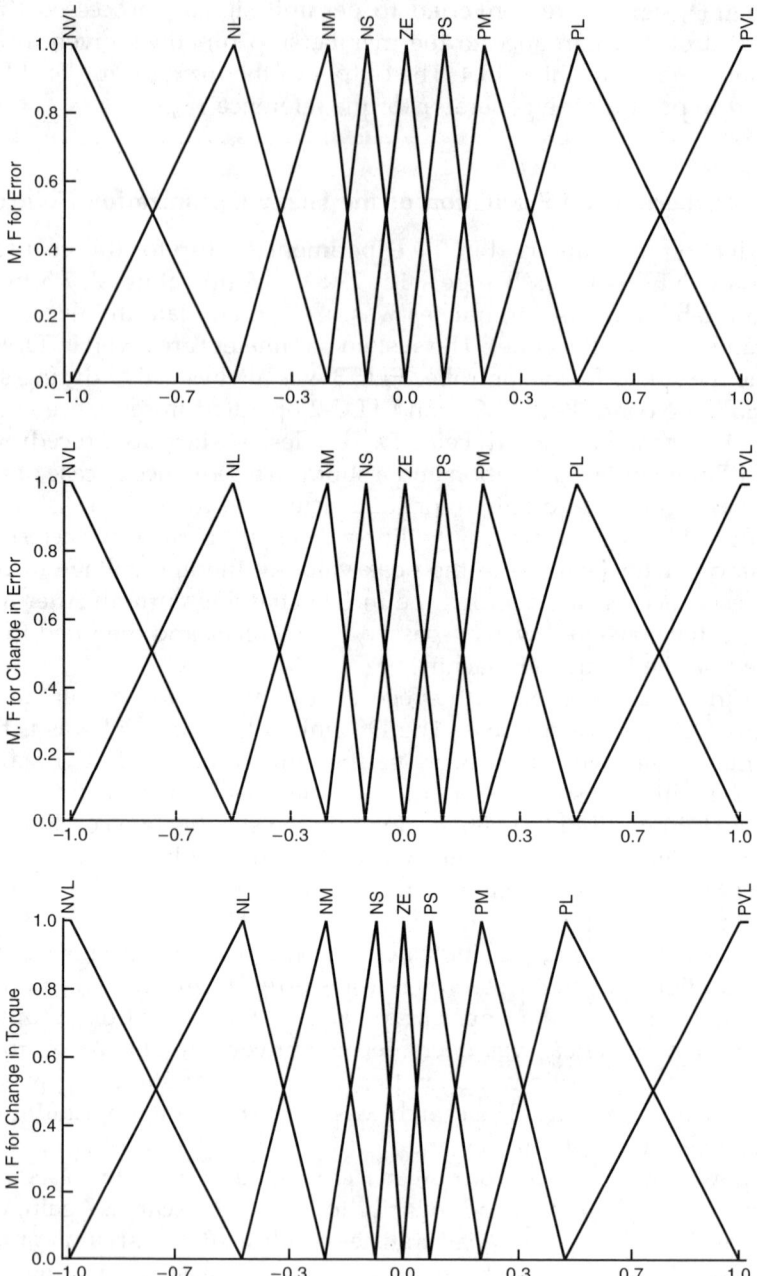

FIGURE 10.27
Membership function for FLC-3.

TABLE 10.4

Rule Table for FLC-3

CE \ E	NVL	NL	NM	NS	ZE	PS	PM	PL	PVL
NVL					NVL	NL	NM	NS	ZE
NL					NL	NM	NS	ZE	PS
NM				NL	NM	NS	ZE	PS	PM
NS			NL	NM	NS	ZE	PS	PM	PL
ZE		NL	NM	NS	ZE	PS	PM	PL	
PS	NL	NM	NS	ZE	PS	PM	PL		
PM	NM	NS	ZE	PS	PM	PL			
PL	NS	ZE	PS	PM	PL				
PVL	ZE	PS	PM	PL	PVL				

TABLE 10.5

Induction Generator and Turbine Parameters

Induction Generator Parameters:

5 hp	230/460 V	13.4/6.7 A
4 poles	1800 RPM	NEMA Class B
Rs = 0.370 Ω		Rr = 0.436 Ω
Lls = 2.13 mH	Llr = 2.13 mH	Lm = 62.77 mH

Turbine Parameters:

A = 0.015	B = 0.03	C = 0.015
	ηGEAR = 5.7	

current or too-high temperature. If the machine inverter tripped, the line-side inverter could easily shut down the system. On the other hand, if the line side inverter tripped, the dynamic break in the DC-link bus would keep the bus voltage in a safe range, while the induction generator decelerated to zero speed. Any fault that occurred during the search procedure transferred the system to the shutdown procedure. The fuzzy controllers were disabled, and a deceleration profile was imposed in the speed control mode. The turbine was mechanically yaw-controlled to permit the wind to pass through the blades, and finally the line circuit breaker was disconnected.

Figure 10.28 shows the static characteristics of the wind turbine at different wind velocity. Basically, these are families of curves for turbine output power, turbine torque, and line-side output power as functions of wind velocity and sets of generator speed. For example, if the generator speed remained constant and the wind velocity increased, the turbine power, turbine torque, and line power would increase and then tend to saturate. The slope of

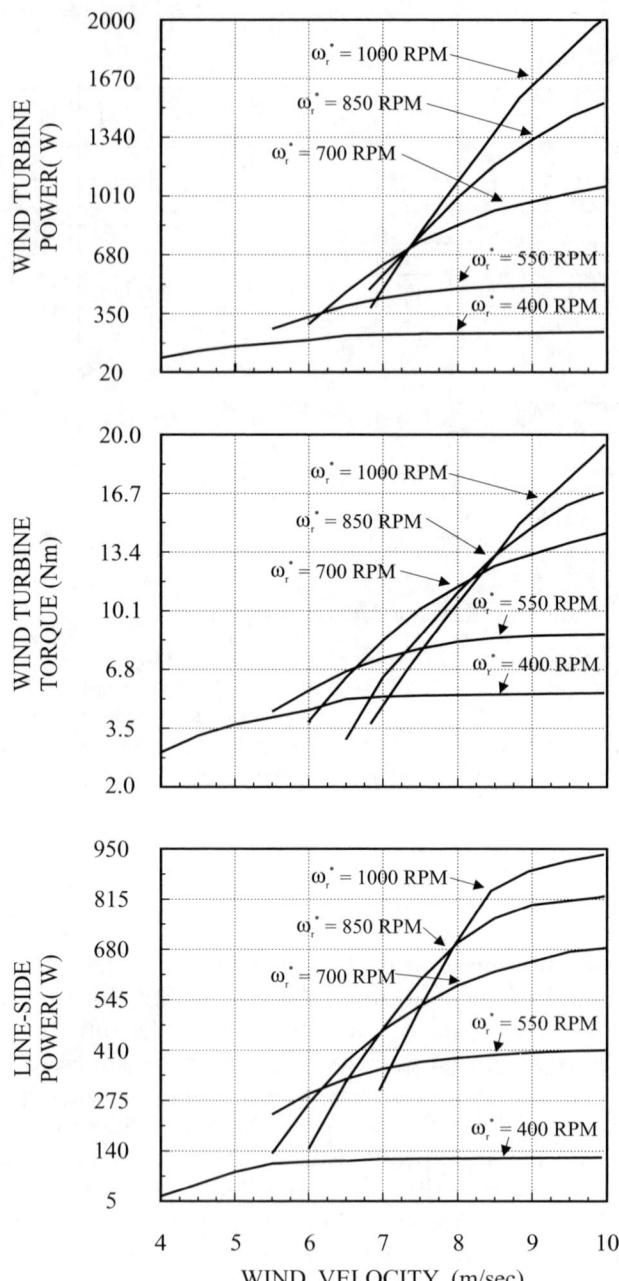

FIGURE 10.28
Static characteristics.

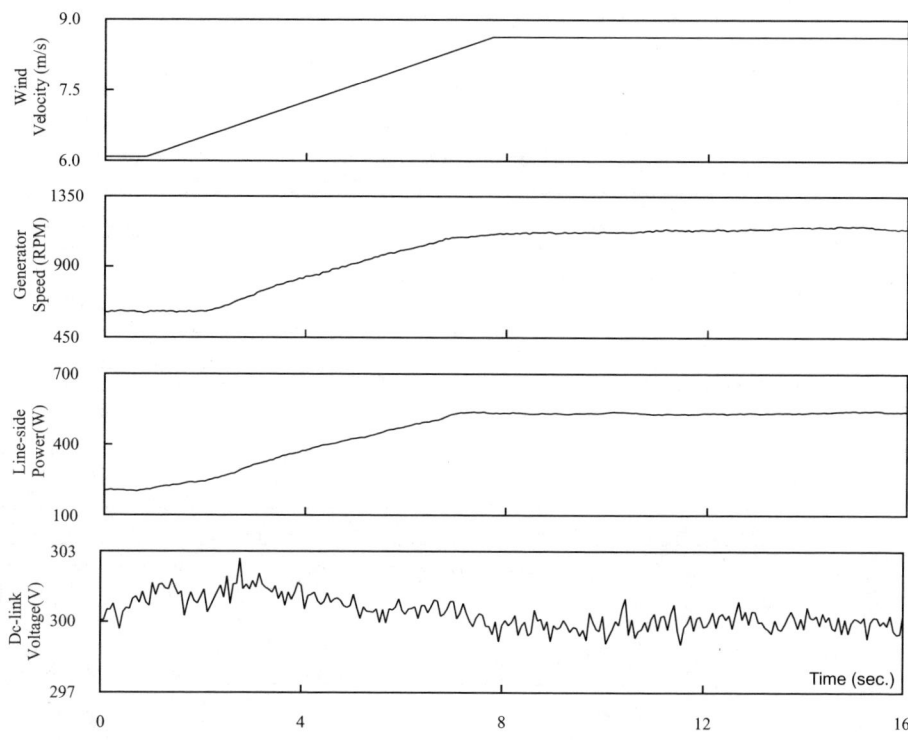

FIGURE 10.29
Aerodynamic optimization with FLC-1.

increase is higher with higher generator speed. For a fixed wind velocity, as the generator speed increased, the torque and power outputs first increased and then decreased.

The effect of fuzzy controller FLC-1 when the wind velocity slowly increased from 6.125 m/s to 8.750 m/s is shown in Figure 10.29. The generator speed tracked the wind velocity, gradually increasing the line-side power, as shown. As the wind velocity settled down to the steady-state value, the generator speed step size decreased until at the optimum power output point oscillated with a small step, as explained before. Note that for a fast transient in wind velocity, the controller FLC-1 remained inoperative. The DC-link voltage is analogous to the level of a low-capacity water tank with input and output water flow pipes, and maintaining its value rigidly constant is extremely difficult. A feedforward power compensator, shown in Figure 10.16, helped reduce the fluctuation. However, as long as the voltage level remained within a tolerance band, the converter system was safe and the performance was satisfactory. The dynamic brake tended to absorb destructive voltage surges in the DC link, as mentioned before.

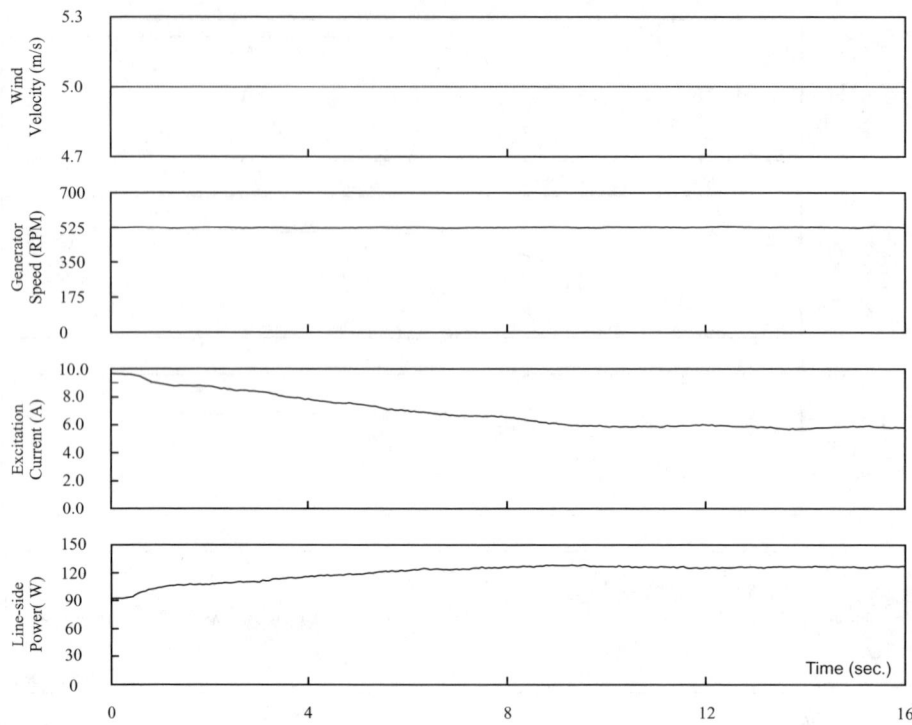

FIGURE 10.30
Flux optimization with FLC-2.

The fuzzy controller FLC-2 started when the operation of FLC-1 ended. Figure 10.30 shows the performance of FLC-2 at a constant wind velocity of 5.0 m/s and constant generator speed of 520 rpm. The generator excitation current gradually decreased from the initial rated value that increased the line-side power because of improved machine efficiency. The step size of the excitation current gradually decreased as the steady state was approached, and then it oscillated around the steady-state point. The increment of power was not very large because the induction generator was operating in a light-load condition.

In order to test the robustness of the fuzzy speed controller FLC-3, the oscillatory torque components were added to the dynamometer wind turbine model. Figure 10.31 shows the smooth speed profile (top) at 900 rpm with oscillatory torque (bottom) that swung from 4.35 Nm to 5.65 Nm with an average value of 5 Nm. The additional pulsating torque introduced by FLC-2 was also highly attenuated by the FLC-3 controller. Figure 10.32 shows an oscillogram of the experimental setup.

Next, the turbine-generator system was operated at constant speed (940 rpm) and the wind velocity was varied. At each operating point, the

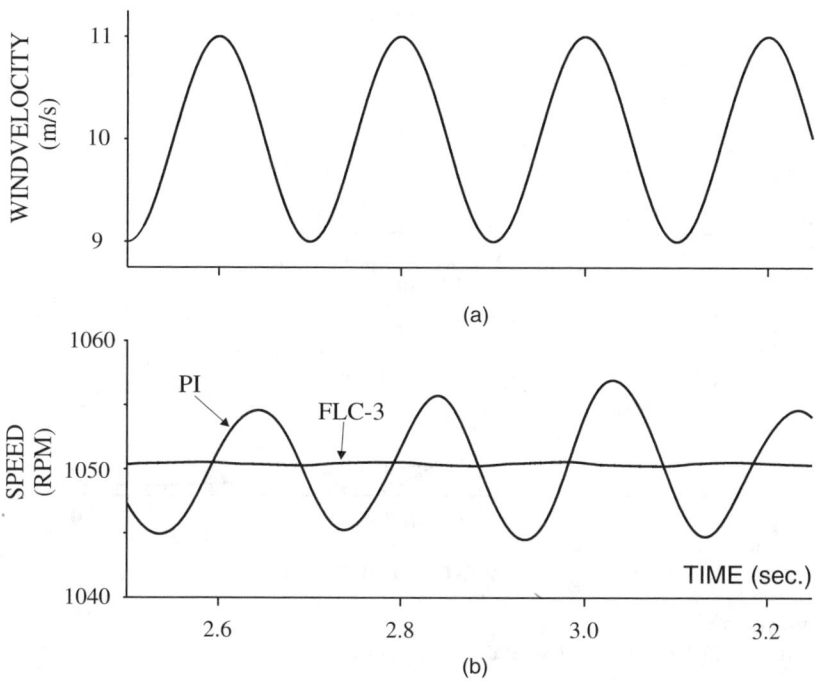

(a)

(b)

FIGURE 10.31
Comparison of speed controller response by PI and FLC-3 for wind vortex (a) wind velocity with vortex and (b) FLC and PI responses.

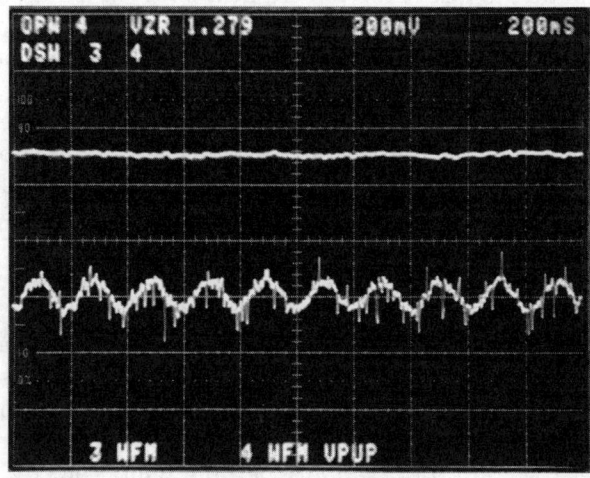

FIGURE 10.32
Speed response compared to turbine oscillatory torque.

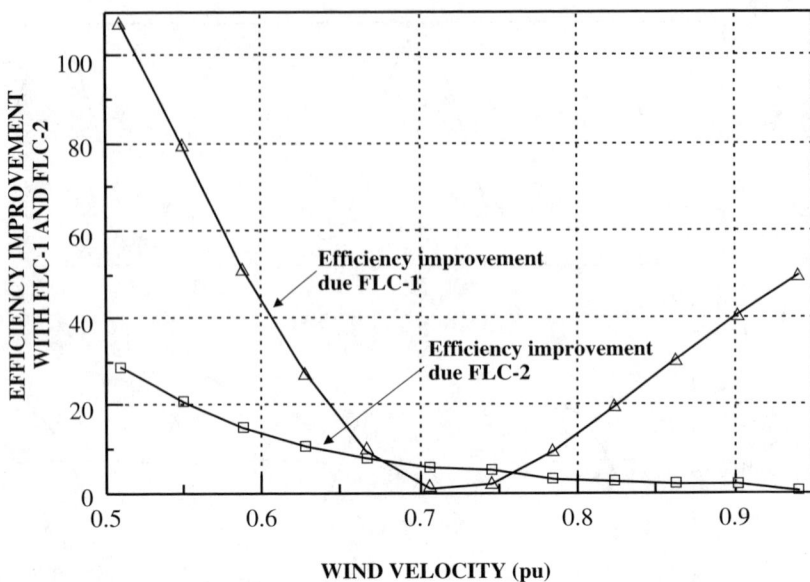

FIGURE 10.33
Wind energy efficiency improvement due fuzzy control.

FLC-1 and FLC-2 controllers were operated in sequence and the corresponding boost of power was observed. From these data, the respective efficiency improvement was calculated and plotted above. Figure 10.33 indicates that the efficiency gain was significant with FLC-1 control compared to that of FLC-2. The gain due to FLC-1 fell to zero near 0.7 p. u. wind velocity where the generator speed was optimum for that wind velocity, and then rose. The gain due to FLC-2 decreased as the wind velocity increased because of higher generator loading.

Figure 10.34 shows the steady-state performance enhancement of control of the wind turbine at several operating points. After start-up, the system was operating at A. As the speed reference command switched from the host communication interface to the fuzzy controller FLC-1, the reference speed increased, and the optimized point at B (for wind velocity of 5 m/sec) was reached. The flux optimization search further enhanced the power generation by boosting the output power to the point C, as shown in Figure 10.34. The figure also shows the output power level, point D, when the wind velocity stepped from 5 m/sec to 7 m/sec. The rated flux was established, and the fuzzy controller FLC-1 was enabled to search the optimum point at E. Again the flux optimization boosted the generated power to F. The path G-H-I indicates the control operation when the wind velocity was stepped down from 7 m/sec to 6 m/sec. The line-side direct vector control kept $i_{ds} = 0$, maintaining unity power factor all the time. Figure 10.35 shows an oscillogram

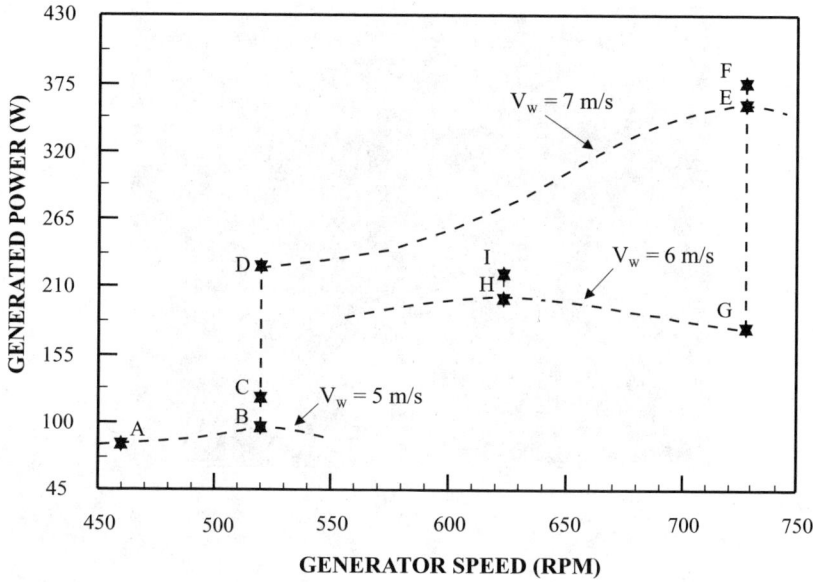

FIGURE 10.34

Fuzzy logic steady state performance enhancement control of wind turbine at several operating points.

with a sinusoidal line current and unity power factor. The out-of-phase current wave indicates that the system was in the generation mode.

10.7 Chapter Summary

This chapter covered principles of hill climbing (HCC) and fuzzy logic control (FLC) and applied them to the energy performance enhancement of induction-generator-based renewable energy systems. HCC is appropriate to implement electronic control by the load, having the load follow the incoming input power, whereas FLC is appropriate to adjust the induction generator speed so as to match the maximum aerodynamic performance, flux programming and robust speed control with three fuzzy controllers. HCC and FLC have also been applied for optimization of other types of renewable energy sources and performed well.[12–15] The advantages of both hill climbing and fuzzy control are that the control algorithms are universal, fast converging, parameter insensitive, and accepting of noisy and inaccurate signals.

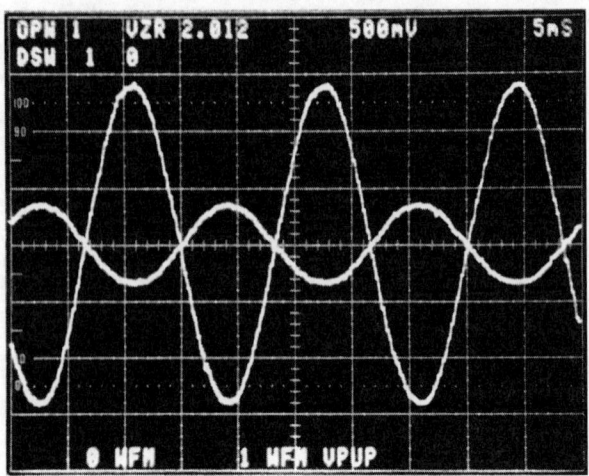

FIGURE 10.35
Line-side voltage and current.

10.8 Problems

10.1 What are the characteristics of optimization methods related to renewable sources of energy? Why may efficiency in Hill Climbing and Fuzzy Control be irrelevant with respect to maximum generated power?

10.2 Give five reasons why optimization in renewable energy systems is important.

10.3 Describe the tracking methodology used in Hill Climbing Control and discuss each method so as to provide a basis for choosing one of them for a particular application.

10.4 Describe the tracking methodology used in Fuzzy Logic Control and compare it with that of Hill Climbing Control.

10.5 Explain under what conditions and why HCC adaptive control may be unstable and why the HCC exponential control tends to be the most stable, although limited.

References

1. Toumiya, T., Suzuki, T., and Kamano, T., Output control method by adaptive control to wind power generating system, *Proceedings of IECON'91*, Vol. 3, 2296–2301, Nov. 1991.

2. Datta, R. and Ranganathan, V.T., A method of tracking the peak power points for a variable speed wind energy conversion system, *IEEE Trans. on Energy Conversion*, 18(1), 163–168, March 2003.

3. Farret, F.A., Pfitscher, L.L., and Bernardon, D.P., An heuristic algorithm for sensorless power maximization applied to small asynchronous wind turbogenerators, *Proceedings of IEEE Industrial Electronics Symposium*, Vol. 1, 179–184, December 2000.

4. Simões, M.G., Bose, B.K., and Spiegel, R.J., Fuzzy logic based intelligent control of a variable speed cage machine wind generation system, *IEEE Transactions on Power Electronics*, Vol. 12, 87–95, Jan. 1997.

5. Simões, M.G., Bose, B.K., and Spiegel, R.J., Design and performance evaluation of a fuzzy-logic-based variable-speed wind generation system, *IEEE Transactions on Industry Applications*, Vol. 33, 956–965, July/August 1997.

6. Bose, B.K. and Simões, M.G., Fuzzy logic based intelligent control of a variable speed cage machine wind generation system, *Environmental Protection Agency EPA/600/SR-97/010*, published by National Technical Information Service (Report # PB97-144851), March 1997.

7. Bose, B.K., *Modern Power Electronics and AC Drives*, Prentice Hall, New York, 2001.

8. Novotny, D.W. and Lipo, T.A., *Vector Control and Dynamics of AC Drives*, Oxford University Press, New York, 1996.

9. Sukegawa, T., Kamiyama, K., Takahashi, J., Ikimi, T., and Matsutake, M., A multiple PWM GTO line-side converter for unity power factor and reduced harmonics, *Conf. records of IEEE Industry Applications Society annual meeting*, Vol. 1, 279–284, Sept./Oct. 1991.

10. Kazmierkowski, M.P., Krishnan and Blaabjerg, R.F., *Control in Power Electronics: Selected Problems*, Academic Press, San Diego, 2002.

11. Duarte, J.L., Van Zwam, A., Wijnands, C. and Vandenput, A., Reference frames fit for controlling PWM rectifiers, *IEEE Trans. on industrial electronics*, 46(3), 628–630, June 1999.

12. Simões, M.G. and Franceschetti, N.N., Fuzzy optimization based control of a solar array system, *IEEE Proceedings Electric Power Applications*, 146(5), 552–558, Sept. 1999.

13. Teulings, W.J.A., Marpinard, J.C., Capel, A., and O'Sullivan, D., A new maximum power point tracking system, *Conf. records of 24th IEEE power electronics specialists conference*, 833–838, June 1993.

14. Senjyu, T. and Uezato, K., Maximum power point tracker using fuzzy control for photovoltaic arrays, *Proceedings of the IEEE International Conference on Industrial Technology*, 143–147, Dec. 1994.

15. Niimura, T. and Yokoyama, R., Water level control of small-scale hydro-generating units by fuzzy logic, *Proceedings of IEEE International Conference. on Systems, Man and Cybernetics*, Vol. 3, 2483–2487, Oct. 1995.

11

Wound-Rotor Induction-Generator Systems

11.1 Scope of This Chapter

In this chapter a theoretical and practical discussion of wound-rotor induction-generator (WRIG) systems will form the basis of their application to high-power renewable energy systems. Since control is tied to the rotor, only slip power is processed where the purpose of control is to synchronize the rotor current with respect to stator reference. Thus the smaller the range of operating slip, the smaller the required power converter. The size of the power converter may strongly affect the initial cost of energy ($/kWh). The utility side of the power converter of the WRIG can be controlled differently from both sides of the power converter. The power factor can be controlled either on the system side or on the utility or rotor side of the WRIG. The wound-rotor generator operates in four quadrants around the synchronous speed. Therefore, it can easily operate in motoring mode, for example for pumped hydro applications.

11.2 Features of WRIG

Before the age of power semiconductors, wound-rotor machines were used with external rotor resistances to control the slip, and consequently torque was lost due to those resistances. The wound-rotor induction generator is not as rugged as the squirrel-cage type, but the brushes have little wear and sparking when compared to DC machines and are the only acceptable alternatives for alternative energy conversion in the megawatts power range.

The fundamentals of slip recovery systems have been known since the beginning of the 20th century. Between 1907 and 1913 Kraemer and Scherbius proposed cascaded connections to the rotor of induction machines using rotary machines. The Kraemer system transformed slip energy back into mechanical energy using a second machine in tandem connected to the shaft,

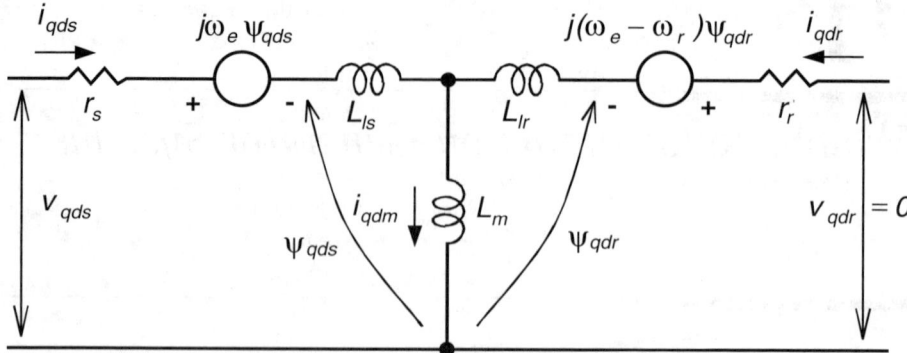

FIGURE 11.1
Complex synchronous dqs equivalent circuit for wound-rotor induction machine.

while Scherbius proposed a second induction machine connected to the rotor to send the slip power back to the line. By using modern power electronic devices it is possible to recover the slip otherwise dissipated in resistances. Therefore, wound-rotor generators use an inverter system connected between the rotor and the grid, while the stator is directly connected to the grid. When the mechanical speed is confined to ±20% of the synchronous speed, the rotor converter is rated only for a portion of the rated stator power, a clear advantage for very high power systems.

Figure 11.1 shows a typical configuration of a doubly fed induction generator with static converters. Compared to the squirrel-cage induction generator (SCIG), the WRIG variable-speed generator has the following advantages:

- The SCIG requires a power converter with full power-processing capability, whereas a WRIG requires a smaller power converter to process the slip power. Thus the smaller the range of the operating slip, the smaller the required power converter. The size of the power converter may strongly affect the cost of energy ($/kWh).
- The WRIG has a simpler control because the magnetizing current is practically constant regardless of the rotor frequency. The purpose of the control is to synchronize the rotor current with respect to stator reference.
- The utility connection to the WRIG can be controlled from both sides of the machine, The power factor can be controlled from the stator or the rotor side of the WRIG.
- The WRIG can be commanded by either of two techniques: (1) active and reactive power from the stator side can be used to control the rotor converter, or (2) the stator voltage can be set up from the rotor side.

- If a wound rotor generator system can operate in four-quadrants around the synchronous speed it can then operate in motor mode, for pumped hydro applications, for example.
- The WRIG has increased power system dynamics and stability permitting the suppression of power-system fluctuations by quickly exchanging energy from the electric system to machine inertia without loss of synchronism, as happens with synchronous machines. The system can also compensate for quick reactive power needs and so can improve overall power system dynamics.

11.3 Sub- and Supersynchronous Modes

The wound-rotor induction generator is usually fed by the stator and by the rotor. That is why it is frequently called the doubly fed induction generator (DFIG) in the literature. Although the term WRIG is more related to the machine itself and DFIG embraces the whole system, both acronyms will be used interchangeably in this book.

A complex synchronous reference frame equivalent circuit as shown in Figure 11.1 can represent a WRIG. The main equations related to this model are

$$\underline{v}_{qds} = r_s \underline{i}_{qds} + j\omega_e \underline{\psi}_{qds} + \frac{d}{dt}\underline{\psi}_{qds} \tag{11.1}$$

$$\underline{v}_{qdr} = r_r \underline{i}_{qdr} + js\omega_e \underline{\psi}_{qdr} + \frac{d}{dt}\underline{\psi}_{qdr} \tag{11.2}$$

$$\underline{\psi}_{qds} = L_s \underline{i}_{qds} + L_m \underline{i}_{qdr} \tag{11.3}$$

$$\underline{\psi}_{qdr} = L_r \underline{i}_{qdr} + L_m \underline{i}_{qds} \tag{11.4}$$

$$T_e = \frac{3}{2}\left(\frac{p}{2}\right)Re\left(j\underline{\psi}_{qds}\cdot\overline{i_{qds}}\right) = \frac{3}{2}\left(\frac{p}{2}\right)Re\left(j\underline{\psi}_{qdr}\cdot\overline{i_{qdr}}\right) \tag{11.5}$$

where $\overline{i_{qds}}$ and $\overline{i_{qdr}}$ are the complex conjugate of the stator-current and rotor-current space vectors and stator and rotor inductances are defined by: $L_s = L_{ls} + L_m$ and $L_r = L_{lr} + L_m$.

The complex torque equation in 11.5 can be resolved in the reference frame $d_e - q_e$ leading to

$$T_e = \frac{3}{2}\left(\frac{p}{2}\right)\left(\psi_{ds}i_{qs} - \psi_{qs}i_{ds}\right) = \frac{3}{2}\left(\frac{p}{2}\right)\left(\psi_{dr}i_{qr} - \psi_{qr}i_{dr}\right) \tag{11.6}$$

In order to find the no-load mechanical speed ω_{r0} for given steady-state voltages $\underline{v}_{qds0}$ and $\underline{v}_{qdr0}$, one can assume that Equations 11.1 and 11.2 have $d/dt(\cdot) = 0$ (steady-state) and $i_{qdr} = 0$ (non-load):

$$\underline{v}_{qds0} = r_s \underline{i}_{qds0} + j\omega_e L_s \underline{i}_{qds0} \tag{11.7}$$

$$\underline{v}_{qdr0} = js\omega_e L_m \underline{i}_{qds0} \tag{11.8}$$

Combining Equation 11.7 with 11.8,

$$\underline{v}_{qds0} = \left(r_s + j\omega_e L_s\right)\frac{\underline{v}_{qdr0}}{js\omega_e L_m} \tag{11.9}$$

For high-power machines the stator resistance can be neglected in Equation 11.9. Since $s = 1 - \omega_{r0}/\omega_e$ the following relation holds for calculation of the no-load rotational speed in accordance with stator and rotor impressed magnitude voltages V_{qdr0} and V_{qds0}.

$$\omega_{r0} = \omega_e\left[1 - \frac{L_s}{L_m} \cdot \frac{V_{qdr0}}{V_{qds0}}\right] \tag{11.10}$$

Equation 11.10 must be interpreted with care due to some restrictions. The phasor voltages in the reference frame are DC quantities, when positive they indicate a positive phase sequence (a-b-c) in the three-phase terminals, whereas negative values indicate a negative phase sequence (a-c-b). Therefore, assuming stator voltage to be always positive $\underline{v}_{qds0}$ the rotor-impressed voltage can be positive $\underline{v}_{qdr0} > 0$ with corresponding rotor power $P_r > 0$ and the rotational speed ω_{r0} lower than the electrical synchronous speed ω_e with the machine operating in subsynchronous mode with positive slip $s > 0$. By commanding only voltage $\underline{v}_{qdr0}$, the machine is naturally stable only in the subsynchronous region. Rotor voltage control leads to motor-only stable operation in subsynchronous mode and to generator-only stable operation in supersynchronous mode. For rotor current control both motor and generator forms are stable in sub- and supersynchronous modes.[1,2]

The operating characteristics of the induction generator from a current source are quite different from those of the induction generator from a voltage source. High-power machines have small stator resistance drops. When under an impressed voltage source the air-gap voltage is close to the stator voltage, but under impressed current, the air-gap voltage and stator voltage vary with shaft loading. Therefore, with current control it is possible to have a negative rotor power $P_r > 0$ for positive slip $s > 0$ by changing the phase sequence of currents in stator and rotor. Equation 11.10 also holds for supersynchronous mode where $v_{qdr0} > 0$ with consequent $s < 0$. Since the machine is stable with rotor voltage control, only reversing the phase sequence of voltages will impose generator mode for rotational speed greater than electrical synchronous speed.

The rotor-injected power P_{ROTOR} coming from a prime mover on the mechanical shaft can be considered as

$$\left(\underline{v}_{qdr} \frac{3}{2} \overline{\underline{i}_{qdr}} \right) = r_s \underline{i}_{qdr} + j s \omega_e \underline{\psi}_{qdr} \cdot \frac{3}{2} \overline{\underline{i}_{qdr}}$$

and therefore, the rotor power can be calculated by

$$P_{ROTOR} = \frac{3}{2} Re\left(\underline{v}_{qdr} \cdot \overline{\underline{i}_{qdr}} \right)$$

leading to

$$P_{ROTOR} = \frac{3}{2} r_r i_r^2 + s \omega_e T_e \qquad (11.11)$$

Since $\omega_e T_e$ is the air-gap power (P_g), neglecting the rotor copper loss, the relation in Equation 11.12 holds, and Equation 11.13 is the torque equation in steady-state Equation 11.14:

$$P_r = s P_g \qquad (11.12)$$

$$T_e = 3\left(\frac{p}{2}\right) I_r^2 \frac{r_r}{s \omega_e} \qquad (11.13)$$

For high-power machines the stator resistance is neglected, and the stator terminal power is P_g. Considering that power flowing out of the machine is negative (generator mode), the induction generator has a power balance in accordance with the torque–slip curve indicated in Figure 11.2. The power distribution for the generator operating in subsynchronous and supersynchronous regions is indicated in the operating region from 0.7 ω_s to 1.3 ω_s. For operation in the subsynchronous region the slip is positive, and, therefore, the rotor circuits receive power from the line, whereas in the supersynchronous region the slip is negative, and the rotor power supplements extra generating power to the grid.

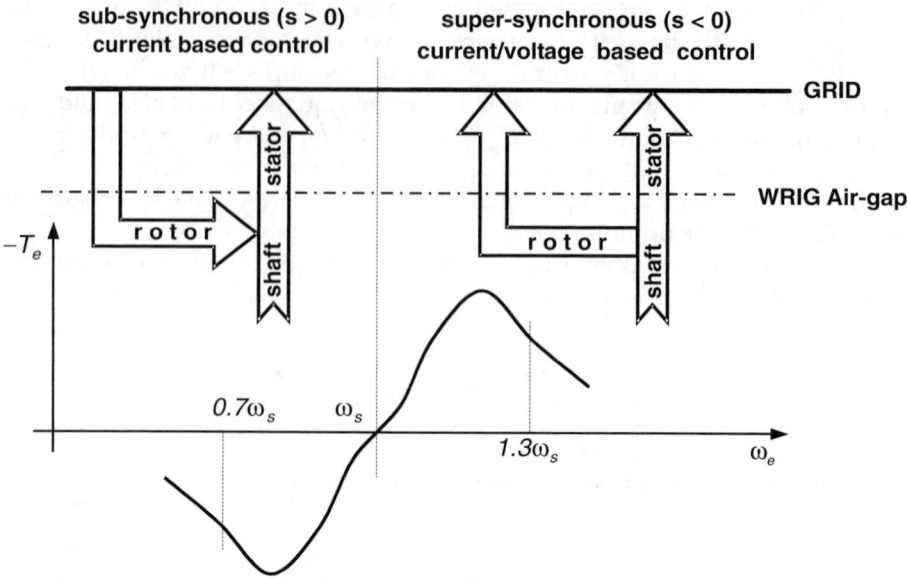

FIGURE 11.2
Torque–slip curve for doubly fed induction generator in sub- and supersynchronous modes.

Figure 11.3 shows the power balance in a WRIG at subsynchronous generation where $s > 0$ and the power flows into the rotor by a current-controlled inverter. A step-up transformer is usually connected between the low-frequency low-voltage requirements and the grid in order to alleviate the rotor converter ratings. Figure 11.4 shows the supersynchronous generating mode where the mechanical speed is greater than the electrical synchronous speed, so the slip is negative ($s < 0$). The rotor voltages will have their phase sequence reversed since $P_g < 0$ and $P_r < 0$ the rotor circuit contributes by generating power to the line with improved efficiency. It is important to note that the shaft incoming power indicates $P_m = (1 + s)P_g$ to show the extra capability of power conversion, but the slip is actually negative. Thus, very efficient generating systems can be achieved using the supersynchronous region. Because the operating region is a limited one, the main drawback is the start-up sequence of the system. One possible way is to use auxiliary resistors in the rotor circuit as indicated in Figure 11.5, then drive the machine in motor mode and just after the cut-in speed plug in the controller that imposes regenerative operation.

11.4 Operation of WRIG

In order to have a variable speed with improved operating covering the operating region $0.7\,\omega_s < \omega_r < 1.3\,\omega_s$ the rotor inverter must be bi-directional

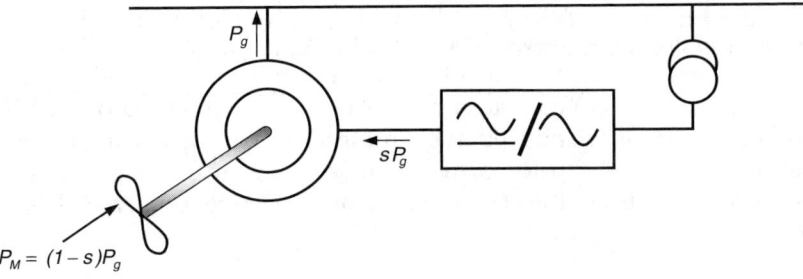

FIGURE 11.3
Subsynchronous generating mode ($s > 0$).

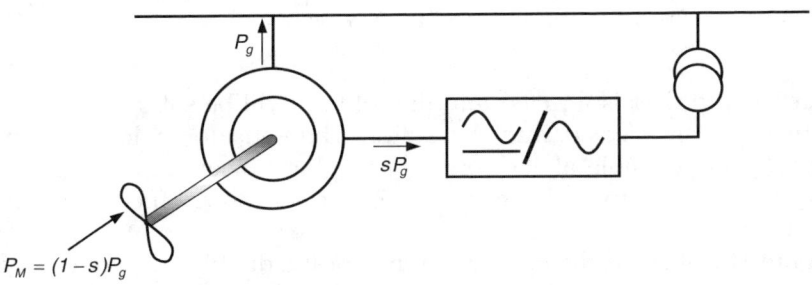

FIGURE 11.4
Supersynchronous generating mode ($s < 0$).

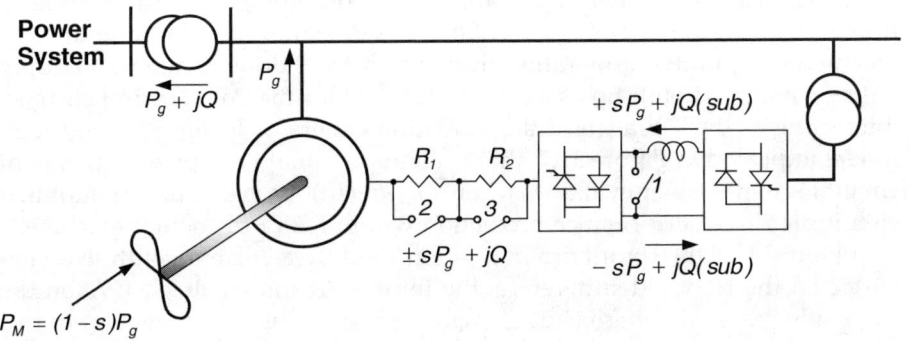

FIGURE 11.5
Wound-rotor IG with a back-to-back double converter.

like the one in Figure 11.5 with a smooth transition through $s = 0$. That is, at synchronous speed the rotor needs DC currents, and the system behaves like a pure synchronous generator. In the past, line-commutated and load-commutated inverters were used in the rotor circuit, but close to zero slip those circuits do not work well. For very-high-power systems cycloconverters are still a very good choice. For medium-power, back-to-back double converters (CSI or VSI) or matrix converters are adequate, provided that a controlled resistor is introduced at the rotor terminals for starting the system in motor mode. If the rotor converter injects the reactive power (jQ) in accordance with Equation 11.14, the stator reactive power can be programmed as indicated by Equation 11.15.[2]

$$Q_{ROTOR} = \frac{3}{2} i_{qdm} \left(\overline{v}_{qdr} \cdot \overline{i}_{qdr} \right) = s\omega_e \frac{3}{2} \left(\left(j\overline{\psi}_{qdr} \cdot \overline{i}_{qdr} \right) \right) \qquad (11.14)$$

$$Q_{STATOR} = \frac{3}{2} i_{qdm} \left(\overline{v}_{qds} \cdot \overline{i}_{qds} \right) = \frac{3}{2} L_{ls} \left(i_{qds} \right)^2 - \frac{Q_{ROTOR}}{|s|} \qquad (11.15)$$

If reactive power is injected into the rotor, it will be subtracted from the reactive power injected into the stator. Theoretically, a leading power factor $Q < 0$ is possible at the cost of a lagging power factor in the rotor. However, as the rotor voltage is very low it is a very difficult solution to implement.

Figure 11.2 showed the operating range of the doubly fed induction generator with a torque–slip curve for sub- and supersynchronous modes. The actual torque can be calculated from a transient d – q model using Equation 11.6. However, a steady-state approach can be useful in the evaluation of the generator torque response. Figure 11.6(a) shows the steady-state per-phase equivalent circuit for a squirrel-cage induction machine (short-circuit in the rotor is considered) where the transformer coupling shows the turn ratio from stator to rotor turns. Figure 11.6(b) considers the wound rotor induction machine with unity turns-ratio where the rotor side is available to apply voltage. Figure 11.6(c) shows a phasor diagram for the voltages and currents interacting at the T-branch of the induction generator leading to the rough model indicated in Figure 11.7: that is an approximate per-phase equivalent circuit for high-power machines where $|(r_s + j\omega_e L_{ls})| \ll \omega_e L_m$; such a simplified circuit provides performance prediction within 5% of the actual machine.

In Figure 11.7 the rotor current I_r is imposed by a current-controlled converter on the rotor side; therefore, the impressed rotor voltage V_r/s on the rotor side has some phase shift $\angle\phi$ with respect to the stator side. From this figure thus the rotor current I_r is related to the stator terminal voltage V_s by Equation 11.16 below. Substituting I_r into Equation 11.13, we get Equation 11.17, which is a function of slip s for constant frequency (ω_e) and supply voltage (V_s). In order to plot the family of curves the slip is extended for subsynchronous ($0 < s < 1$) and supersynchronous ($-1 < s < 0$) for fixed synchronous rated stator frequency ($\omega_e = \omega_s$) as indicated in Figure 11.8.

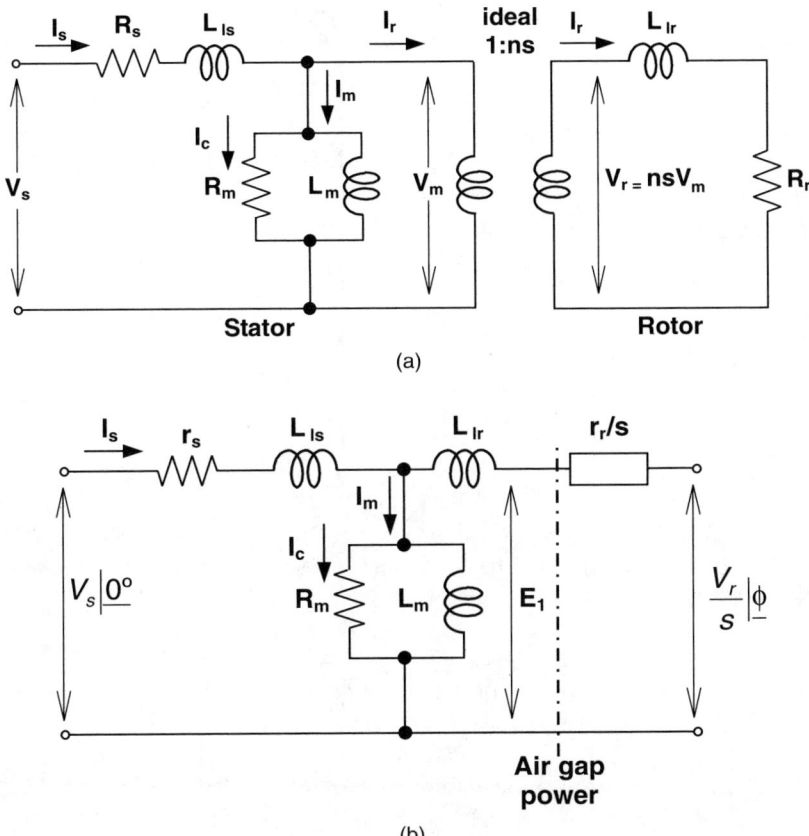

FIGURE 11.6
Per-phase equivalent model of an IG: (a) equivalent circuit with transformer coupling, (b) equivalent circuit with respect to the stator, (c) T-branch of an IG and correspondent phasor diagram for rotor current control.

$$\bar{I}_r = \frac{V_s \angle 0^\circ - \dfrac{V_r}{s} \angle \phi}{\sqrt{\left(r_s + \dfrac{r_r}{s}\right)^2 + \omega_e^2 (L_{ls} + L_{lr})^2} \angle \tan^{-1}\left(\dfrac{\omega_e (L_{ls} + L_{lr})}{r_s + \dfrac{r_r}{s}}\right)} \tag{11.16}$$

$$T_e = 3\left(\frac{P}{2}\right)\frac{r_r}{s\omega_e} \frac{\left(V_s - 2\dfrac{V_s V_r}{s}\cos\phi + (\dfrac{V_r}{s})^2\right)}{\left(r_s + \dfrac{r_r}{s}\right)^2 + \omega_e^2 (L_{ls} + L_{lr})^2} \tag{11.17}$$

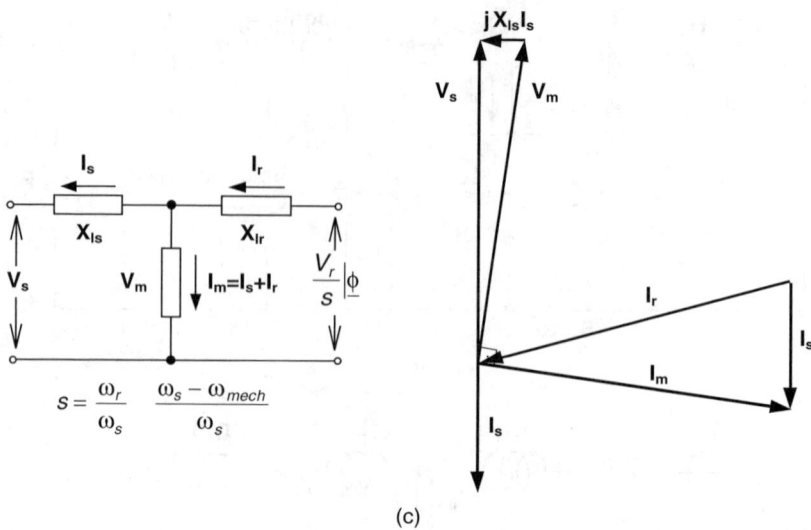

FIGURE 11.6 (continued) (c) T-branch of an IG and correspondent phasor diagram for rotor current control.

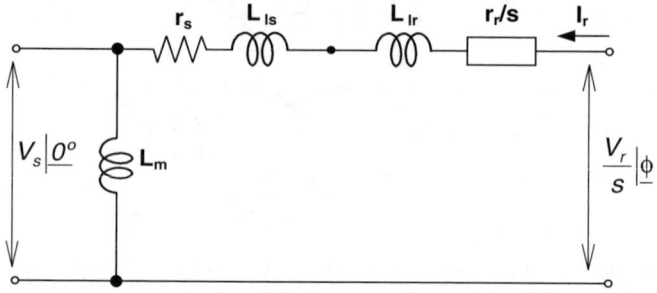

FIGURE 11.7
Approximate high-power equivalent circuit.

Figure 11.8 shows a family of curves where the phase shift between stator voltage and current will define the position and shape of the torque–slip curve. The rotor side voltage is assumed to be nearly constant and the phase shift would be varied to match the required constant air-gap flux. Figure 11.8 shows the torque–slip characteristic for short-circuited rotor windings (passive rotor power flowing in a squirrel-cage type machine). As negative rotor power is extracted from the rotor the machine increases speed to the super-synchronous region in the curve operating at point a. The machine can have positive rotor power injected, causing a decrease in speed to the subsynchronous region and operates at point b. It is clear that although supersynchronous operation is preferable to generation by both stator and rotor, a stable position for the same torque level requires higher slip s and more losses incurred at the rotor side. Therefore, it is advisable to implement management

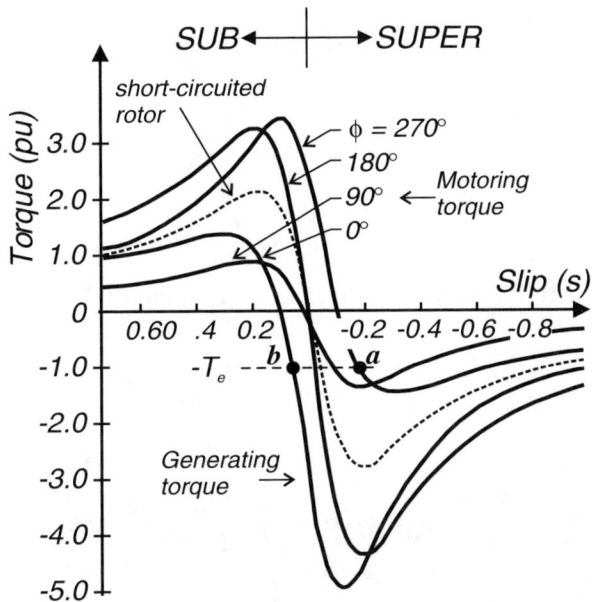

FIGURE 11.8
Effect of injected slip rotor power on the torque–slip characteristics of a DFIG.

of the optimum operating point that minimizes the losses on the supersynchronous mode.

Several converter topologies can be used to control a DFIG and many control approaches have been studied and implemented. Nonetheless it is possible to classify two broad approaches to control the DFIG: (1) a complete controller where the prime mover active power and the line-side reactive power are the set points for a rotor frame controller for a grid-connected system, and (2) a simplified current controller with a line voltage control for stand-alone applications that do not require full active/reactive power management.

Both approaches are currently implemented using vector control theory. The converter connected to the rotor side will receive the set points from the vector-control-based system. Depending on the topology (cycloconverters, matrix converters, back-to-back double CSI or VSI converters) an inner control for a possible DC-link (for the back-to-back double converters) will be required or there will be a natural bi-directional operation (for the cycloconverters and matrix converters). The following approach will consider that the rotor side converter receives set points for the rotor currents $i_{ra}^*, i_{rb}^*, i_{rc}^*$. For a grid-connected DFIG it is important to administer the injected generator active and reactive power written as Equations 11.18 and 11.19.

$$P_s = \frac{3}{2} Re\left(\underline{v}_{qds} \cdot \overline{\underline{i}_{qds}}\right) = \frac{3}{2}\left(v_{qs}i_{qs} + v_{ds}i_{ds}\right)$$ (11.18)

$$Q_s = \frac{3}{2} Im\left(v_{qds} \cdot \overline{i_{qds}}\right) = \frac{3}{2}\left(v_{qs} i_{ds} - v_{ds} i_{qs}\right) \tag{11.19}$$

Considering that

$$\overline{i}_{qds} = \frac{1}{L_s} \overline{\psi}_{qds} - \frac{L_m}{L_s} \overline{i}_{qdr}$$

the following active and reactive power equations hold.

$$P_s = \frac{3}{2}\left[\frac{1}{L_s}\left(v_{qs}\psi_{qs} + v_{ds}\psi_{ds}\right) - \frac{L_m}{L_s}\left(v_{qs}i_{qr} + v_{ds}i_{dr}\right)\right] \tag{11.20}$$

$$Q_s = \frac{3}{2}\left[\frac{1}{L_s}\left(v_{qs}\psi_{ds} - v_{ds}\psi_{qs}\right) - \frac{L_m}{L_s}\left(v_{qs}i_{dr} - v_{ds}i_{qr}\right)\right] \tag{11.21}$$

By simplifying and considering stator resistance to be $r_s \approx 0$, there is alignment of active power with q_e axis and alignment of reactive power with d_e axis. Therefore, stator flux will be also aligned with d_e axis and $\psi_{qs} = 0$, consequently $d/dt\ \psi_{qs} = 0$. The steady-state values for the following simplified power equations may be used.

$$P_s = -\frac{3}{2}\frac{L_m}{L_s}|\psi_s|\omega_e i_{qr} \tag{11.22}$$

$$Q_s = \frac{\omega_e}{L_s}\left[|\psi_s|^2 - L_m i_{qr}|\psi_s|\right] \tag{11.23}$$

In the coordinate system of the stator voltage, the P_s and Q_s powers are given by

$$P_s = -\frac{3}{2}\frac{L_m}{L_s}V_s i_{qr} \tag{11.24}$$

$$Q_s = \frac{3}{2}\left(\frac{V_s^2}{L_s\omega_e} - \frac{L_m}{L_s}V_s i_{qr}\right) \tag{11.25}$$

Considering the stator flux orientation of the DFIG, the stator current i_{qds} can be considered to have a projection i_{sT} responsible for a hypothetical active torque, which is

$$T_T = \frac{3}{2} Re\left(j\underline{\psi}_{qds} \cdot \overline{i_{qds}} \right);$$

by the same reasoning, a reactive torque i_{sQ} might be defined as

$$T_Q = \frac{3}{2} Im\left(j\underline{\psi}_{qds} \cdot \overline{i_{qds}} \right).$$

The equations for the hypothetical active and reactive torques will help to define the feedback approach. The equations for T_T and T_Q are similar to the electromagnetic torque, but in this case they are measuring the active and reactive power flow at the stator side and do not need any correction for pair of poles.

$$T_T = \frac{3}{2}\left(\psi_{ds} i_{qs} - \psi_{qs} i_{ds} \right) \tag{11.26}$$

$$T_Q = \frac{3}{2}\left(\psi_{qs} i_{qs} + \psi_{ds} i_{ds} \right) \tag{11.27}$$

Thus, the current phasor can be defined as $i_{qds} = i_{sT} - j\left(i_{sQ} \right)$, where i_{sT} is generated from a closed loop control on T_T and i_{sQ} is generated from a closed loop control on T_Q. Figure 11.9 shows how the control law for the induction generator can be evaluated for this system; when current-control is impressed, only the T-branch of the generator indicated in Figure 11.9 needs to be evaluated. The rotor current i_{qdr} must supply the required stator current i_{qds} subtracted from the magnetizing current i_{qdm}. By assuming that magnetizing current is fully aligned with the reactive stator current component i_{sQ} we get the following:

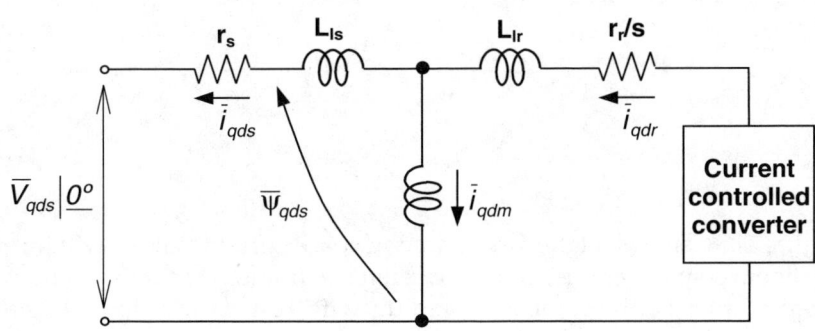

FIGURE 11.9
Distribution of rotor, magnetizing and stator current.

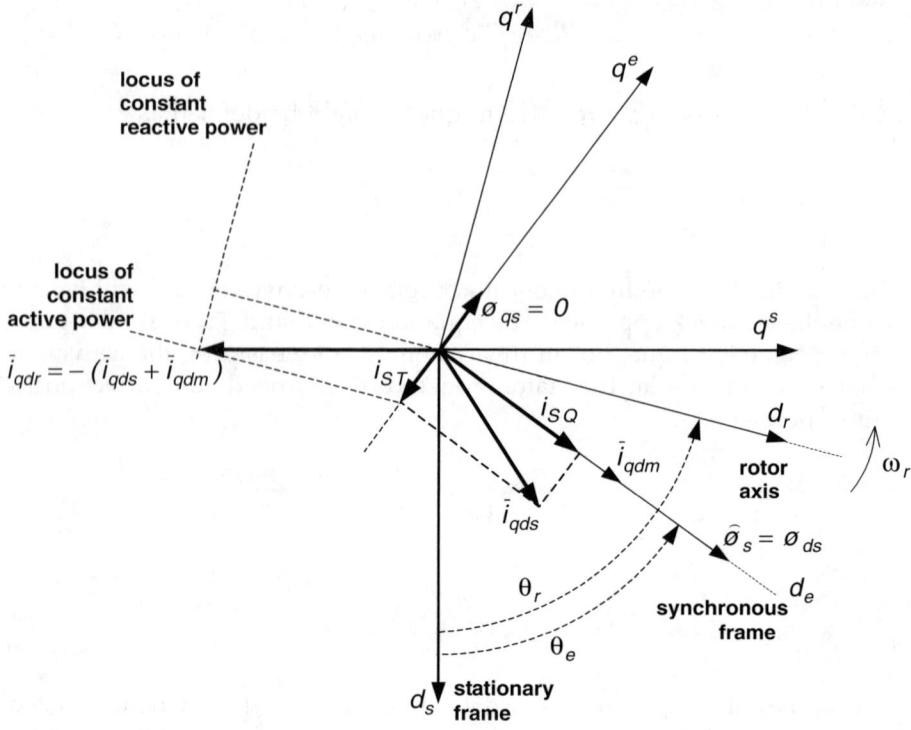

FIGURE 11.10
Diagram of space current vectors for supersynchronous operation.

$$i_{qdr}^{*} = -\left(i_{qds}^{*} - i_{qdm}^{*}\right) = -i_{sT} - j\left(i_{m} - i_{SQ}\right) \tag{11.28}$$

$$i_{qr}^{*} = -i_{sT} \tag{11.29}$$

$$i_{dr}^{*} = i_{m} - i_{SQ} \tag{11.30}$$

$$i_{m} = \frac{|\psi_{s}|}{L_{s}} \tag{11.31}$$

Figure 11.10 shows relationships among space current vectors with respect to stationary, synchronous, and rotor reference frames for supersynchronous operation, that is, the rotor axis leading with respect to d_{e} axis. The rotor current is negative providing stator current (with active and reactive components) plus the magnetizing current. The rotor current has a locus of constant active power that keeps a prescribed i_{ST} and a locus of constant reactive power that keeps a prescribed i_{SQ}. The calculation of the magnetizing current makes the reactive component very sensitive to stator voltage integrations, so that variations in the grid AC voltage and frequency are taken into consideration. However, the saturation of the generator is not considered by Equation 11.31. The diagram in Figure 11.10 can be also drawn for subsynchronous operation. When the rotor axis is lagging the d_e axis, the slip becomes negative and the vector rotation that takes the field coordinates to the rotor frame automatically reverses the phase sequence, and the rotor current is injected in the rotor.

Figure 11.11 shows the complete block diagram for the active-reactive-based power control of a doubly fed induction generator. The three-phase currents and voltages are acquired and converted to a two-phase stationary frame ($d_s - q_s$). Because the frequency of operation is high, a low-pass-filter performs the function of integrator to obtain the stationary stator flux projections ψ_{qs}, ψ_{ds} and the stator flux amplitude $|\psi_s|$. The hypothetical active and reactive torques T_T, T_Q are computed and compared to the required values of active and reactive power P_s, Q_s by the division with ω_e. Such error signals E_T, E_Q drive the PI controllers to generate the reference values for stator current i_{sT}^*, i_{sQ}^*. The reactive component is subtracted from i_m, and after limiting blocks the reference values for rotor currents i_{qr}^* and i_{dr}^* are provided in synchronous coordinates. Therefore, the vector rotation must consider the angle for transformation as the rotor position in field coordinates—that is, the subtraction of the mechanical angle θ_r (compensated by the number of poles) from the stator electrical frequency angle θ_e. A two-phase to three-phase transformation brings the reference values for the three-phase rotor currents i_{ra}^*, i_{rb}^*, i_{rc}^*. More sophisticated techniques to control the inner loops of the bi-directional AC/AC converter are outside the scope of this chapter.

It is possible to generate the rotor voltage set-points v_{qr}^*, v_{dr}^* and through synchronous PI controllers or space-vector or direct torque control modulate the rotor currents. Although the inner management of such bi-directional power flow of the rotor side converter is not discussed here, we provide several references as discussed in Chapter 9, on vector-controlled drive systems that are helpful in designing such controllers. In addition, sensorless techniques so well developed for squirrel-cage induction generators can also be applied for enhanced performance of DFIG.

The active-reactive power control for a DFIG presented in Figure 11.11 requires the full management of power. Active power can be available from information from the shaft—for example, wind speed can be measured and a look-up table may generate the active power available on the shaft to be converted for the electric grid. Reactive power requires communication with the grid and commands from neighboring power systems must request lagging/leading power factor.

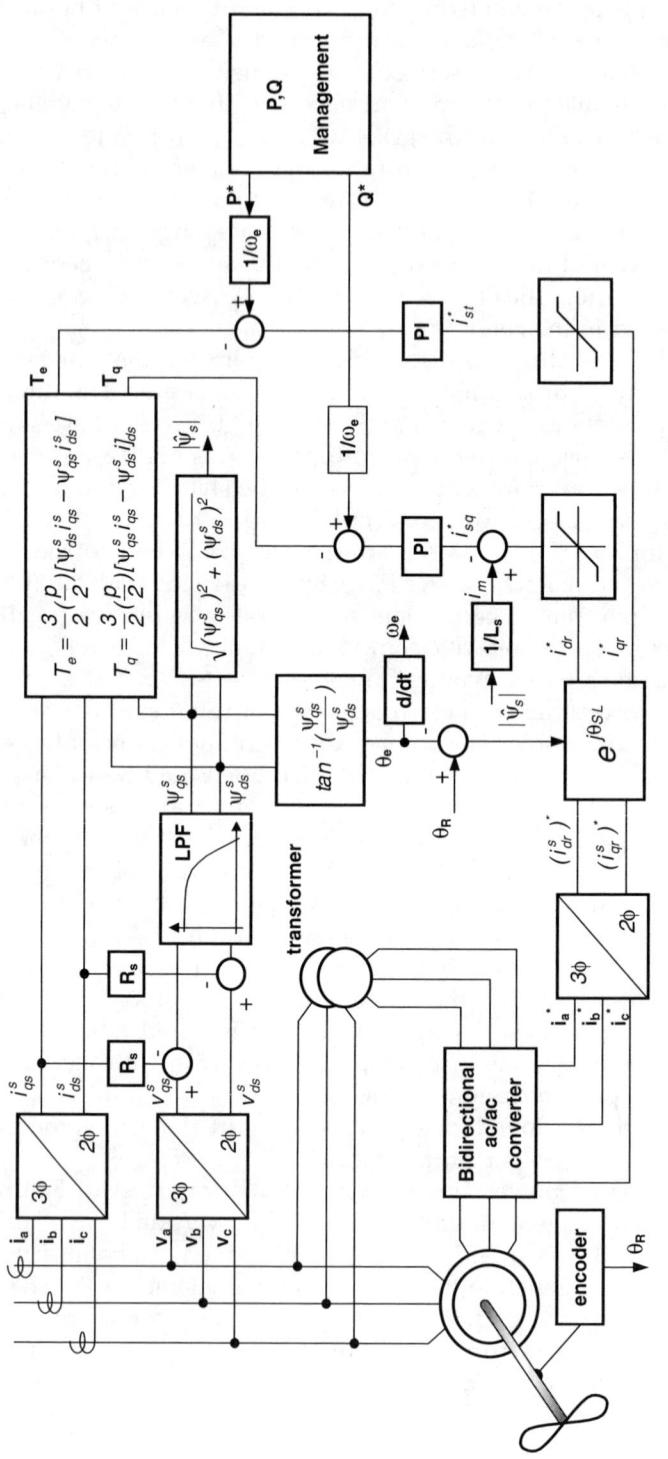

FIGURE 11.11

Active-reactive power control for a doubly fed induction generator.

Stand-alone applications can be simplified, since they do not need this sort of management, and a simplified current controller with a line-voltage control suffices. Figure 11.9 shows the T-model relating the stator current i_{qds}, magnetizing current i_{qdm}, and the rotor current i_{qdr}. From Equation 11.3 and considering that $L_s = L_{ls} + L_m$, the following $d_s - q_s$ equations hold.

$$\psi_{qs} = L_{ls} i_{qs} + L_m \left(i_{qs} + i_{qr} \right) \tag{11.32}$$

$$\psi_{ds} = L_{ls} i_{ds} + L_m \left(i_{ds} + i_{dr} \right) \tag{11.33}$$

The stator flux will be aligned with d_s axis and, therefore, $\psi_{qs} = 0$, and consequently $d/dt \, \psi_{qs} = 0$. The following control law of Equation 11.34 must be implemented. The equations 11.35 to 11.37 are valid for these flux-aligned conditions.

$$i_{qr}^* = -\frac{\left(L_m + L_{ls} \right)}{L_m} i_{qs} \tag{11.34}$$

$$v_{qs} = r_s i_{qs} + \omega_e \psi_{ds} \tag{11.35}$$

$$v_{ds} = r_s i_{ds} + \frac{d}{dt} \psi_{ds} \tag{11.36}$$

$$\left| \psi_s \right| = \psi_{ds} = L_{ls} i_{ds} + L_m \left(i_{ds} + i_{dr} \right) \tag{11.37}$$

The reference i_{qr}^* commands the injection of active power. The reactive power required for the generator can be commanded by a closed loop control on the output voltage. The stator voltage is calculated by

$$\left| v_s \right| = \sqrt{\left(v_{ds} \right)^2 + \left(v_{qs} \right)^2}$$

and a PI controller is used to generate the reactive current set-point i_{dr}^* as indicated in Figure 11.12. The set points i_{qr}^* and i_{dr}^* are on the synchronous rotating reference $d_e - q_e$ frame and the inverter needs commands on the rotor frame $d_r - q_r$ as previously considered for the DFIG active-reactive power control system of Figure 11.11, with a subtraction of the mechanical angle θ_r (compensated by the number of poles) from the stator electrical frequency angle θ_e, which is generated internally since this system is in stand-alone operation mode.

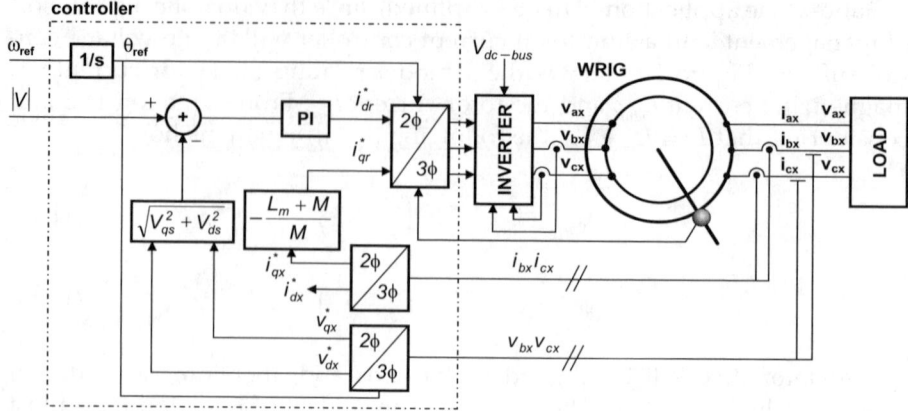

FIGURE 11.12
Stand-alone simplified control of DFIG.

11.5 Problems

1. Do research to find applications for doubly fed induction generators and write a summary report. Focus your analysis on power rating, economic reasons, control approach, and flexibility of such systems.

2. A transformer is frequently used to connect the rotor side to the grid. What will the effect be if a lower transformer ratio is selected?

3. Why is DFIG not considered for motor applications for the speed range from zero to rated speed?

4. Equation 11.17 relates stator voltage, rotor voltage, and torque. Use a spreadsheet program to develop a macro to plot a family of curves in relation to input parameters.

References

1. Boldea, I. and Nasar, S.A., *Electric Drives: CD-ROM Interactive*. CRC Press, Boca Raton, 1998.
2. Azaza, H. and Masmoudi, A., On the dynamic and steady state performances of a vector controlled DFM drive. *IEEE International Conference on Systems Man and Cybernetics*, 6(6–12), 2002.
3. Bose, B.K., *Modern Power Electronics and AC Drives*. Prentice Hall, New York, 2001.

4. Cadirici, I. and Ermis, M., Double-output induction generator operating at subsynchronous and supersynchronous speeds: steady-state performance optimization and wind-energy recovery, in *Electric Power Applications, IEEE Proceedings B Electric Power Applications*, 139 (5), 429–442, 1992.

5. Brown, G.M., Szabados, B., Hoolbloom, G.J., and Poloujadoff, M.E., High-power cycloconverter drive for double-fed induction motors, *IEEE Trans. on Ind. Electronics*, 39(3), 230–240, 1992.

12

Simulation Tools for the Induction Generator

12.1 Scope of This Chapter

Modern control of induction machines is a very challenging area of electrical engineering, especially when control is based on detailed machine physical and mathematical models. The correct transient state representation of an induction generator is a very complex matter because of the various electromagnetic and electromechanical phenomena involved in a detailed equivalent circuit. This complexity is mostly related to the dependence of the parameters on the frequency, transient mechanical and electromagnetic states, and degree of accuracy of the approximations included in the transient model.

The study of distributed stray inductances and capacitances in the machine windings and core, the magnetic path, and the exact core geometry representations are some of the major difficulties that make this one of the most complex fields in electrical engineering. Nevertheless, the exact representation of so many minute details can lead to important findings in terms of reduction of electrical and mechanical stresses and improvement of the electric field distribution. Most of these details have been resolved in daily practice and experience accumulated over the years. For these reasons, many simplifications are usually made throughout the kinds of practical representations in Chapters 2, 3, 4, and 5. Industrial simulations of such scalar and vector fields use finite elements and are beyond the scope of this book. For most practical purposes an extremely simplified model is sufficient to represent the main variations of mechanical speed, frequency, torque, power, voltage, and current. Peaks of torque, current, and voltage approximate the real ones encountered in practice, though electromagnetic interference and shaft oscillations are still not quite clear.

This chapter discusses the application of computer tools to the induction generator. The design of a small wind power plant and an appreciation of overall machine behavior are proposed as applications for MATLAB®, PSpice®, the Pascal computer language, the C computer language, Excel®, and PSim®. Each of these computer tools has features more or less suitable for applications in the field of renewable sources of energy that lead to precious increments of efficiency and durability of equipment under operating conditions. Furthermore, electrical and mechanical stresses can be minimized with knowledge of the causes of torque, voltage, and current peaks or the voltage and current intensity of switching surges. There are simulation companion files available for download from the publisher's Web site, http://www.crcpress.com/. Although it is not possible for the authors to give technical support for the simulations based on those files, the reader is encouraged to provide feedback for possible improvements.

12.2   Design Fundamentals of Small Power Plants

When designing a small power plant, the first thing to be done is to determine the energy potential of the primary source and the potential load. During project planning and development several factors are taken into consideration:

- Primary energy resources and siting
- Financial resources
- Local primary energy analysis
- Consideration of the consumer and environment
- Project planning and analysis
- Project development
- Economical operation of the power plant
- Expansion planning

The higher the final costs of the whole installation are expected to be, the more accurate the studies leading up to it need to be. In large power plants (power plants of more than 10 MW), minute details of the natural behavior of the primary source, the distance to the main consumers, and the quality of the energy and voltage levels of transmission and distribution will be very relevant and decisive issues because they will be closely related to every watt-hour generated.

On the other hand, in micropower plants (power plants of less than 100 kW), generally installed at distant sites where there is no other viable alternative for energy supply, most of those restrictions will not apply or the load will be so sophisticated and self-regulated that quality of energy and other constraints may be made irrelevant by the filtering, storing, and regulating capabilities of the load itself, as long the average energy is available. Typically, in the present state of the art, an induction generator does not supply more than 1 MW of energy, and so it should be considered only for micro and mini power plants (power plants up to 100 kW). Particularly, the self-excited induction generator is definitely only suitable for micropower plants, since it cannot supply viably more than 20 kW without using advanced power electronics and digital controls of speed and voltage.

Useful indicators of the primary energy potential are:

- Data already available for surrounding areas
- Topographical evidence (mountains, valleys, sea surface)
- Other visual site phenomena (like those suggested by Beaufort tables)
- Social and cultural identification
- Measurement and recording of the primary source intensity

As this book is mostly about induction generators, our main interest is directed to the types of primary sources related to rotating primary movers such as turbine, internal combustion engines (ICE), and micro-turbines. Worldwide, the best known primary sources of energy for rotating conversion are hydro and wind power.

12.3 Simplified Design of Small Wind Power Plants

This section will present a brief discussion to motivate an appreciation of the induction generator models introduced in other chapters. A typical application of an induction generator in a small wind power plant with variable speed will be studied. Obviously there is a minimum amount of available wind energy required to have a viable small power plant. This minimum is based on the cubic wind power density of the site, since the wind power is proportional to the cube of the wind speed. In practice, most of these studies base their final decision on sites where the speed of the wind is at least 3 m/s (11 mph); below that it is almost impossible to have a viable power plant.

Several factors affect wind speed, some natural, some caused by civilization (cities, wildfires, roads, forest devastation). Since our source of energy changes as the cube of the wind speed changes, the output power of the generator is very sensitive to these changes. A synchronous source of conversion is recommended. The main factors affecting these data usually considered are: weather systems, local terrain geography, and height above the surrounding area. Such factors cause periodic changes in the wind speed that are hard to be predicted accurately. Some studies recommend 10 years or more of statistical data, which for mini and micropower plants could be well beyond the budget. In that case, a year's data from a nearby site can be used to estimate the long-term annual wind speed. During one year most

TABLE 12.1

Usual Formulas for Wind Speed, Root Mean Cubic Power and Energy Density

Average Speed	Root Mean Cubic Speed	Cubic Power Density	Cubic Energy Density
$V_{av} = \dfrac{1}{n}\sum\limits_{i=1}^{n} v_i$	$V_{rmc} = \sqrt[3]{\dfrac{1}{n}\sum\limits_{i=1}^{n} v_i^3}$	$P_{rmc} = \dfrac{1}{2n}\sum\limits_{i=1}^{n}\rho_i v_i^3$	$E_{yw} = \dfrac{8{,}760}{2n}\sum\limits_{i=1}^{n}\rho_i v_i^3$
m/s	m/s	W/m²	Wh/m²/y

$\rho_0 = 1.225 \text{ kg/m}^3$

v_i = wind speed sample at the instant i in m/s
V_{av} = average wind speed in m/s
V_{rmc} = root mean cubic wind speed in m/s
P_{rmc} = root mean cubic power in W/m²

E_{yw} = root mean cubic yearly energy in Wh/m²/y
ρ_i = air density at the instant i in kg/m³
n = number of samples

variations in wind speed happen; minor modifications can be introduced. Statistically, such wind speed variations can be described by a probability function known as Rayleigh distribution. The formulas usually used for this purpose are given in Table 12.1.

The output power in the turbine shaft including all losses and geometrical details can be approximated by

$$P/A = 0.25 V_{rmc}^3$$

where
    P is the electrical generator output power
    A is the area swept by the turbine blades in m²

It is important to observe at this point that the normal speed of high winds is not high enough to deliver sufficient speed directly to the electrical generator shaft so as to produce electrical energy. The required speeds are related to the rated synchronous speeds (3600 rpm, 1800 rpm, 900 rpm and 450 rpm, respectively for commercial generators of two, four, six, and eight poles), which are very high for the usual wind speed patterns (equivalent to less than 600 rpm). Therefore, it is always necessary to use a speed multiplier with inherent losses.

Once the turbine size is selected for its output power, the next step is to determine the electrical generator size. Though there is no universal standard procedure to obtain the rated speed of a turbine, a good starting point is to use the European manufacturers' recommendation called specific rated capacity (SRC), the ratio of the generator electrical capacity to the rotor swept area. If this recommendation is followed, then it is easy to determine the generator size once the wind turbine is selected.

We can relate the blade tip speed to the wind speed known as $TSR = \lambda$ by

$$\lambda = v/V_w = \omega R/V_w$$

where

    v is the blade tangent tip speed
    V_w is the instant wind speed
    R is the length of each blade
    ω is the shaft angular speed

Using the blade tangent tip speed we obtain the angular speed. From basic electromechanical theory, we can relate the turbine-generator common shaft power to the electrical generator torque as

$$P_{mec} = kT_{electrical}\omega$$

where k is the speed ratio of the mechanical multiplier and its inherent losses.

The load on the electrical generator should be matched to the rotor speed of the turbine using a good control strategy, since the maximum torque of the turbine always happens before the maximum power with respect to the speed increase. The electrical generator selected should be one that always runs at the turbine maximum power.

In order to maximize the turbine output power, a very important definition in wind turbine design is the power coefficient or the rotor efficiency expressed by Betz as

$$C_p = \left(1 - \frac{v_2^2}{v_1^2}\right)\left(1 + \frac{v_2}{v_1}\right) \Big/ 2$$

where v_1 and v_2 are, respectively, the wind speed immediately before and soon after reaching the rotor blades.

The maximum theoretical value of C_p is $16/27 = 0.5926$. So, the aim of any wind turbine control system should be to reach this maximum value, which is strongly dependent on the wind speed. This suggests that for maximum output power at any time the rotor speed must be adjusted in such a way to have the correct v_2/v_1 ratio. Any modern control system operates at variable rotor speed.

After the generator has been selected, the power system components must be sized on the peak capacity of the electrical generator. The control system is in charge of keeping these limits within the right boundaries, which usually follows the bands given in Figure 12.1.

The components of a power system are: circuit breakers, lightning arresters, switch gears, transformers, and transmission lines; they are dealt with in many books of power systems design. Also, a protection system is always embedded in any generating systems.

An important matter to consider is the steady-state matching of both generator-load and turbine-generator connections. In addition, transient analysis must be performed. In steady-state operation, the maximum load should not ever go much beyond the nominal electrical ratings of the generator for the

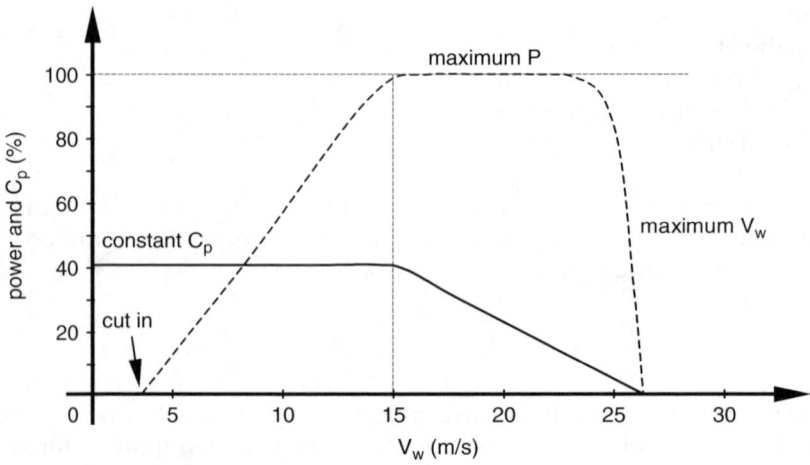

FIGURE 12.1
Speed control bands for wind turbines.

sake of its physical integrity. In electromechanical transient states the high generated currents will create intense magnetic forces and intense vibrations on the generator windings against the metallic machine core and bearings. As a result, mechanical and electrical stresses are produced over the machine's useful life. Furthermore, electromagnetic noise is also generated, which can have devastating effects on sophisticated control and communication circuits.

Maximum power and torque speeds are not usually the same, as can be observed in Figure 12.2; power speed is higher than torque speed. So the electrical load of the generator should always try to match the maximum power of the turbine. As wind speed changes randomly the maximum power point will be changing all the time. For this reason, a well designed speed control should allow for changes in rotor speed so as to keep a maximum value of the power coefficient, C_p.

For injection of energy into the public network it is always expected that the generator load remains constant or within a controllable range, since the public network is considered to be a machine of infinite size and almost constant voltage and frequency. In stand-alone operation it must be assured that the load size would not represent an impact to the generator rotation.

12.4 Simulation of the Self-Excited Induction Generator in PSpice

This example of simulation of the induction generator under transient state uses an approach based on the development of an equivalent triple-T circuit

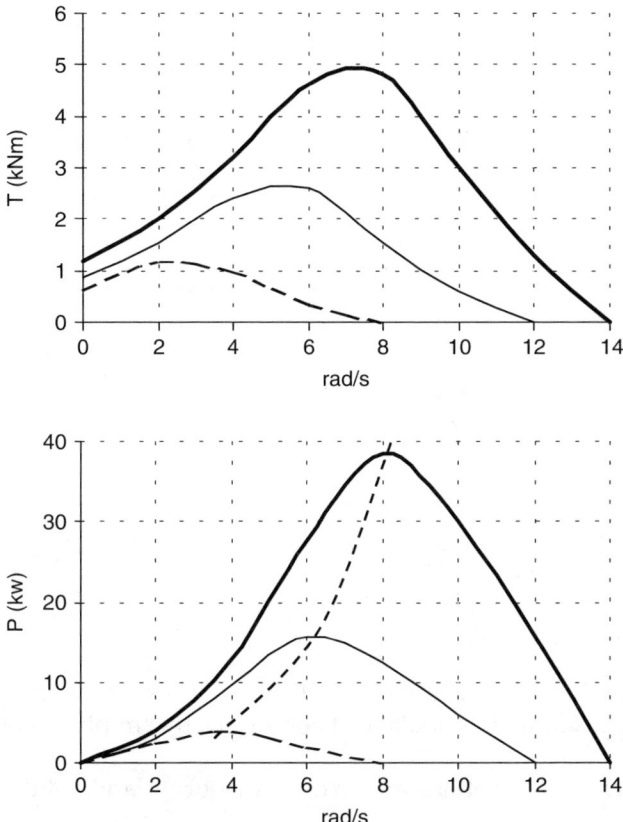

FIGURE 12.2
Torque and power characteristics of a small wind turbine.

referred to a three-axis stationary reference frame as presented by Szczesny and Ronkowski,[7] adapted by Marras and Pomilio,[8] and called the αβγ induction machine model (see Figure 12.3). This model has proved to be accurate and quite handy for simulating the aggregation of three terminal induction machines with another device connected to its stator terminals like, for example, a three-phase inverter. The αβγ transformation does not affect the stator parameters and likewise the parameters of any other devices connected to the stator.

The circuit parameters of the αβγ model are derived from the no-load and blocked rotor tests, in the same way they can be obtained for the single-phase equivalent T circuit. As the αβγ induction machine model provides an immediate circuit representation, it is especially suitable for carrying out simulations assisted by circuit symbolic interface simulation programs, such as PSpice and PSim. Furthermore, the quality of the representations of line stator currents and voltages obtained from these simulations has to be the same as those measured in the physical machine. An outstanding advantage

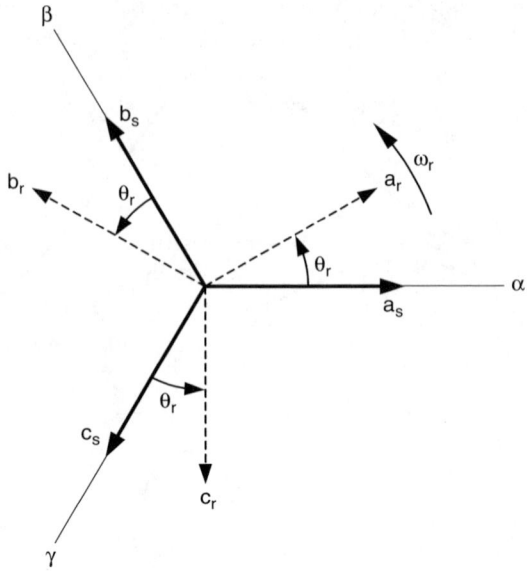

FIGURE 12.3
$\alpha\beta\gamma$ stationary reference frame.

of the $\alpha\beta\gamma$ model, when compared with the two-axis conventional models, is the ability to simulate unbalanced conditions by simply changing parameter values.

The following assumptions are made in the formulation of the $\alpha\beta\gamma$ model.

- Rotor variables are referred to the stator.
- All magnetic and mechanical losses are ignored.
- The machine is three-wire fed and three-phase.
- The magnetomotive force (mmf) has a sinusoidal pattern distributed along the stator (harmonic free).

The third of these assumptions imposes the following constraints to the stator and rotor currents:

$$i_{as} + i_{bs} + i_{cs} = 0 \tag{12.1}$$

$$i'_{ar} + i'_{br} + i'_{cr} = 0 \tag{12.2}$$

The induction machine model in the $\alpha\beta\gamma$ stationary reference frame is derived from the projection of the stator and rotor quantities in the $\alpha\beta\gamma$ axis, respectively positioned in the magnetic axis of the stator a, b and c windings of a virtual two-pole machine,[7] as shown in Figure 12.3.

Symbols ω_r and ω_{mr} stand for, respectively, the rotor electrical and mechanical speed in rad/s in relation to the reference axis; p is the number of poles; and θ_r is the rotor angular position in electrical radians.

$$\omega_r = \frac{p}{2}\omega_{mr} \tag{12.3}$$

The cage rotor induction machine model written in the abc reference frame with machine variables leads to rotor flux and stator flux dependence on the rotor position θ_r as shown in the equations that follow.[2]

$$\left[v^a_{abcs}\right] = r_s\left[i^a_{abcs}\right] + \frac{d\left[\lambda^a_{abcs}\right]}{dt} \tag{12.4}$$

$$0 = r'_r\left[i'^a_{abcr}\right] + \frac{d\left[\lambda'^a_{abcr}\right]}{dt} \tag{12.5}$$

$$\left[\lambda^a_{abcs}\right] = (L_{ls} + M)\left[i^a_{abcs}\right] + \left[L_{sr}\right]\left[i'^a_{abcr}\right] \tag{12.6}$$

$$\left[\lambda'^a_{abcr}\right] = (L'_{lr} + M)\left[i'^a_{abcr}\right] + \left[L_{sr}\right]^T\left[i^a_{abcs}\right] \tag{12.7}$$

$$[L_{sr}] = L_{ms}\begin{bmatrix} \cos\theta_r & \cos\left(\theta_r + \frac{2\pi}{3}\right) & \cos\left(\theta_r - \frac{2\pi}{3}\right) \\ \cos\left(\theta_r - \frac{2\pi}{3}\right) & \cos\theta_r & \cos\left(\theta_r + \frac{2\pi}{3}\right) \\ \cos\left(\theta_r + \frac{2\pi}{3}\right) & \cos\left(\theta_r - \frac{2\pi}{3}\right) & \cos\theta_r \end{bmatrix} \tag{12.8}$$

$$M = \tfrac{3}{2}L_{ms} \tag{12.9}$$

where
  $[v]$ is phase voltage vector
  $[i]$ is line current vector
  $[\lambda]$ is flux linkage vector
  L_{sr} is the stator and rotor mutual inductance matrix
  r is resistance
  L is inductance

The reference frame transformation for the $\alpha\beta\gamma$ induction machine model M is defined as

$$\left[f_{\alpha\beta\gamma}\right] = \left[K^a_\alpha\right]\left[f_{abc}\right] \tag{12.10}$$

where

[$f_{\alpha\beta\gamma}$] and [f_{abc}] are variables in the αβγ stationary and abc reference frames, respectively.

[K_α^a] is the abc to αβγ reference frame transformation matrix

Matrix [K_α^a] may be divided into a matrix for stator transformation [$K_{\alpha s}^a$] and another one for rotor transformation [$K_{\alpha r}^a$].[7] Matrixes [$K_{\alpha s}^a$] and [$K_{\alpha r}^a$] are defined as follows.

$$[K_{\alpha s}^a] = diag\begin{bmatrix} 1 & 1 & 1 \end{bmatrix} \tag{12.11}$$

$$[K_{\alpha r}^a] = \begin{bmatrix} \cos\theta_r + \frac{1}{2} & \cos\left(\theta_r + \frac{2\pi}{3}\right) + \frac{1}{2} & \cos\left(\theta_r - \frac{2\pi}{3}\right) + \frac{1}{2} \\ \cos\left(\theta_r - \frac{2\pi}{3}\right) + \frac{1}{2} & \cos\theta_r + \frac{1}{2} & \cos\left(\theta_r + \frac{2\pi}{3}\right) + \frac{1}{2} \\ \cos\left(\theta_r + \frac{2\pi}{3}\right) + \frac{1}{2} & \cos\left(\theta_r - \frac{2\pi}{3}\right) + \frac{1}{2} & \cos\theta_r + \frac{1}{2} \end{bmatrix} \tag{12.12}$$

Application of Equation 12.10 to the machine model described by Equations 12.4 to 12.9, taking Equations 12.1 and 12.2 into account, yields

$$[v_{abcs}] = r_s[i_{abcs}] + \frac{d[\lambda_{abcs}]}{dt} \tag{12.13}$$

$$0 = r_r[i'_{\alpha\beta\gamma r}] + \frac{d[\lambda'_{\alpha\beta\gamma r}]}{dt} + \frac{\omega_r}{\sqrt{3}}[\lambda'_{\alpha x}] \tag{12.14}$$

$$[\lambda_{abcs}] = (L_{ls} + M)[i_{abcs}] + M[i'_{\alpha\beta\gamma r}] \tag{12.15}$$

$$[\lambda'_{\alpha\beta\gamma r}] = (L_{lr} + M)[i'_{\alpha\beta\gamma r}] + M[i_{abcs}] \tag{12.16}$$

$$[\lambda'_{\alpha x}] = \begin{bmatrix} (\lambda'_{\beta r} - \lambda'_{\gamma r}) \\ (\lambda'_{\gamma r} - \lambda'_{\alpha r}) \\ (\lambda'_{\alpha r} - \lambda'_{\beta r}) \end{bmatrix} \tag{12.17}$$

The notation for stator variables (voltage, current, and flux) are kept as originally used for the abc reference frame, apart from the omission of the superscript a. The αβγr subscript indicates rotor variables referred to the αβγ stationary frame. The [$\lambda_{\alpha x}$] vector comes out as a consequence of the abc-to-αβγ frame transformation.

The mechanical system equation is

$$T_e = (\tfrac{2}{P})J\frac{d\omega_r}{dt} + (\tfrac{2}{P})B_m\omega_r + T_L \qquad (12.18)$$

where
  J is the rotor inertia in kg · m²;
  B_m is the rotational friction constant in kg · m²/s;
  T_e is the electromagnetic torque and
  T_L is the load torque in *Nm*.

Torque T_e may be defined as

$$T_e = \sqrt{3}(\tfrac{P}{2})M\left(i_{as}i_{\gamma r} - i_{cs}i_{\alpha r}\right) \qquad (12.19)$$

The αβγ stationary frame induction machine model is governed by Equations 12.13 to 12.19.

The magnetizing current of the induction machine $[i_{abcm}]$ is the sum of the stator and the rotor currents given as

$$\left[i_{abcm}\right] = \left[i_{abcs}\right] + \left[i_{\alpha\beta\gamma r}\right] \qquad (12.20)$$

The saturation of the air-gap magnetizing inductance *M* is taken into account, considering that the machine is magnetically and electrically symmetrical, and hence the saturation, caused by the air-gap flux, affects all windings equally and is independent of the direction of the flux at any instant and dependent only on its magnitude.[3] The magnetizing current space vector, as defined in Equation 12.21, is the quantity that determines the saturation degree; that is, it defines the value of *M* from

$$\bar{I}_m = \tfrac{2}{3}\left(i_{am} + i_{bm}e^{j\frac{2\pi}{3}+} + i_{cm}e^{-j\frac{2\pi}{3}}\right) \qquad (12.21)$$

When considering saturation, the derivatives present in the second member of Equations 12.13 and 12.14 yield

$$\frac{d\left[\lambda_{abcs}\right]}{dt} = L_{ls}\frac{d\left[i_{abcs}\right]}{dt} + M\frac{d\left[i_{abcm}\right]}{dt} + \left[i_{abcm}\right]\frac{dM}{dt} \qquad (12.22)$$

$$\frac{d\left[\lambda'_{\alpha\beta\gamma r}\right]}{dt} = L'_{lr}\frac{d\left[i'_{\alpha\beta\lambda r}\right]}{dt} + M\frac{d\left[i_{abcm}\right]}{dt} + \left[i_{abcm}\right]\frac{dM}{dt} \qquad (12.23)$$

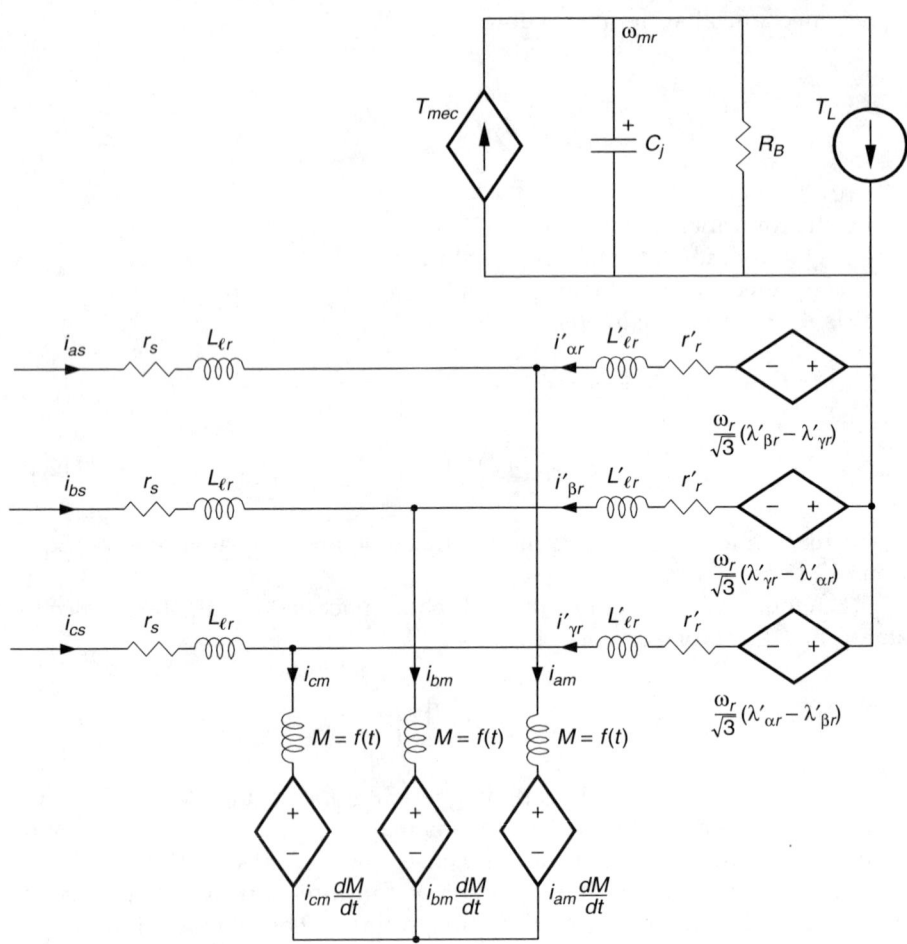

FIGURE 12.4
Equivalent triple-T circuit for the $\alpha\beta\gamma$ machine model.

Notice in Equations 12.22 and 12.23 that the leakage inductances are assumed constant and the saturation inclusion leads to a new element in the magnetizing branch of the equivalent T circuit, which is proportional to the derivative of M with time.

The theory above is fundamental to the $\alpha\beta\gamma$ machine model represented by the equivalent triple-T circuit presented in Figure 12.4. The mechanical system Equation 12.18 is represented by an equivalent circuit as shown in Figure 12.5, referring to the same grounding point as the circuit in Figure 12.4. In this circuit T_e is obtained from Equation 12.19, C_j represents J, R_B represents the inverse of B, and T_L is the mechanical load torque.

The $\alpha\beta\gamma$ model described by Equations 12.13 to 12.19 can be simulated by the implementation of the circuits in Figure 12.4 and Figure 12.5 in PSpice graphical language. As can be seen in that figure, the stator currents, rotor

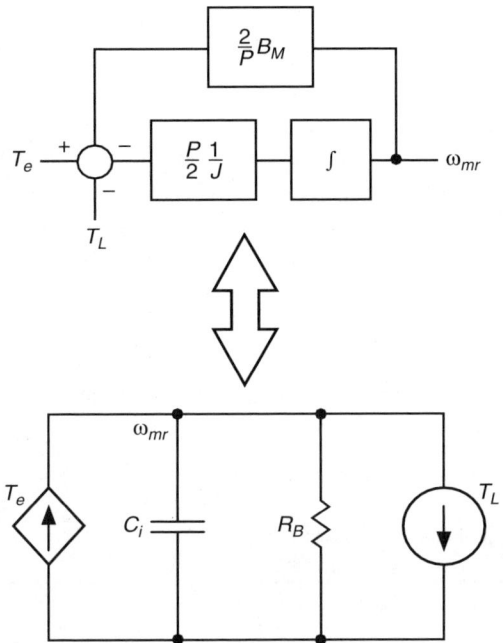

FIGURE 12.5
Mechanical system equivalent circuit.

currents, and magnetizing currents are sampled with a current-controlled voltage source (H device).

The value of M is determined by the output of a PSpice® ETABLE block whose entry is the magnitude of the magnetizing current space vector. This ETABLE block comprises values of M as a function of the magnitude of the magnetizing current space vector, assessed from the machine magnetizing curve, experimentally obtained. The output of the ETABLE block is the input of the three variable inductance ZX blocks, which are the inductances M of each phase of the magnetizing branches. The output of the ETABLE block is also an entry to a DIFFER block to determine the value of dM/dt. The output value of the DIFFER block is then multiplied by each one of the magnetizing currents and used to control three voltage-controlled voltage sources, which represent the sources in the magnetizing branch of Figure 12.4.

The voltage sources present in the rotor circuit, relative to the $[\lambda'_{ax}]$ vector are calculated using ABM blocks whose entries are the sampled rotor and stator currents M and the rotor electrical speed ω_r. The value of ω_r is obtained from the mechanical system circuit by the multiplication of ω_{mr} and $p/2$. The electromagnetic torque T_e at the mechanical system equivalent circuit in Figure 12.5 is calculated with an ABMI block whose entries are the rotor and stator currents and M.

In this example, the machine represented in the simulation is a four-pole, ½-hp induction machine whose parameters and magnetizing curves were

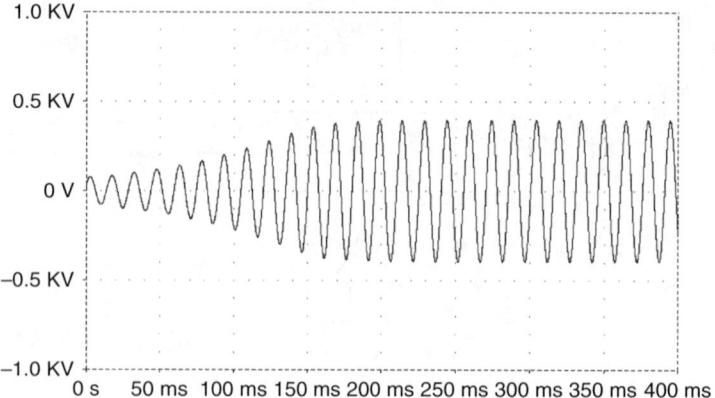

FIGURE 12.6
IG self-excitation line voltage obtained by simulation.

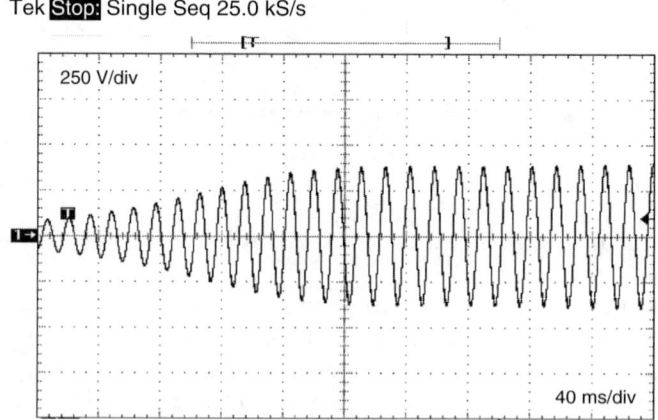

FIGURE 12.7
IG self-excitation line voltage obtained experimentally.

obtained from experimental tests. The machine self-excitation as an induction generator was simulated using a Y-connected $37\mu F$ capacitor bank and sinusoidal sources, which were disconnected at 15 ms, to represent the residual flux effects. The parameters of the induction machine were $r_s = 4.4\ \Omega$, $r_r = 5.02\ \Omega$, $L_{\ell s} = L'_{\ell r} = 15.6$ mH, and $J = 6.10^{-4}$ kg $\cdot$ m^2. The self-excitation was carried out at $\omega_{mr} = 2000$ rpm. The line voltage across the machine terminals obtained from simulation is presented in Figure 12.6 and similar voltage was experimentally obtained as shown in Figure 12.7. The satisfactory agreement between simulation and experimental results attests to the model reliability.

Although the $\alpha\beta\gamma$ model is suitable for representing average saturation effects, the assumption that the machine remains symmetrical when saturated does not allow for representation of the harmonic phenomena in currents and voltages or effects caused by machine asymmetries.

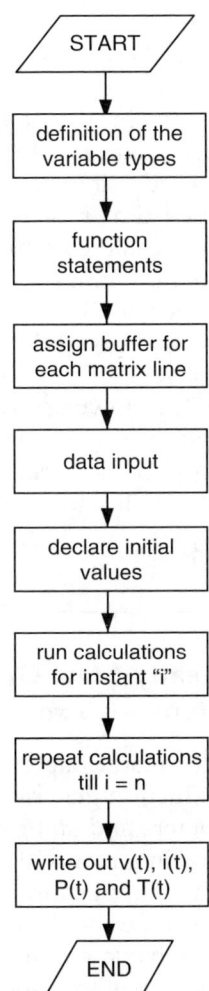

FIGURE 12.8
Pascal flowchart for the transient calculations of a SEIG.

12.5 Simulation of the Self-Excited Induction Generator in Pascal

Pascal is popular with students, researchers, and teachers because it is a simple, easily understood language, though this simplicity makes it less flexible than other more powerful languages like C and C++. For study of less complex models, it can be a quite acceptable and fast computer language for engineering problems where high speed and machine interactions are not the main issue. Iterative numerical processes and on-line fast control schemes are the main features of these computer languages.

Two high-level computer languages commonly used in engineering, Pascal and C, can be of extreme value in simulating the transient phenomenon of self-excitation. In a few minutes, details of the power, torque, voltage, and current transient states can be easily simulated and plotted for several conditions of load and speed of the prime mover. The flowchart in Figure 12.8 can be used to develop a fast dedicated computer program in either of these languages. The code of this program for a stand-alone self-excited induction generator is given below in Appendix B.

Details of the power, torque, voltage, and current transient states can be easily simulated and plotted for several conditions of load and speed of the prime mover in few minutes. The flowchart of Figure 12.8 can be used to develop a dedicated fast computer program in either of these languages.

As Pascal differs from C only in the form of statements, the computer program and the output results are left to be discussed in Section 12.10.

12.6 Simulation of the Steady-State Operation of an Induction Generator Using Microsoft Excel

To understand this section better, read Chapter 4 first. That chapter says that in many applications of the induction generator it is necessary to establish relationships among its parameters and static characteristics in such a way that the designer is able to know in advance if a specified generator is going to be suitable or not for a particular application. That is the case, for example, when an induction motor is considered to be a self-excited induction generator. The self-exciting capacitor bank has to be calculated and sized for extreme conditions of load, and it has to respond to real-world conditions. Preliminary tests can be done based on an Excel spreadsheet, as displayed in the block diagram shown in Figure 12.9 with the theory discussed in Chapter 4.

Table 12.2 displays a fitted Microsoft Excel spreadsheet with all the necessary data for the steady-state calculation of an induction generator in steady-state operation. Table 12.3 defines each variable for the calculations in Table 12.2. An example of the results of these calculations is given in Figure 12.10 where the terminal voltage of the induction generator is plotted versus its output power to the load.

Another example of the use of an Excel spreadsheet is the calculation of the averages and values shown in Table 12.1. This is a very helpful tool that saves a lot of time on tedious calculations for every new estimation of a wind power or hydro power plant performance. The statistical averages of

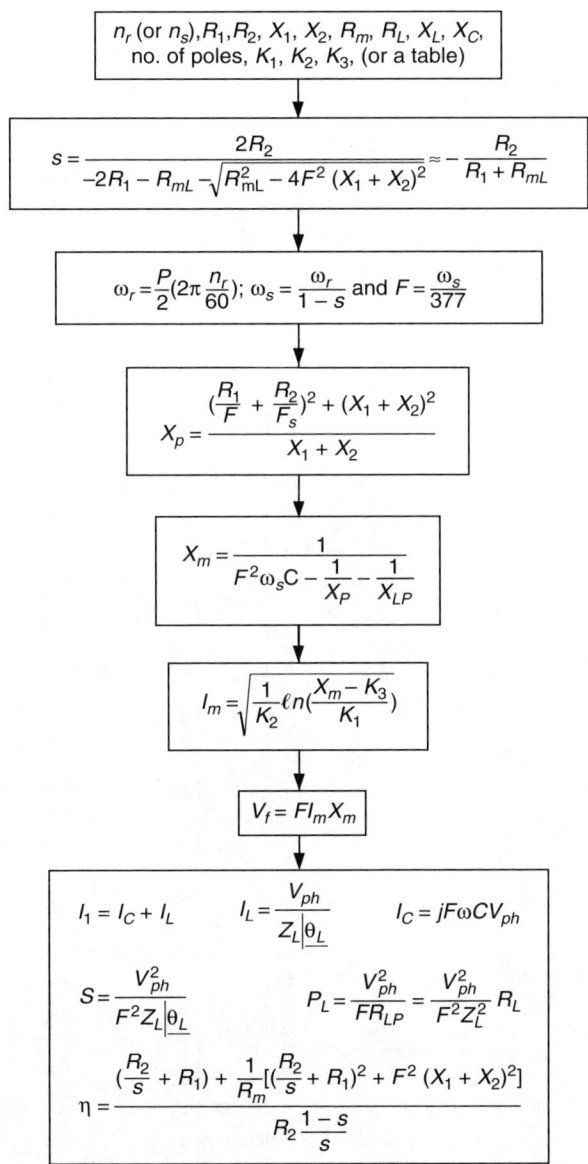

FIGURE 12.9
Excel flowchart for the steady-state calculation of a SEIG.

wind speed distribution or water flow estimation and the discrete integration processes involved are a straightforward task for such spreadsheets. The main limitation of most spreadsheets is that they cannot perform iterative processes unless helped by a lengthy user data transfer.

TABLE 12.2

Example of IG Data Fitting in Excel Spreadsheet for Steady-State Calculations

A P_L	B Vf	C Im	D Xm	E RmL	F Rp	G Xp	H s	I F	J ws	K nr	L wr	M R1	N R2	O X1
3.573516	267.2788	25.16681	10.62521	918.3069	862016.9	322807.1	-0.00044	0.999537	376.8254	1800	376.9911	0.575	0.404	1.3074
71.38949	267.0765	25.1253	10.63878	490.4483	246036.3	92149.94	-0.00082	0.999154	376.6812	1800	376.9911	0.575	0.404	1.3074
355.1414	266.1861	24.93403	10.70182	165.5928	28125.34	10541	-0.00243	0.997551	376.0768	1800	376.9911	0.575	0.404	1.3074
1387.328	262.2742	23.9844	11.02729	47.53132	2345.553	881.2441	-0.0084	0.991649	373.8515	1800	376.9911	0.575	0.404	1.3074
1717.418	260.7511	23.58812	11.16917	38.40436	1538.384	578.4512	-0.01036	0.989719	373.1239	1800	376.9911	0.575	0.404	1.3074
2247.662	257.9207	22.84026	11.44641	29.09348	890.3077	335.214	-0.01362	0.986543	371.9265	1800	376.9911	0.575	0.404	1.3074
2393.875	257.0348	22.60647	11.53557	27.20874	780.6757	294.0474	-0.01454	0.985644	371.5879	1800	376.9911	0.575	0.404	1.3074
2473.964	256.5259	22.47268	11.58711	26.26352	728.4258	274.4247	-0.01505	0.985147	371.4004	1800	376.9911	0.575	0.404	1.3074
2559.254	255.9637	22.32548	11.64426	25.31638	677.9009	255.4478	-0.0156	0.984613	371.1991	1800	376.9911	R_{s1}0.575	R_{s1}0.404	R_{s1}1.3074
2650.198	255.339	22.16283	11.70794	24.36733	629.1119	237.1208	-0.0162	0.984038	370.9822	1800	376.9911	R_{s1}0.575	R_{s1}0.404	R_{s1}1.3074
2747.279	254.6407	21.98229	11.77925	23.41635	582.0695	219.4476	-0.01684	0.983416	370.7479	1800	376.9911	R_{s1}0.575	R_{s1}0.404	R_{s1}1.3074
2851.005	253.8546	21.78089	11.85958	22.46344	536.7848	202.4325	-0.01754	0.982743	370.4942	1800	376.9911	R_{s1}0.575	R_{s1}0.404	R_{s1}1.3074

TABLE 12.2 (continued)

Example of IG Data Fitting in Excel Spreadsheet for Steady-State Calculations

P X2	Q X1+X2	R Rm	S XC	T COS(θ_L)	U [Z_L]	V RLp	W XLp	X P	Y K1	Z K2	AA K3	AB SQRT(R)	AC h	AD f_s
1.3074	2.6148	962.5	10.61499	0.99999	20000	20000.2	4472147	4	11.88	-0.0022	7.68	918.292	4.585827	59.97363
1.3074	2.6148	962.5	10.61906	0.99999	1000	1000.01	223607.4	4	11.88	-0.0022	7.68	490.4204	48.94519	59.95067
1.3074	2.6148	962.5	10.63613	0.99999	200	200.002	44721.47	4	11.88	-0.0022	7.68	165.5102	82.30469	59.85448
1.3074	2.6148	962.5	10.69944	0.99999	50	50.0005	11180.37	4	11.88	-0.0022	7.68	47.24275	93.12882	59.50031
1.3074	2.6148	962.5	10.7203	0.99999	40	40.0004	8944.294	4	11.88	-0.0022	7.68	38.04663	93.60563	59.38451
1.3074	2.6148	962.5	10.75481	0.99999	30	30.0003	6708.221	4	11.88	-0.0022	7.68	28.61961	93.79725	59.19395
1.3074	2.6148	962.5	10.76461	0.99999	28	28.00028	6261.006	4	11.88	-0.0022	7.68	26.70144	93.77368	59.14005
1.3074	2.6148	962.5	10.77005	0.99999	27	27.00027	6037.399	4	11.88	-0.0022	7.68	25.73759	93.75043	59.11021
1.3074	2.6148	962.5	10.77589	0.99999	26	26.00026	5813.791	4	11.88	-0.0022	7.68	24.77035	93.71838	59.07817
1.3074	2.6148	962.5	10.78219	0.99999	25	25.00025	5590.184	4	11.88	-0.0022	7.68	23.79953	93.67648	59.04365
1.3074	2.6148	962.5	10.789	0.99999	24	24.00024	5366.577	4	11.88	-0.0022	7.68	22.82491	93.62354	59.00637
1.3074	2.6148	962.5	10.79639	0.99999	23	23.00023	5142.969	4	11.88	-0.0022	7.68	21.84622	93.55815	58.96598

Note: Rated values of the induction generator: Power = 3.0 kW; Phase voltage = 220 V; V_{ph} = 60 Hz

TABLE 12.3

Excel Formula Definitions for Table 12.2 Spreadsheet Cells

Cell	Variable	Formula / Value
A	PL	= B2*B2/(I2*V2)
B	Vf	= I2*C2*D2
C	Im	= SQRT((1/Z2)*LN((D2-AA2)/Y2))
D	Xm	= 1/((I2*I2/S2)-(1/G2)-(1/W2))
E	RmL	= R2*V2/(R2+V2)
F	Rp	= ((M2+(N2/H2))*(M2+(N2/H2))/(I2*I2)+Q2*Q2)/(M2+(N2/H2))/(I2*I2)+Q2*Q2)/Q2
G	Xp	= ((M2+(N2/H2))*(M2+(N2/H2))/(I2*I2)+Q2*Q2)/Q2
H	s	= -N2/(M2+E2)
I	F	= J2/377
J	ws	= L2/(1-H2)
K	nr	1800
L	wr	= PI()*X2*K2/60
M	R1	0.575
N	R2*	0.404
O	X1	1.3074
P	X2 — A	1.3074
Q	X1+X2 — B	= O2+P2
R	Rm — C	962.5
S	XC — D	= 1/(J2*250*0.000001)
T	COS(QL) — E	0.99999
U	[ZL]	20000
V	RLp	= U2/T2
W	XLp	= U2/(SQRT(1-T2*T2))
X	P	4
Y	K1	= Y2
Z	K2	-0.002202
AA	K3	7.68
AB	SQRT(R)	= SQRT(E2*E2-4*Q2*Q2)
AC	h	= 100*H2*((M2+(N2/H2))+((M2+(N2/H2))*(M2+(N2/H2))+(I2*(O2+P2))*(I2*(O2+P2)))/R2)/(N2*(1-H2))
AD	fs	= J2/(2*PI())

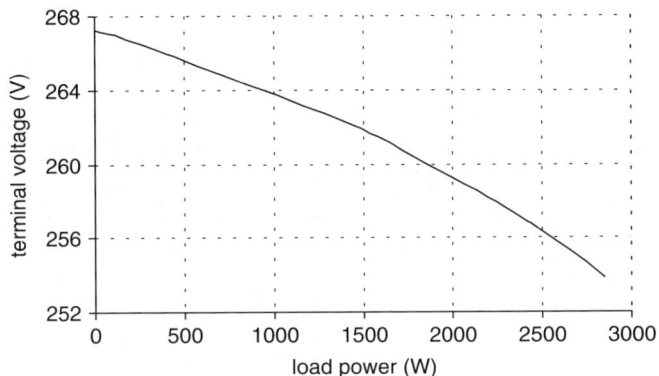

FIGURE 12.10
Example of the voltage regulation Excel plot of an induction generator.

12.7 Simulation of Vector-Controlled Schemes Using MATLAB/Simulink

Simulink is a subprogram in the MATLAB environment. It uses a graphical user interface and enables a system to be simulated by block diagrams and equations. For all of the programs in this section, the versions that are used are as follows:

- MATLAB 6.1.0.450 (R12.1)
- MATLAB Toolbox 6.1
- Simulink 4.1
- DSP Blockset 4.1

The DSP blockset is used to run the Butterworth filters in the system. If it is not available in this blockset, then the Butterworth filters can be replaced by transfer functions that represent low-pass filters.

Shown in Figure 12.11 is a typical block diagram of one of the models that can be run in Simulink. The system contains all of the equations necessary for an induction machine as described in Chapter 9. The reader is also referred to the literature on vector control, as referenced in Chapter 9. Subsystem #1 in Figure 12.11 is the controller for the model; it changes based on which model is running. Subsystem #2 in Figure 12.11 is the three-phase inverter block, which changes the reference voltages from the controller to the phase voltages through a PWM converter. Subsystem #3 in Figure 12.11 is the induction machine block, which contains all of the equations necessary for an induction machine. The block takes in the phase voltages and calculates speed, currents, fluxes, and developed torque. Subsystem #4 in Figure 12.11

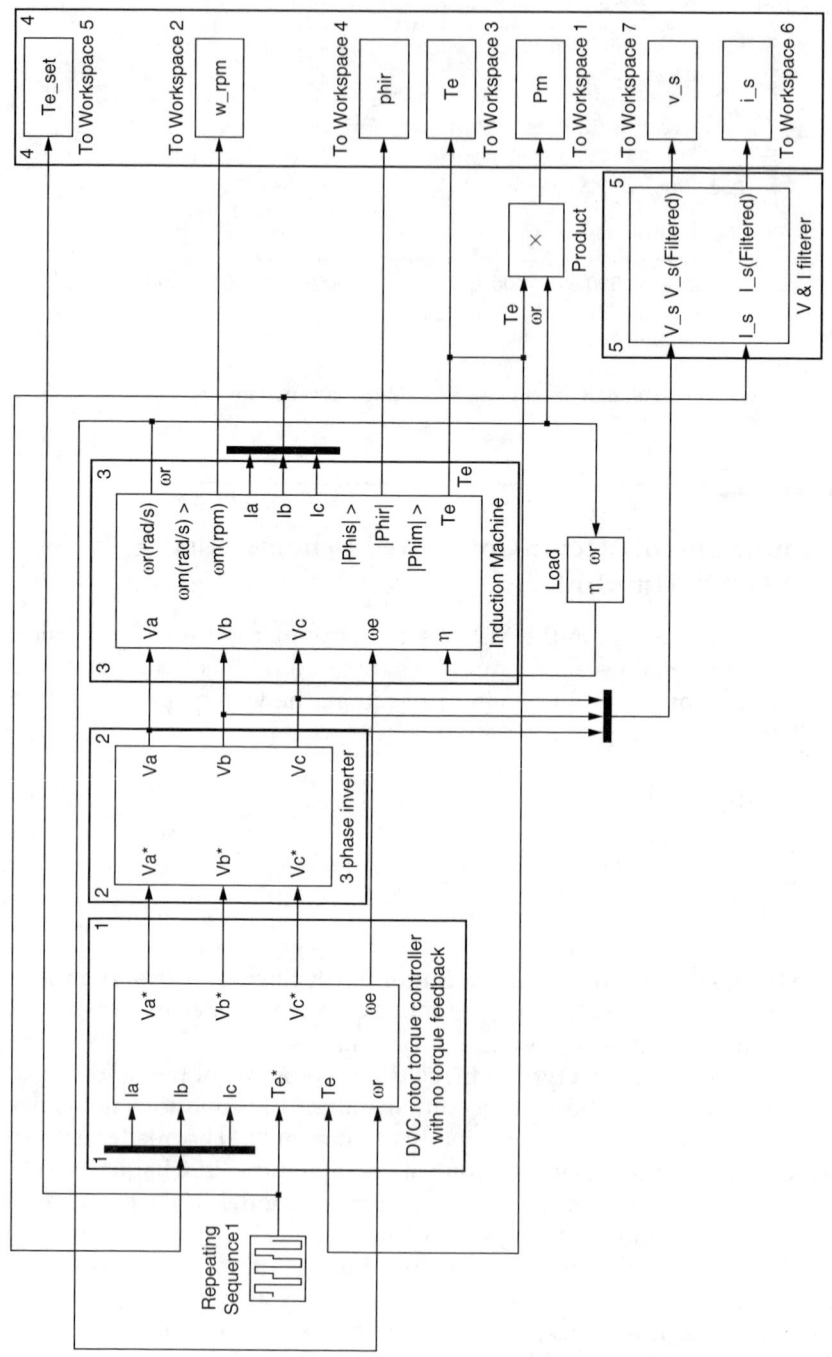

FIGURE 12.11
Simulation diagram for vector controlled schemes.

TABLE 12.4

Inputs for Simulink-Based Vector Control Scheme

Repeating sequence	Applies to all of the block diagrams and is the reference torque that the machine is trying to follow.
J	This is the inertia on the shaft, usually 0.05.
Kw	This value is the constant to determine rotational losses in the machine when running. $T_L = K_w \omega^2$, usually 0.00047502.
freqt	This value is the frequency that the PWM operates at, usually at 2000 Hz.
Vdc	This is the value of the DC bus, usually at 300 V.
Tstop	This value is how long the simulation runs, usually 10 s.
t_J_large	After this time value, the inertia of the machine will be changed to a very large value so the speed of the machine doesn't change with torque, usually at 4 s.
fc*	All these values fci_dvc, fcv, etc. are used to control the filtering in the output, controllers and the machine. Lower filtering values can create smoother signals, but could also create instabilities in the system.

TABLE 12.5

Output for Simulink-Based Vector Control Scheme

Te_set	Torque command, which derives directly from the repeating sequence in Nm.
w_rpm	Mechanical speed of the shaft in rpm.
phir	Rotor flux within the machine in Wb.
Te	Torque developed by the machine in Nm.
Pm	Mechanical power created by the machine in W.
v_s	Va, Vb, and Vc, filtered to 1000 Hz in V.
i_s	Ia, Ib, and Ic, filtered to 1000 Hz in A.

takes the outputs of the model where the matrices are sent to the MATLAB workspace as the value listed. The time index ("tout") is also output to the workspace. Subsystem #5 in Figure 12.11 is a filtering block that filters both the input voltages and output currents with an eighth-order low-pass Butterworth filter.

12.7.1   Inputs

Before any of the models can be simulated, the *consts.m* file must be run to initialize the variables within the machine and also the controller constants. Table 12.4 displays the inputs to the model, which can be changed; all of the values are included in *consts.m* except for the repeating sequence.

12.7.2   Outputs

In terms of the outputs of the model, every time the model is run it creates the matrices given in Table 12.5.

There are several scopes inside of the blocks to show instantaneous variables such as i_{qs} and i_{ds} and the reference voltages v_{qs}^* and v_{ds}^*. To obtain any other values within the machine or the controllers, the reader can add either a scope connected to that point or to a workspace block. To run the model, follow these steps.

1. Change the MATLAB directory to where the files are.
2. Type *consts* in the MATLAB command window. MATLAB will display:
   - The current revision
   - "Constants have been initialized"
3. Open one of the models to be run.
4. Click on the icon for simulation at the top of the screen. The time will be displayed at the bottom of the screen and will go to Tstop.

After the simulation runs to the end of the time, all of the output variables will be available to use in the MATLAB workspace. For example, to display the currents on a graph with time, use the following command: *plot(tout,i_s)*, which will plot all three currents in a figure.

All of the simulations run with the following sequence. A low (1 Nm) reference torque is applied for 2 s (0 s–2 s). This allows the fluxes within the machine to stabilize. A high (5 Nm) reference torque is applied for 4 s (2 s–6 s). While this is happening, the machine is running as a motor and the speed will increase until the rotational losses are equal to the applied torque. A very high inertia (10,000 J) is applied at time 4 s. This makes the speed of the machine constant so that power can be generated with a negative torque. Without this change, when a negative torque is applied, the machine will reverse rotation and spin in the other direction. A high negative (-5 Nm) reference torque is applied for 4 s (6 s–10 s). Since the speed is locked due to the high inertia, the motor is now a generator generating power.

The next three sections outline how to run the models for the specific controllers and give typical outputs for these models.

12.7.3   Indirect Vector Control

Files needed:

- consts.m
- ivc_rotor_model_torque_no_feed.mdl
- ivc_rotor_model_torque_w_feed.mdl

In the indirect-vector torque-control system, both of the IVC rotor models find the electrical speed from the speed of the shaft and the calculated slip angle. The controller imposes the restraint that the rotor flux, Phir, is aligned

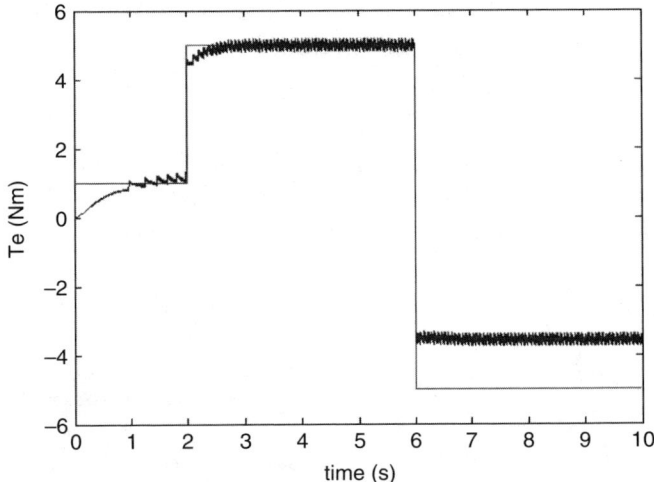

FIGURE 12.12
Torque response for indirect vector control.

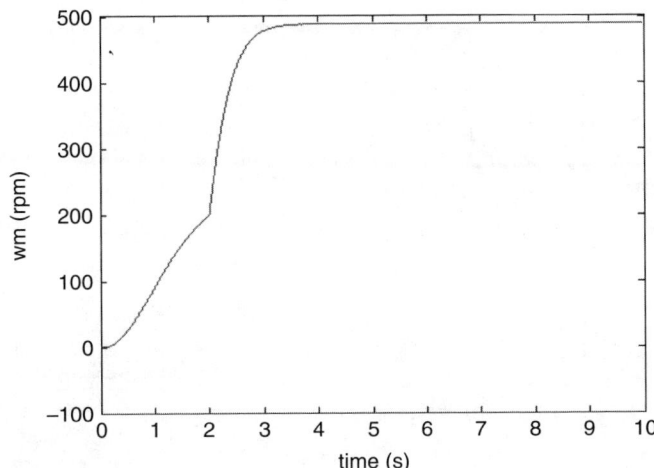

FIGURE 12.13
Speed response for indirect vector control.

with the ds-axis. Since I_{qs} is directly proportional to torque, the current I_{qs} is controlled to follow the reference torque applied to the block.

The model without feedback uses no measurement from torque to control the torque of the motor, relying on the proportionality of I_{qs} and torque. The torque response of the system is shown in Figure 12.12, and the torque follows the reference torque, except that when the torque is negative, there is a DC offset. Shown in Figure 12.13 is the speed response of the machine and in Figure 12.14 is the power to the machine. As can be seen from those figures,

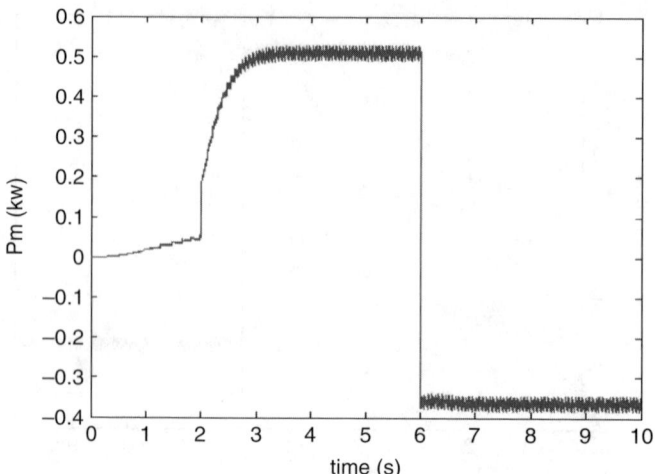

FIGURE 12.14
Power response for indirect vector control.

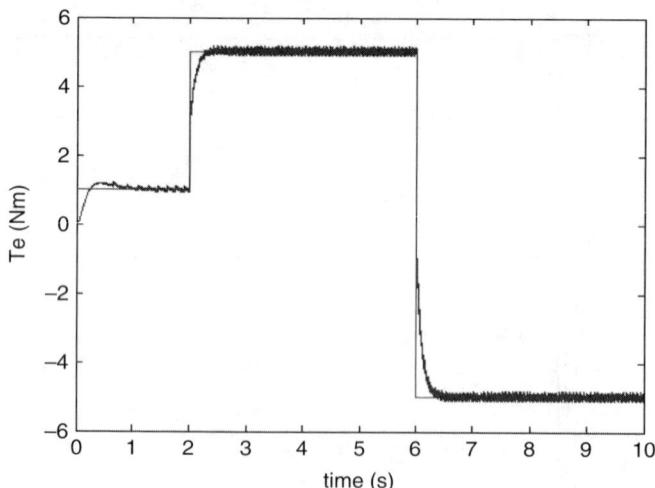

FIGURE 12.15
Closed loop torque response for indirect vector control.

from time 0s to 6s, the machine is in motoring mode, and from 6s to 10s, the motor is generating power due to a positive speed and a negative torque.

The feedback model assumes that there is a torque sensor on the shaft and therefore the reference I_{qs} is based on the difference between the reference torque and the actual torque, commanded through a PI controller. This provides significantly better response on the torque, as shown on Figure 12.15, Figure 12.16, and Figure 12.17. Except for the better torque response, this system is identical to the system without feedback.

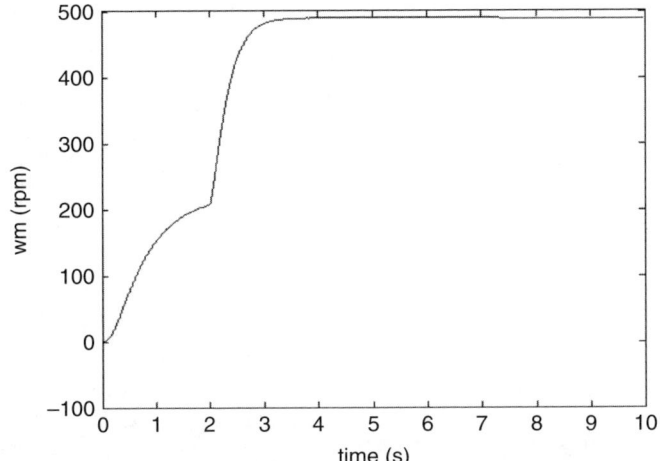

FIGURE 12.16
Closed loop speed response for indirect vector control.

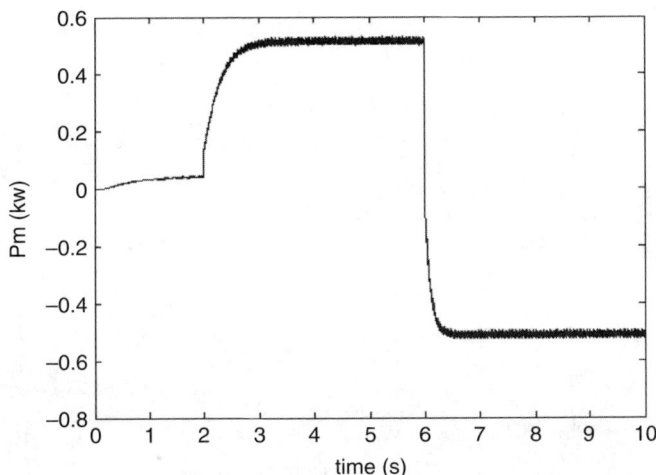

FIGURE 12.17
Closed loop power response for indirect vector control.

12.7.4 Direct Vector Control with Rotor Flux

Files needed:

- consts.m
- dvc_rotor_model_torque_no_feed.mdl
- dvc_rotor_model_torque_w_feed.mdl

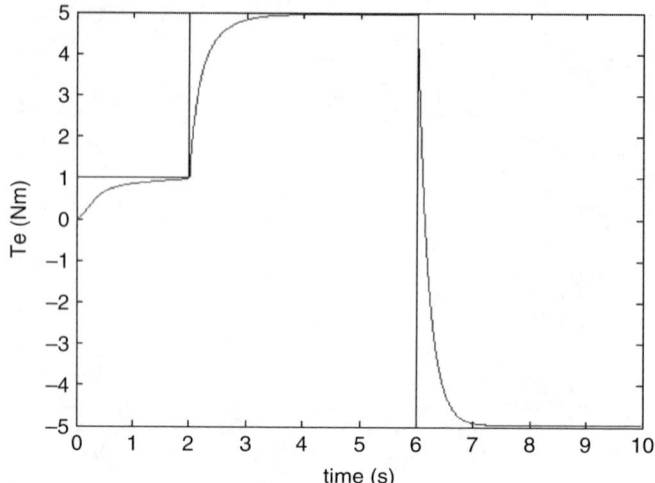

FIGURE 12.18
Torque response for direct vector control with rotor flux.

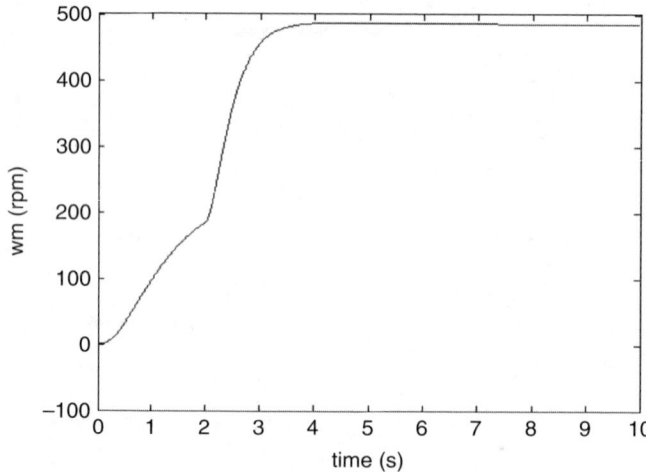

FIGURE 12.19
Speed response for direct vector control with rotor flux.

In the direct-vector-control system with rotor flux, both of the DVC rotor models find the electrical speed from value of the rotor flux on the stationary axis. The controller imposes the restraint that the rotor flux is aligned with the ds-axis. Since I_{qs} is directly proportional to torque, the current reference I_{qs} is proportional to the reference torque applied to the block.

The model without feedback uses no measurement from torque to control the torque of the motor, relying on the proportionality of I_{qs} and torque. The torque response of the system is shown in Figure 12.18, and the torque

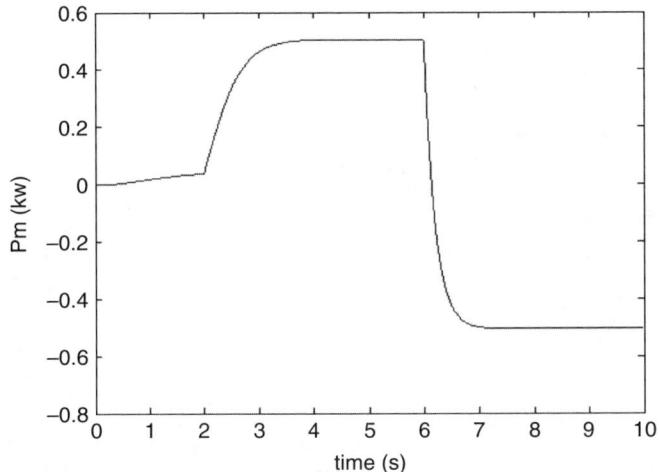

FIGURE 12.20
Power response for direct vector control with rotor flux.

follows the reference torque, with a slow response. Shown in Figure 12.19 is the speed response of the machine, and on Figure 12.20 is the power to the machine. As can be seen from those figures, from time 0 s to 6 s, the machine is in motoring mode, and from 6 s to 10 s, the motor is generating power due to a positive speed and a negative torque.

The feedback model assumes that there is a torque sensor on the shaft and therefore the reference I_{qs} is based on the difference between the reference torque and the actual torque, commanded through a PI controller. This provides significantly better response on the torque, as shown in Figure 12.21, Figure 12.22 and Figure 12.23. Except for the better torque response, this system is identical to the system without feedback.

12.7.5 Direct Vector Control with Stator Flux

Files needed:

- consts.m
- dvc_stator_model_torque_no_feed.mdl
- dvc_stator_model_torque_w_feed.mdl

In direct vector control with stator flux, both of the DVC stator models get the electrical speed directly from the currents and voltages in the machine. The controller imposes the restraint that the stator flux in the machine is aligned with the ds-axis. Since the torque produced by the machine is proportional to the current I_{qs}, this current is used to control the torque developed by the machine.

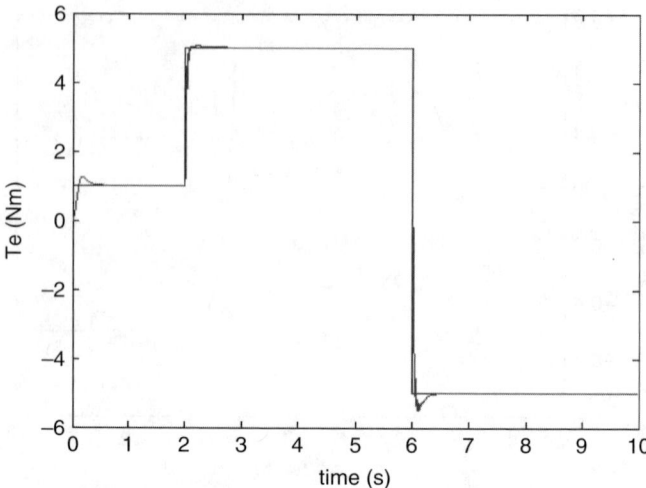

FIGURE 12.21
Closed loop torque response for direct vector control with rotor flux.

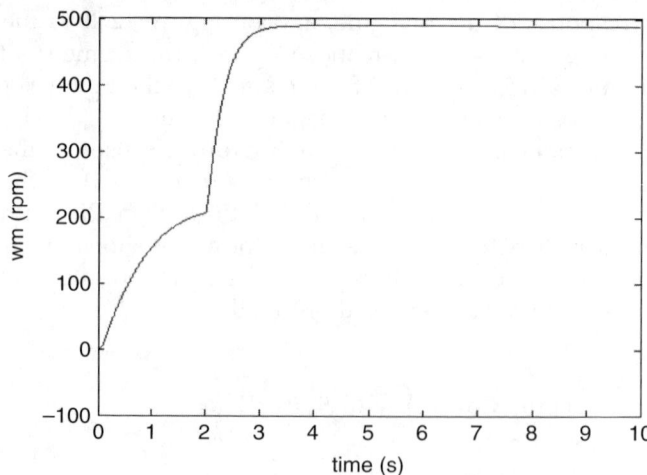

FIGURE 12.22
Closed loop speed response for direct vector control with rotor flux.

The model without feedback uses no measurement from torque to control the torque of the motor, relying on the proportionality of I_{qs} and torque. The torque response of the system is shown in Figure 12.24, and follows the reference torque very closely.

This is the best controller in terms of response, especially considering that there is no feedback from the torque being developed by the machine. Shown in Figure 12.25 is the speed response of the machine, and in Figure 12.26 is the power to the machine. As can be seen from the figures, from time 0 s to

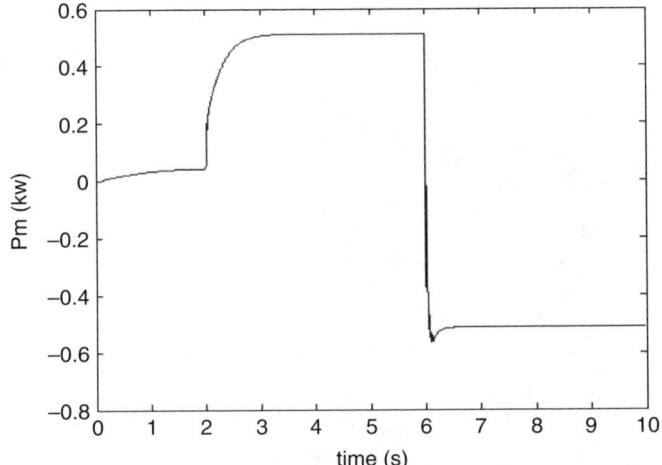

FIGURE 12.23
Closed loop power response for direct vector control with rotor flux.

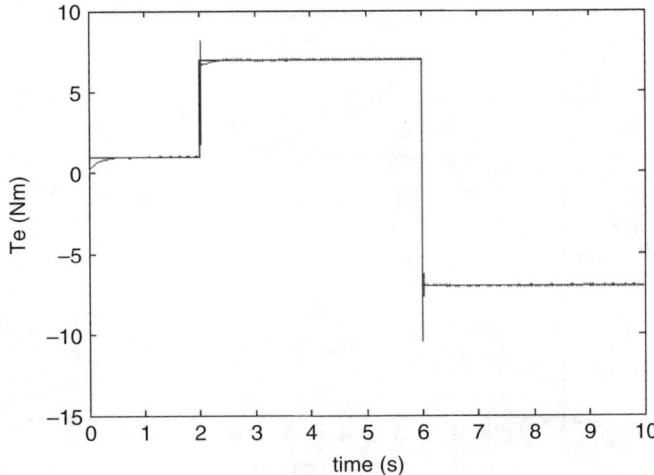

FIGURE 12.24
Torque response for direct vector control with stator flux.

6 s, the machine is in motoring mode, and from 6s to 10 s, the motor is generating power due to a positive speed and a negative torque.

The feedback model assumes that there is a torque sensor on the shaft and therefore the I_{qs} is based on the difference between the reference torque and the actual torque, commanded through a PI controller. Adding this feedback does not make the controller act that much better, as the model without the feedback is very good at following the reference torque. The responses with feedback are shown on Figure 12.27, Figure 12.28 and Figure 12.29.

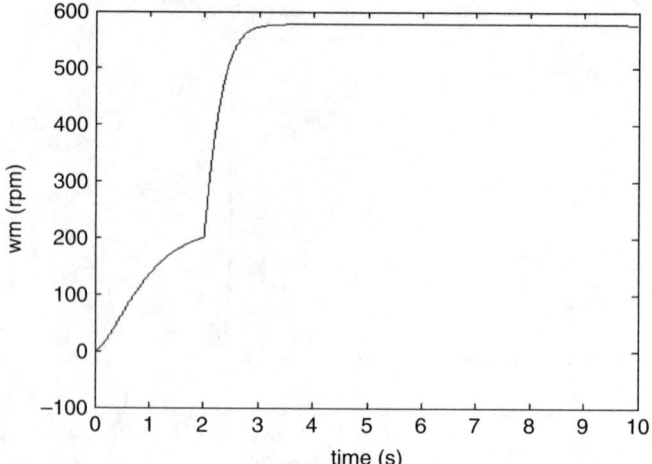

FIGURE 12.25
Speed response for direct vector control with stator flux.

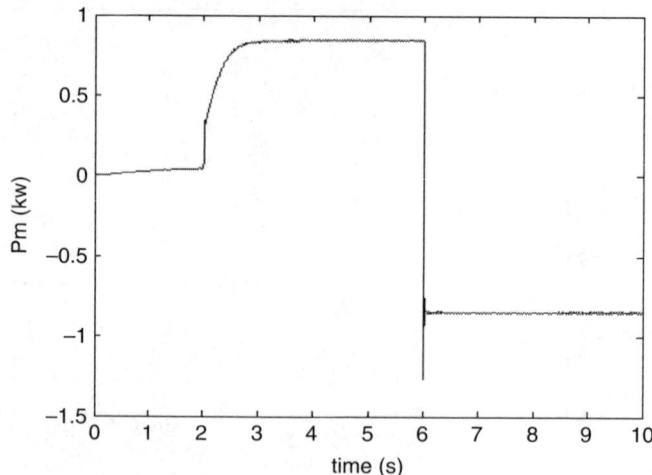

FIGURE 12.26
Power response for direct vector control with stator flux.

12.7.6   Evaluation of the MATLAB/Simulink Program

Simulink is a very powerful program that can be used to simulate a wide variety of systems. The graphical user interface is very easy to use, and scopes can be placed at any junction to observe the current state of the system. And the ability of Simulink to output variables to the MATLAB environment is a large plus when dealing with complex systems. Simulink

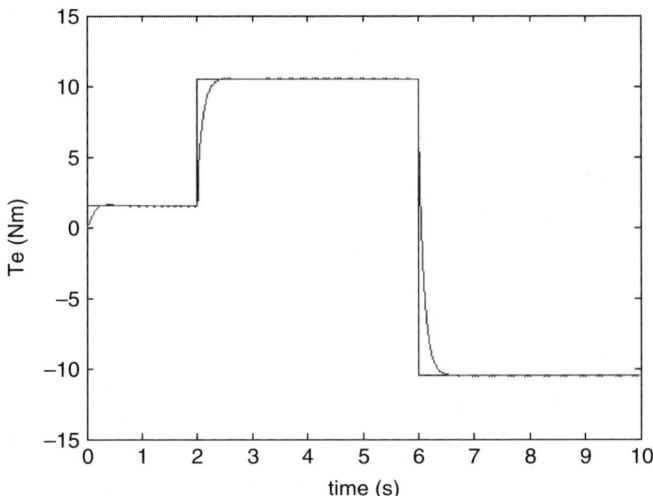

FIGURE 12.27
Closed loop torque response for direct vector control with stator flux.

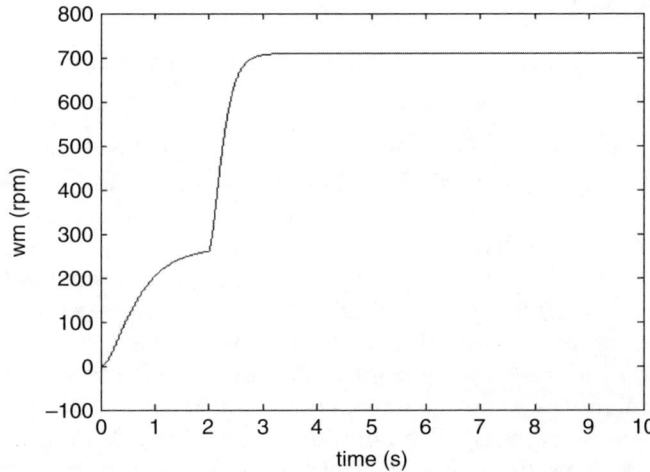

FIGURE 12.28
Closed loop speed response for direct vector control with stator flux.

can also interface with hardware and perform calculations in real time to control systems.

However, the graphical user interface imposes problems in terms of simulation time. The models that are shown in this section take 5 minutes to run a 10-second simulation on a Pentium 3 1.2 GHz processor with 512 MB RAM. When running, the program will consume over 200 MB of RAM; you need a powerful system to run these simulations.

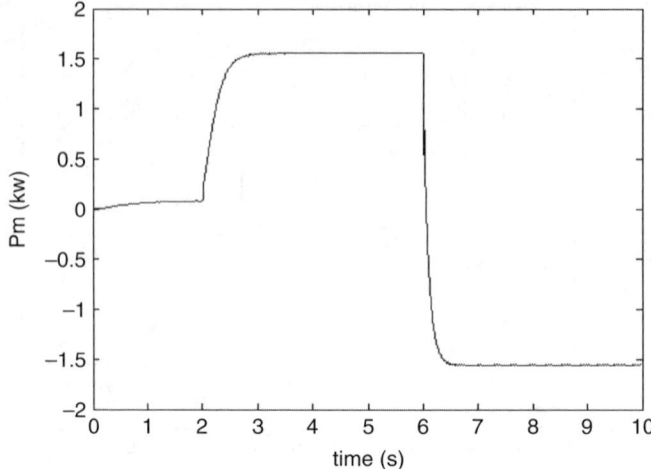

FIGURE 12.29
Closed loop power response for direct vector control with stator flux.

12.8 Simulation of a Self-Excited Induction Generator in PSim

PSim is a commercial program that enables speedy analysis of an induction generator. Readily available tools and electrical and electromechanical components make this program an easy tool to observe phenomena that would otherwise only be observable in a good power system laboratory. PSim is an internationally accepted computer program used to demonstrate practical results in many high-level international congresses and magazines.

Before reading this section, the student should have first read Chapter 4. In this section we present a short introduction to the PSim block that can be used to simulate the stand-alone operation of a wound-rotor self-excited induction generator (SEIG). The PSim block is password protected by Powersim Inc. However, the parameters of the model can be changed by highlighting the model—it is a subcircuit—or going to *Subcircuit/Edit Subcircuit* and clicking on *Subcircuit Variables*. The induction generator parameters in the PSim SEIG block are as follows. (All parameters in Figure 12.30 refer to the stator side.)

- R_s stator resistance, Ohms
- $L_{\ell s}$ stator leakage inductance, H
- R_r rotor resistance, Ohms
- $L_{\ell r}$ rotor leakage inductance, H
- P no. of poles

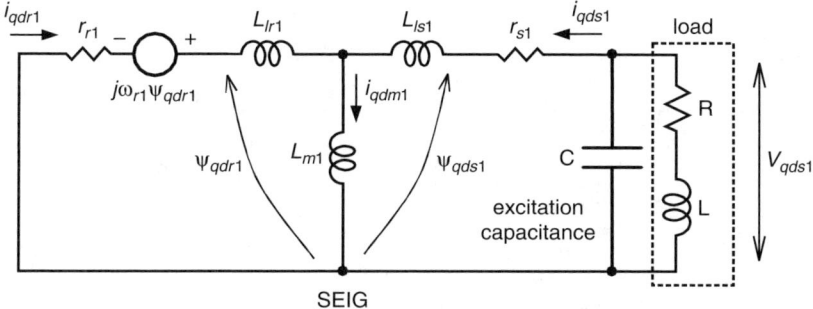

FIGURE 12.30
Induction generator d – q model.

- J moment of inertia, J
- F_{rated} rated operating frequency, Hz
- V_{rf} residual flux vector, V
- T_{start} time at which the initial flux is applied, s
- T_{end} time at which the initial flux is removed, s

The nonlinear relationship between the magnetizing current I_m and the magnetizing inductance X_m is expressed in the form of a lookup table and is stored in the file *file_Lm_Im.tbl*. The lookup table name must be named as above; no other names are allowed. The lookup table has the following format:

0	66.4097
1.0000	66.2097
2.0000	65.7694
3.0000	65.0458

Column 1 is the magnetizing current I_m, in A, and Column 2 is the magnetizing inductance L_m, in mH. The PSim model applies the residual flux voltage in the same way as the MATLAB/C simulation. That is, a pulse with amplitude of V_{rf} is applied to both the d-axis and q-axis circuit for a short period of time. The circuit presented in Figure 12.31 is used to simulate the transient self-excitation and the load response of a stand alone SEIG whose parameters are shown in Table 12.6.

The load parameters are initialized to simulate an open circuit. The induction generator might fail to self-excite if started on a heavy load. A suitable value for self-excitation capacitance should be chosen from the magnetization curves to self-excite the generator up to the rated voltage. Figure 12.32 presents some simulation results from the circuit in Figure 12.31.

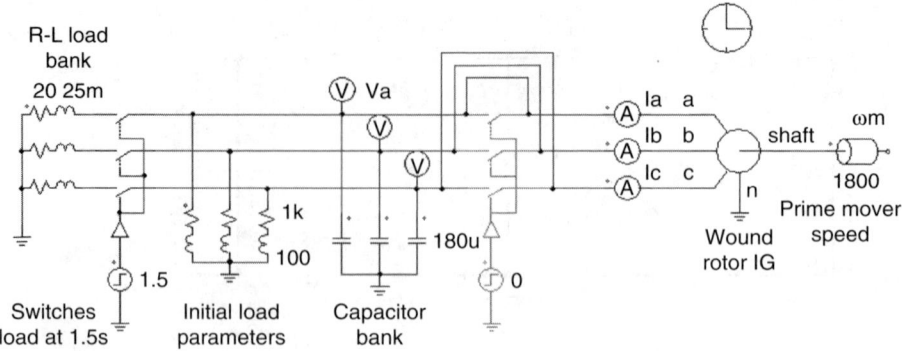

FIGURE 12.31

PSim SEIG simulation with linear load.

TABLE 12.6

Parameters Used for the SEIG Simulated
in PSim®

Simulation Parameters		Open Circuit Load Parameters	
R_s	0.262 Ω	R	1000 Ω
$L_{\ell s}$	1.679 mH	L	100 H
R_r	0.447 Ω	Load switched at	t = 1.5 s
$L_{\ell r}$	3.899 mH	R	20 Ω
P	4	L	20 mH
J	0.05 J	Shaft speed	1800 rpm
f_{rated}	60 Hz	Tsim	2 s
V_{rf}	10 V	DeltaT	0.0001 s
T_{start}	0		
T_{end}	0.001 s		
C_1	180µF		

The load is switched on at t = 1.5 s after the generator has completely excited. Computation time is 25 s for a simulation time of 2 s. PSim is fast compared to a similar program in MATLAB but slower than C. The transient in PSim is somewhat different from that in MATLAB, although the overall trend is the same as will be discussed in the next section. The steady-state values in PSim and in MATLAB are very close.

12.9 Simulation of a Self-Excited Induction Generator in MATLAB

This is a short introduction to the MATLAB program that simulates the stand-alone operation of a self-excited induction generator (SEIG). The flow-chart in Figure 12.32 describes the step-by-step simulation procedure. The

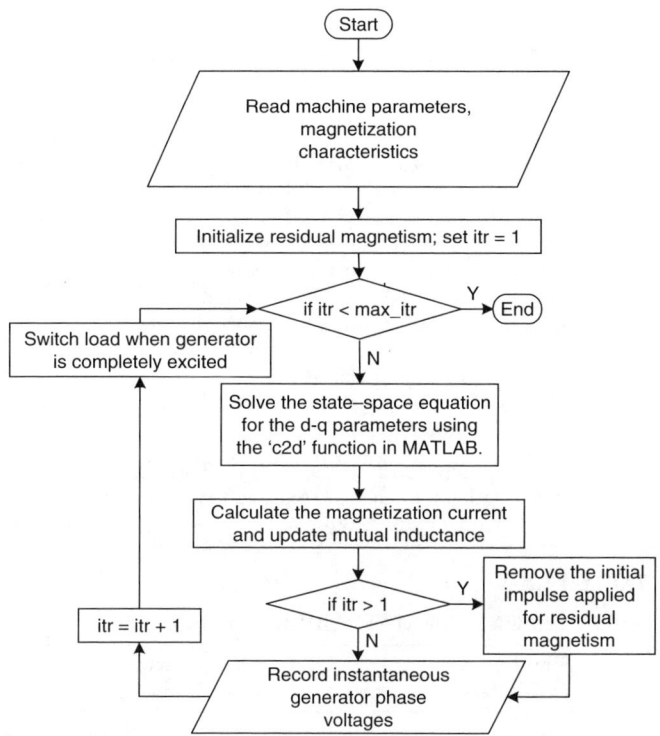

FIGURE 12.32
Flow chart for MATLAB and C simulations.

inputs to the program are the machine parameters, magnetization characteristic, residual magnetism, and self-excitation and load parameters as presented in Table 12.7. A step of 1 ms can produce accurate results without sacrificing execution speed.

An impulse function is used as well to represent the remnant magnetic flux in the core. The load parameters are initialized with large values for load resistance and capacitance to simulate an open circuit or no load condition. On the other hand, the induction generator might fail to self-excite if started on a heavy load. A suitable value for self-excitation capacitance should be chosen from the magnetization curves to self-excite the generator at the rated voltage. The load parameters can be modified after the generator has been completely excited. Mutual inductance is computed at every step using the relation obtained from experimental data. To change the magnetization characteristic, the relationship between mutual inductance and magnetization current has to be entered as shown in Figure 12.30.

The program solves the generator differential equations for the d – q axis voltages and currents using the *c2d* function in MATLAB. The d – q variables are then converted back to a-b-c variables using the d – q transformation. Table 12.8 gives some SEIG parameters for MATLAB simulation results.

TABLE 12.7

SEIG Parameters for MATLAB Simulation

Parameters

R_{s1}	Stator resistance, Ohms
X_{s1}	Stator reactance, Ohms
R_{r1}	Rotor resistance, Ohms
X_{r1}	Rotor reactance, Ohms
M_1	Mutual Inductance, Ohms
u_1	Residual magnetism vector
C_1	Self-excitation capacitance, F
R	Load resistance, Ohms
L	Load inductance, H
w_{r1}	Prime mover speed, rad/sec
Tisim	Simulation time, s
DeltaT	Sampling time for discretization in s

Note: (All parameters are referred to the stator side, Figure 12.30).

TABLE 12.8

SEIG Parameters for MATLAB and C Simulations

Simulation Parameters		Open Circuit Load Parameters	
R_{s1}	0.262 Ω	R	1000 Ω
X_{s1}	0.633 Ω	L	100 H
R_{r1}	0.447 Ω	Load switched at	t = 2 s
X_{r1}	1.47 Ω	R	20 Ω
M_1	0.25 H	L	0.02 H
u_1	[10 10 0 0] V	w_{r1}	377 rad/sec
C_1	180 μF	Tsim3	3s
$M_1 = 0.423e^{-0.0035I_m^2} + 0.0236$		DeltaT	0.001 s

The load is switched on at t = 2 s after the generator has completely excited. Computation time is 8 s for a simulation time of 3 s. The prime mover model has not been included in this program. The prime mover speed w_{r1} is assumed to be constant in the program, but any function of time can be assigned to the variable based on load variations or prime mover input. MATLAB is relatively slow compared with similar programs in C or Pascal.

12.10 Simulation of a Self-Excited Induction Generator in C

To better understand these simulations, the student should first read Chapter 4. This is a short introduction to the C program that simulates the stand-alone operation of a self-excited induction generator (SEIG).

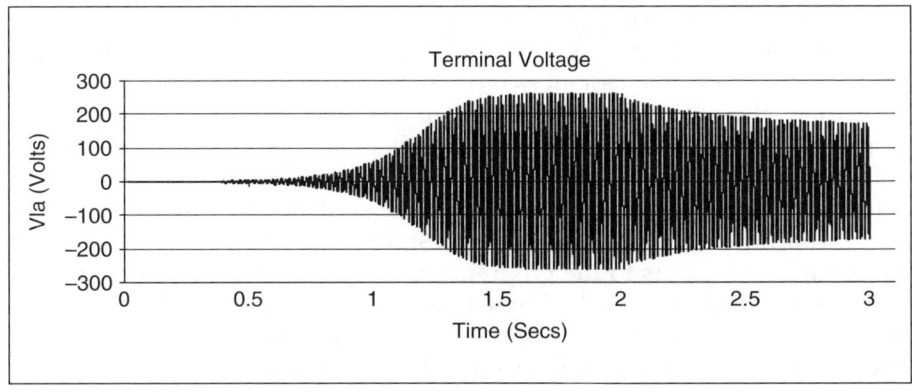

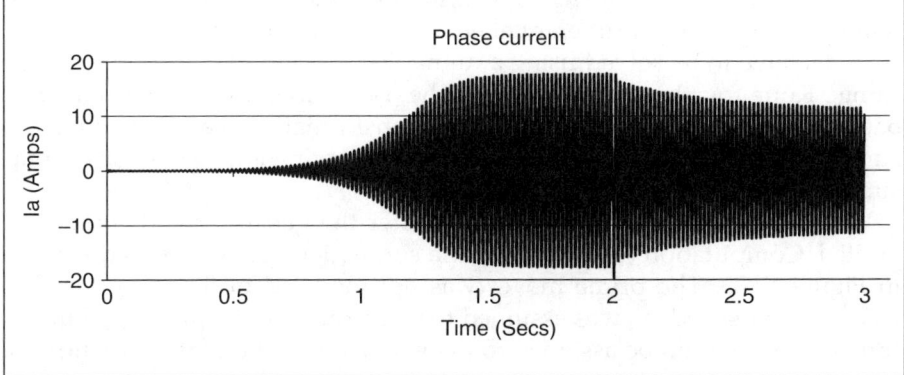

FIGURE 12.33
Self-excitation and load response in C language of a stand-alone SEIG.

C has three modules: the compiler, the assembly, and the debugger. The compiler converts a program statement into object code. When you write a C or C⁺⁺ program the filename of your source code usually ends with an extension *.c* or *.cpp*. Examples of that are *counter.c* and *satura.cpp*.

Assembly language is a low-level structured language that allows you to write a computer program as close as possible to the real machine code of the computer. Low level means that its language symbols are very close to the machine's physical connections. Writing assembly language can be a very complicated and laborious task. The advantages of assembly language are speed and direct full control over the hardware circuitry. Assembly language programs can be written separately with an editor and later assembled and linked altogether. The role of a compiler is to convert program code into object code. An object code filename usually ends with an extension *.asm*. Examples of that are *seig.asm* and *charact.asm*.

It is difficult to write a first version of a computer program without making mistakes of two kinds: syntax errors and programming errors. Syntax errors are due to a wrong compilation or assembly of the source codes, which can cause abortion of the executable computer program. As a rule, syntax errors are easier to correct than programming errors. Programming errors are mistakes in programming logic. Most of the time, they are not detected in compilation or in code assembly but will show up as unreasonable results. Sometimes these errors are very difficult to detect. The debugger program exists to help the user detect errors in general by giving logical hints or general rules to be followed to shortcut the search for mistakes.

Program simulations in C are developed similarly to MATLAB simulations. The flowchart in Figure 12.32 describes the step-by-step simulation procedure. The inputs to the program are the machine parameters, magnetization characteristic, residual magnetism, self-excitation, and load parameters listed in Table 12.7. Notice that a discretization sampling time of 0.1 ms can produce very accurate results without sacrificing execution speed.

The main difference between the C solution and the MATLAB solution is that in C the generator differential equations for the d – q axis voltages and currents have to be solved using a numerical method like, for example, the Runge-Kutta fourth order procedure. The d – q variables are then converted back to a-b-c variables using the d – q transformation. Parameters for this simulation are also given in Table 12.8. The following is a discussion of some simulation results.

The load was switched on at t = 2 s after the generator had completely excited. Computation time was 2 s for a simulation time of 3 s, as displayed in Figure 12.33. The prime mover was not modeled in this program. The prime mover speed ω_{r1} was assumed to be constant in the program, but any function of time can be assigned to the variable based on load variations or prime mover input. The C program calculations are also very fast compared to a similar program in MATLAB.

12.11   Problems

1. Collect data for a wind power site being considered for the installation of a small power plant. Develop a computer program to estimate the average speed, the root mean cubic speed, the cubic power density, and the cubic energy density.

2. Do the same as in Problem 1 for a small hydro power plant.

3. Develop a program to estimate induction generator performance as an efficient and qualified source of energy for an industrial enterprise.

4. Give good technical and economic reasons for recommending a computer program from among the ones you know to estimate the torque and speed of an induction generator relating it only to its electrical parameters. What would be the best program to recommend as an equivalent circuit for these calculations?

5. Use the MATLAB/Simulink block diagram shown in Figure 12.11 to simulate a vector control for a small hydropower plant of 100 kW so that voltage and frequency control does not exceed a 5% tolerance range.

References

1. Vosburgh, P.N., *Commercial Applications of Wind Power*, Van Nostrand Reinhold Co., New York, 1983.
2. Patel, M.R., *Wind and Solar Power Systems*, CRC Press, Boca Raton, 1999.
3. Messenger, R. and Ventre, J., *Photovoltaic Systems Engineering*, CRC Press, Boca Raton, 2000.
4. Kusko, A., *Emergency/Standby Power Systems*, McGraw-Hill Book Company, New York, 1989.
5. Lawrence, R.R., *Principles of Alternating Current Machinery (Máquinas de Corriente Alterna H.A.S.A.)*, Editora Hispano Americana S.A., Buenos Aires, Argentina, 1953.
6. Chapman, S.J., *Electric Machinery Fundamentals*, Third Edition, McGraw-Hill International Edition, New York, 1999.
7. Szczesny, R. and Ronkowski, M., A new equivalent circuit approach to simulation of converter induction machine associations, *Proceedings of the European Conference on Power Electronics and Applications (EPE'91)*, p. 4/356–4/361, Firenze, Italy, 1991.
8. Krause, C., *Analysis of Electric Machinery*, McGraw-Hill, New York, 1986.
9. Kovács, P.K., *Transient Phenomena in Electrical Machines*, Elsevier, London, 1984.

13

The Economics of Induction Generator-Based Renewable Systems

13.1 Scope of This Chapter

The vision of the interaction of values, principles, and resource dynamics, of policies and social effects constitute the paradigm of economics, the environment, and distributed generation. Such a paradigm brings together the views of environmentalists, economists, and end users. Environmentalists are interested in maintaining the equilibrium, renewal, and regeneration of ecosystems; their main perspective is embodied in the sustainability of natural systems. Economists and market decision makers are interested in how natural resources or natural services can be used for human benefit. From this perspective, only things useful to human beings have any value. Standard economic analysis so far does not consider the ecosystem to have any intrinsic value, whereas the ecological view suggests that natural systems must be protected independently of their use or value to humans. The end user is interested in converting the available resources into products and services. If electric energy is good for home and industrial needs, the end user is interested in how electrical energy will be priced and how reliable its availability will be.

Public utilities sell power to users at tariffs based on historic load demand and investments made to bring electricity to the end user. Public utilities also regulate how end users are tied into the distribution system and monitor power quality, reliability, and harmonic pollution affecting the end user. If the end user is interested in investing in distributed generation because of reliability, harmonic pollution, or power quality concerns, he or she may not be concerned about the ecological impact. On the other hand, if the end user is investing in distributed generation of community or small-scale power, he or she may be interested in keeping up with the environmental issues involved in the broader impact of such installation. This chapter covers the fundamental economic ideas that support decision-making criteria for renewable energy systems, particularly those that need induction generators to convert the energy source into electricity.

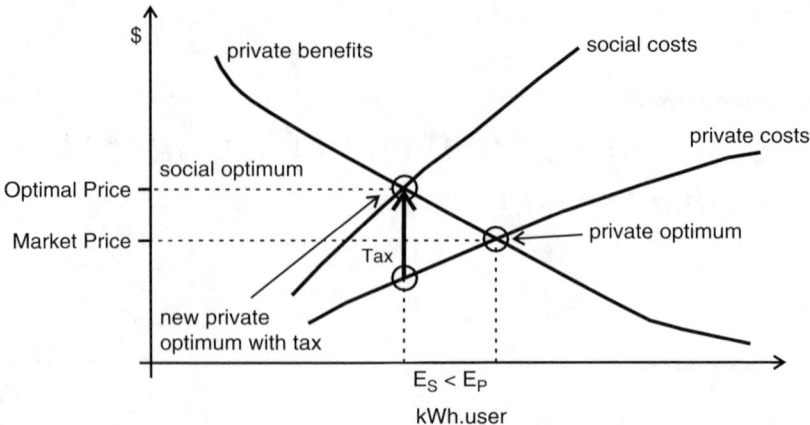

FIGURE 13.1
Optimal and market price of energy in terms of private benefits and social costs.

13.2   Optimal and Market Price of Energy in a Regulatory Environment

A comprehensive understanding of the interaction of the environment and electric energy users through an economic analysis sometimes binds incompatible perspectives. Electric energy is not found in nature directly (except in lightning and statics). Electric energy has to be converted from fossil fuels (oil, coal, and natural gas), nuclear power, hydropower (large scale renewable) and small-scale renewable sources like wind and solar. Nonrenewable sources have a very high energy density at the expense of generating wastes and polluting by-products. The costs associated with pollution and resource depletion must be taken into consideration when balancing the social costs of energy production with the social benefits.

Figure 13.1 shows the private benefit of a given electric power infrastructure: demand for a kWh of energy consumed by users. The private cost of supply is indicated in the figure with a private optimum at energy E_p. However, there are costs not included under utilities perspectives: some losses may occur that affect people. Perhaps the landscape will be transformed to make room for a coal thermal plant, or pollution will transform the perceived impact of the environment on people. Accumulated carbon dioxide may increase due to loss of land area including beaches and wetlands, thinning of species and forest area with heat waves, droughts, and disruption of water supplies. Anything related to climate change is hard to quantify as a social cost but surely represents a curve to be considered to find an optimal social cost for an energy investment. If all private benefits are captured by social benefits, the new intersection E_s represents a new

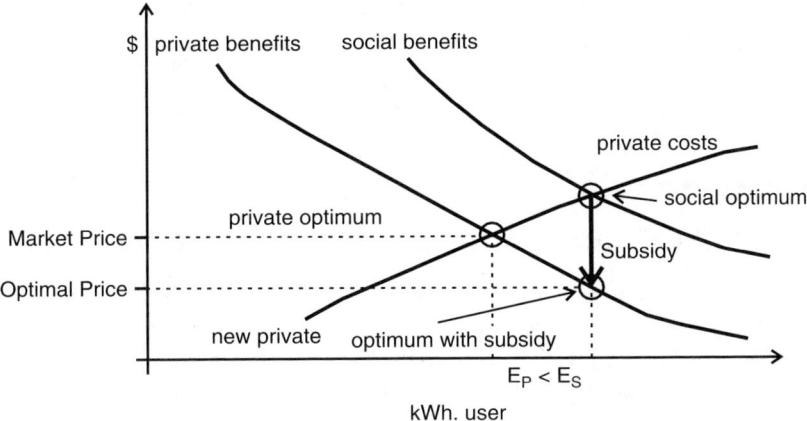

FIGURE 13.2
Effects of subsidies on social optimal market price.

equilibrium point representing a compromise between the need for energy and the desire to keep a resilient ecosystem.

One way to have an ecosystem-friendly market-oriented policy is to raise the price of energy with a pollution tax, as indicated by the arrow in Figure 13.1. A pollution tax looks like an easy way to make consumers buy less of the "pollution source" at tight regulatory government guidelines. On the other hand, a subsidy could also be implemented to put projects forward that create external benefits. Figure 13.2 shows such a situation where green energy would be fostered by encouraging private investors in small distributed energy — for example, small, more efficient hydropower plants, combining heat-power cycles (cogeneration), and combining agricultural processing machines with local energy production. Provisions for hybrid systems incorporating hydro, solar, and wind installations will make each technology capable of complementing or circumventing the gaps in power generation, physical limitations, or economic efficiencies of the other sources. The arrow in Figure 13.2 indicates a subsidy that lowers the costs, through a tax rebate or offers payments for green power investments and the purchase of on-site generated power.

With environmental policies, the government acts to modify market outcomes through taxes, subsidies, and regulations that will consider major economic benefits. For example, hydroelectric power can bring a stable water supply for irrigation and flood control at the cost of flooding existing farmlands and creating distress for wildlife and local communities. Economic theory is generally in favor of removing subsidies. However, there are perceived social benefits from shifting to a renewable fuel economy and subsidizing the development of new renewable and energy efficient technologies.

Policy conditions for a shift to renewable energy include a mix of free market competition and regulation with environmental taxes correcting distortions and temporary subsidies to support the market entry of these renewable energies and the removal of hidden subsidies to conventional sources.

13.3   World Climate Change Related to Power Generation

The atmosphere is a global public good into which individuals and companies release gases and particulates. Despite two conferences dealing with this issue, at Rio de Janeiro, Brazil in 1992 and at Kyoto, Japan in 1997, progress on combating global climate change has been slow.

World consumption of energy since the Industrial Revolution has affected the concentration of greenhouse gases. In addition to increased burning of fossil fuels like coal, oil, and natural gas, emissions of man-made chemicals like chlorofluorocarbons (CFCs), methane, and nitrous oxide from agriculture and industry contribute to the deterioration of the atmosphere.

Primitive man used to burn 200 kcal/day. Energy consumption grew in one million years to almost 250,000 kcal/day per person.[1] This enormous growth of per-capita energy consumption is due to

- Increased use of coal as a source of heat and power in the 19th century
- Use of internal combustion engines with massive use of petroleum and its derivatives
- Electricity generation from thermoelectric plants
- Degree of comfort demanded by the modern society.

In 1990, world average yearly per-capita energy consumption was about 1.5 TOE (tons of oil equivalent); that is, the average energy consumption was 30,000 kcal/day. With a current population of almost 6 billion people, this could lead to an overall consumption of approximately 8×10^9 TOE. There is a huge gap between the per-capita energy consumption of the industrialized countries where 25% of the world's people live and the less developed countries where the remaining 75% live. The U.S. alone, with 6% of the world's population, consumes 35% of the world's energy, contributing proportionally to carbon (CO_2) emitted per unit of energy.

Most economists agree that action is necessary, though the options considered differ drastically depending on government interests. More than 2500 economists, including eight Nobel Laureates, endorsed a report for the Intergovernmental Panel on Climate Change in favor of a global approach to climate change, recommending unified environmental, economic, social, and geopolitical decisions. Economic studies have found that there are many potential policies that would reduce greenhouse-gas emissions for which the total benefits outweigh the total costs.

For the United States in particular, sound economic analysis shows that there are policy options that would slow climate change without harming American living standards, and these measures may in fact improve U.S. productivity in the long run. The most efficient approach to slowing climate change is through market-based policies. In order for the world to achieve its climatic objectives at minimum cost, a cooperative approach among nations is required — for example, an international emission trading agreement. The United States and other nations can most efficiently implement their climate policies through market mechanisms, such as carbon taxes or auction of emissions permits. The revenues generated from such policies can effectively be used to reduce the deficit or to lower existing taxes.

Carbon emissions C can be related to energy consumption E and economy measured as gross domestic product GDP and population as shown by Kaya.[2]

$$E = \left(\frac{C}{E}\right)\left(\frac{E}{GDP}\right)\left(\frac{GDP}{P}\right)(P) \tag{13.1}$$

where

$\dfrac{C}{E}$ is the index of carbonization, measured in tons of carbon per TOE

$\dfrac{E}{GDP}$ is the energy intensity, the energy needed to produce one unit of GDP, measured in TOE per GDP

$\dfrac{GDP}{P}$ is the per capita gross domestic product

P is the population

The rate of increase of carbon emissions $\Delta C/C$ is given as the sum of four factors:

$$\frac{\Delta C}{C} = \frac{\Delta \dfrac{C}{E}}{\dfrac{C}{E}} + \frac{\Delta \dfrac{E}{GDP}}{\dfrac{E}{GDP}} + \frac{\Delta \dfrac{GDP}{P}}{\dfrac{GDP}{P}} + \frac{\Delta P}{P} \tag{13.2}$$

Stabilization of CO_2 can be evaluated by identities (Equations 13.1 and 13.2) suggesting that $\Delta C/C = 0$ could be reached by (1) reducing population growth, (2) reducing per capita income, and (3) reducing the rate of energy intensity. However, the only long range policy is to actually increase the rate of decarbonization by massive investments in clean energy solutions, renewable or fusion technologies.

Hydrogen can facilitate the transition from fossil fuels to renewable energy sources, playing an essential role in the decarbonization of the global energy

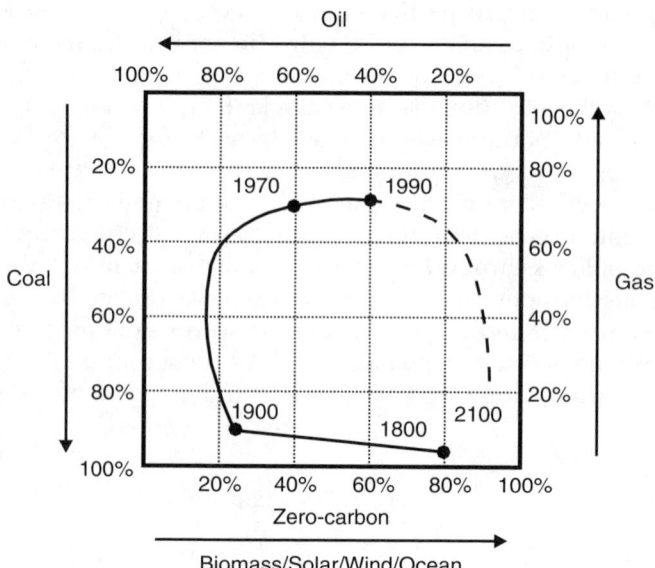

FIGURE 13.3
Evolution towards a hydrogen-based economy.

system, ultimately improving the outlook with regard to climate change. Hydrogen can provide a major hedge by enabling decentralized micropower plants, vehicles based on efficient fuel cells, integration with solar and wind technologies, and the use of fuel-cell-powered vehicles in home and rural residences.

Figure 13.3 shows historical data of global share[4] of and estimated evolution towards a hydrogen-based economy. The first step toward a hydrogen-based economy would probably be the use of natural gas as the main fuel that will be reformed to supply hydrogen for fuel cell power plants. By avoiding burning natural gas at its maximum efficiency of 35% through the thermodynamic cycle but converting the energy through the electrochemistry cycle at an efficiency of 70%, the reserves of fossil fuels would be extended with a corresponding decrease in carbon emissions.

The second step would be complete generation through renewable primary sources like hydro, wind, solar, and tidal. Although solar is primarily reliant on static converters, induction generators are the major devices for the other sources and could act as flywheel-based storage devices.

A third effort could be concentrated on the direct conversion of electricity from carbon, without burning and tied to biomass systems. Fusion technology, depending on international agreements and regulatory issues, is still a promise that would require massive investments. It has been noted that valuation and cost-benefit analysis can be applied to an environmental economic analysis, providing insights for policy making. Cline[3] suggested that limiting global warming would require aggressive action, which would appear to have higher costs than benefits for several decades, but a long-term effect with a very low discount rate would sway from "what is better for the economy" to "what is better for the environment." This means that climate stabilization should be the goal, rather than economic optimization of costs and benefits.

Balancing the valuation of human life, intrinsic value of ecosystems, and ethical considerations under the pressures of population and economic growth requires a system perspective. Therefore, energy resources and power conversion systems play an important role in reducing greenhouse emissions. Cogeneration of heat and power improves overall efficiency, and fuel cells can implement a sustainable paradigm integrating economic needs and environmental constraints through distributed generation.

Hydrogen can currently be produced with a diversified mix of technologies: steam reforming of natural gas, gasification of biomass, and efforts for dry reforming using ultra-high-temperature solar reactors, but those technologies are still in the research and development phase. Electrolysis from renewable sources (wind and solar) is already competitive today, and reversible fuel cells capable of combining electrolysis (electrical energy to H_2 and O_2), H_2 fuel cell mode (power from stored gases), and propane/air fuel cell mode (power from propane and air) are in the initial stages of research.

There are two types of measures to address climate change: preventive measures which tend to lower or mitigate the greenhouse effect, and reactive measures dealing with the consequences of the greenhouse effect and trying to minimize their impact. Preventive measures include:

- Reducing emissions of greenhouse gases, either by reducing the level of economic activities responsible for it or by shifting to more energy-efficient technologies that allow the same level of economic activity at a lower levels of CO_2 emissions.
- Enhancing greenhouse gas sinks. Since forests recycle CO_2 into oxygen, maintaining forested areas intact and implementing significant campaigns of reforestation would reduce the concentration of CO_2 in the atmosphere.

An economic approach suggests that we should apply cost-effectiveness analysis in considering such policies. This differs from cost-benefit analysis in having a more modest goal. Rather than attempting to decide whether a policy should be implemented, cost-effectiveness analysis asks what is the most efficient way to reach a policy goal. In general, economists favor approaches that work through market mechanisms to achieve their goals. Market-oriented approaches are considered cost-effective. Rather than attempting to control market actors directly, they shift incentives so that individuals and firms change their behavior to take into account external costs and benefits. Pollution taxes and transferable permits are potentially useful tools for greenhouse gas reduction. Other relevant economic policies include measures to create incentives for the adoption of renewable energy sources and energy-efficient technology.

13.4 Appraisal of Investment

As a basis for financing a renewable energy technology project, a feasibility analysis is required, particularly due to the need for bank assistance and financial commitments. A scenario must be constructed for this appraisal including the following topics.

- Does a renewable energy system make sense where the grid is present?
- Is there a demand for the project?
- Is there an administrative capacity to carry out the project?
- Is there information about the market?
- Is there a historical study about environmental conservation?
- What are the capital, operating, maintenance, and replacement costs of the renewable energy options in comparison to conventional energy sources?
- What are the advantages of the natural resources to be harnessed?
- Is it possible to obtain sufficient technical data to define the equipment needs?
- What are the technical options and what can they do?
- Does the country (if public sector) or the company (if private sector) have the financial capacity or access to sufficient funds to carry out the project?
- If private sector, how would this investment contribute to overall profit?
- If public sector, how would this project contribute to the economic growth of the country?
- Who benefits and by how much? Who pays and how much? Are there other social objectives to be achieved (stabilization, diversification), and what disadvantages do renewable energy technologies have relative to fossil fuel-based options?

The feasibility analysis can initially be viewed as an internal project appraisal effort. Given a particular load profile and range of available energy sources, the final choice of energy system will be influenced by the initial capital costs, operating costs, and maintenance costs. Technical alternatives should be compared using life-cycle costs, not just initial capital costs. The following factors impact the cash flow and the payback from an induction-generator-based project.

- Electricity demand
- Rate of interest and structure of the loan
- Utilization factor of the generator site
- Incentives for external benefits
- Economic and financial conditions
- Sustainability
- Population growth
- Financing
- Environment (local, regional, global)
- Socioeconomic factors (for example, employment)
- Security and diversity of fuel supply (import dependency)
- Integration (including decentralized versus centralized supply)
- Purchase price for the generated power
- Tax liability and savings from depreciation

Several factors will determine the interaction of the investor with the utilities. The utility will be avoiding fuel costs for contracted power. Buyback rates reflect the relative costs of providing power by the utility, with power injected into the grid during peak-load periods earning higher rates than power generated in off-peak periods. A meter to record exported and imported energy in the various periods and enables accurate billing to occur.

Another approach used by some utilities for small-scale domestic buyback rates is that of "net metering." In this scheme, the utility pays the same rate for power generated as it charges for power consumed by the consumer. In effect, the meter runs backwards when the system is exporting power to the grid. This is a simple approach, as a new meter is not necessarily required. However, a utility may still install an additional meter to get an indication of the amount of energy exported.

For larger systems the price paid for exported power is generally set by negotiation. The electricity tariffs charged to consumers are quite complex. These can be broadly split into time-of-use tariffs and general tariffs. The general tariffs base the cost on energy consumption only. A consumer on this tariff may find it advantageous to install a renewable energy source to reduce total energy consumption but still maintain the reliability of the grid system.

Wind energy uses the force of moving air to power generators connected to the shaft of turbines. Electricity is produced from the rotational energy in units consisting of a tower, rotor (with blades, hub, and shaft), generator, control equipment, and power conditioning and protection equipment. A gearbox is usually used to match the generator speed with the natural impressed velocity of the wind. Some sort of braking is used to protect the system against high winds and severe weather conditions. Combining wind power with other sources of generation like photovoltaic and sometimes diesel generators form more reliable hybrid systems.

In the past few years, wind energy has become competitive on a large scale on a cost-per-kilowatt-hour basis. The typical lifetime of a wind generator is assumed to be 20 to 25 years for a modern turbine design, during which time capital costs will be amortized. Economies of scale apply to capital costs, though the price increases as the turbine size increases. A rule of thumb for costs is US$1000 per kW capacity on average, with the marginal tower cost at US$1600 per meter. Installation costs include foundations, transportation, road construction, utilities, telephones and other communications, substation, transformer, controls, and cabling. Depending on soil conditions and distance to power lines, a safety margin must be applied. Operation and maintenance costs generally increase throughout the life of a wind power project.

Electricity costs may range from about US$0.008 to US$0.014 per kWh generated (although some authors claim even lower numbers), and increase at a rate of about 2.5% per year. The high end (US$ 0.014/kWh) would include a major overhaul of equipment such as rotor blades and gearboxes, which are subject to a higher rate of wear and tear. Other operating costs include plant monitoring and semiannual inspections.

The potential income from wind energy is based on annual energy production, which in turn is dependent on average annual wind speed. Wind speed varies globally, regionally, and locally following seasonal patterns. The duration and force of the wind is critical to having a lucrative operation of wind turbines to pay the investments on the equipment, maintenance, and operational costs. Large-scale wind turbines require at least an annual average wind speed of 5.8 m/s (13 mph) at 10 m height, while small-scale wind turbines require 4 m/s (9 mph).

Energy production is also a function of turbine reliability and availability. Even though the induction generator is a small fraction of all the required investments, it is indeed the unit that converts mechanical energy from the shaft to electricity. Therefore, even tiny improvements in the efficiency of the induction generator pay off. Better electrical and magnetic design associated with insulation, thermal, and mechanical design will make the energy converter unit extract more reliable power. Controllers capable of putting the induction generator in the best operating condition, programming the optimum magnetic flux level, producing required lagging/leading unit power factor energy contribute to faster amortization and a productive investment.

Hydropower converts the potential and kinetic energy of water conveyed through a pipeline or canal to the turbine into electrical energy. The energy in the water enters at high pressure, rotates the generator shaft, and leaves at lower pressure. Power generation can be amplified by increases in water elevation, river or stream flow, or the size of the watershed. Small hydropower generators are usually run-of-the-river systems that use a dam or weir for water diversion but not for reservoir storage. The amount of pressure at the turbine is determined by the actual flow of the stream or river and the head or height of the water above the turbine.

Controlling the flow of a river can have uses other than power generation including irrigation, flood control, municipal and industrial water supply, and recreation. Those parameters affect the financial return from such projects and sometimes have long-term benefits that may interest local officials. Hydropower is characterized by extremely high up-front costs, low operating costs, and long life cycles. This makes projects sensitive to financial variables, construction timing, bank interests, and discount rates.

Hydro projects usually have long lead times, typically taking 10 years from analysis to deployment. Such projects have acerbic requirements on resource assessment and environmental and social considerations such as water use, displacement of homes, and distance to transmission lines. Small power plants on the order of 10 to 200 kW for rural applications in distributed energy systems do not have the same requirements as larger projects (greater than 1 MW). However, those small-scale applications for a target cost of $2000/kW will need a 35% investment in civil engineering, 40% for plant installations and commission, 7% for overall design and management, 8% for electrical engineering, and contingencies on the order of 10%. The induction generator may need a peak power tracking control, a dummy load, or any type of storage system, and pumping-back schemes are sometimes used in conjunction of water flow needs.

A cost-benefit analysis can help in the construction of a list of economic, environmental and social indicators. Economic figures are naturally quantified, but social and environmental indicators are generally hidden impacts and may be viewed either as external costs to the environment, local or global, or as external benefits, for example, job creation. An appraisal must be undertaken in order to provide a basis for selection or rejection of projects by ranking them in order of profitability or social and environmental benefits and ensuring that investments are not made in projects that earn less than the cost of capital, generally expressed as a minimum rate of return.

Cost comparisons in industry often assume that the annual cost of capital is a fixed percentage of the investment. However, a long-term renewable energy project needs a cash flow analysis to assess the difference between generated revenue and ongoing expenses. Some of the analytical tools that can be used are as follows.

- Benefit-cost ratio
- Net present value (or discounted cash flow)
- Internal rate of return
- Payback period
- Least cost analysis
- Sensitivity analysis

13.4.1   Benefit-Cost Ratio

The benefit-cost ratio (BCR) is the ratio of discounted total benefits to total costs. For a project to be acceptable, the ratio must have a value of 1 or greater. Among mutually exclusive projects, the rule is to choose the project with the highest benefit-cost ratio. The disadvantages of the BCR are that it is typically sensitive to the choice of discount rate, and can provide incorrect analysis if differences in size or scale of the various projects being compared are great.

13.4.2   Net Present Value (or Discounted Cash Flow)

The net present value (NPV) approach (also referred to as the discounted cash flow approach) measures the present value of money exclusive of inflation. NPV uses the time value of money to convert a stream of annual cash flow generated by a project to a single value at a chosen discount rate. This approach also allows one to incorporate income tax implications and other cash flows that may vary from year to year. The discounted cash flow or net present value method takes a spread of cash flows over a period and discounts the cash flows to yield the cumulative present value. When comparing alternative investment opportunities, the project with the highest cumulative NPV is the most attractive one. The only serious limitation with this approach is that it should not be used to compare projects with unequal time spans.

13.4.3   Internal Rate of Return

The internal rate of return (IRR) and net present value approaches are very similar. As outlined, the NPV determines today's values of a future cash flow at a given discount rate. On the other hand, the IRR approach one seeks to determine the discount rate (or interest rate) at which the cumulative net present value of the project is equal to zero. This means that the cumulative NPV of all project costs would exactly equal the cumulative NPV of all project benefits if both were discounted at the internal rate of return. In the private sector, this computed financial internal rate of return (FIRR) is compared to the company's actual cost of capital. If the FIRR exceeds the company's cost of capital, the project is considered financially attractive. The higher the IRR compared to the cost of capital, the more attractive the project. On the other hand, if the IRR is less than the company's cost of capital, then the project is not considered financially attractive.

For projects financed in whole or in part by the public sector, the discounted cash flow may need to be adjusted to account for social benefits or economic distortions such as taxes and subsidies, economic premiums for foreign exchange earnings that accrue from the project or employment benefits. The resulting statistic would be the economic internal rate of return

(EIRR) and would be compared with the country's social opportunity cost of capital. If the EIRR exceeds the social opportunity cost of capital the project would provide economic benefits to the society.

13.4.4   Payback Period

Payback period is the easiest and most basic measure of financial attractiveness of a project. The payback period reflects the length of time required for a project's cumulative revenues to return its investment through the annual (non-discounted) cash flow. A more attractive investment is one with a shorter payback period. In development settings, however, there is little reason to assume that projects with short payback periods are superior investments. Also, the criterion has a bias against long gestation projects such as renewable energy.

13.4.5   Least Cost Analysis

The least-cost analysis method is used to determine the most efficient (least costly) way to perform a given task, reach a specified objective, or obtain a set of benefits measured in terms other than money. For example, the objective might be to supply a fixed quantity of potable drinking water to a village. The alternatives might be wind pumping, run-of-river off-take, or impoundment. One would calculate all costs, capital and recurrent, to achieve the objective, apply economic adjustments and discount the resulting stream of costs for each alternative examined. The one with the lowest NPV would be the most efficient one.

13.4.6   Sensitivity Analysis

Sensitivity analysis refers to the testing of key variables in the cash flow pro forma balance sheet to determine the sensitivity of the project's NPV to changes in these variables. For example, in a renewable energy project proposal, one may increase fuel costs or fuel transport costs, remove import restrictions on solar panels, lower labor costs, and increase land acquisition at different rates to determine the corresponding impact on the NPV. It is useful to test in the cash flow pro forma a variable that appears to offer significant risk or probability of occurring. The analysis becomes another useful tool when combined with others to improve the decision making process.

While each index provides data needed to make efficient decisions, multicriteria analysis (combining one or more tools with other project data and benefits) can be helpful in evaluating future financial performance. For example, other criteria might include the distribution of benefits, ease and speed of implementation, and replicability that might be combined with one of the quantifiable indices illustrated above.

13.5   Concept Selection and Optimization of Investment

In induction-generator-based renewable energy systems the energy source is typically a low-speed prime mover: hydropower or wind power. The investments required for such power plants are basically in three areas: (1) a power generation section, consisting of the turbine, gear box, generator and all structural elements to support them, (2) a protective and control operational center capable of withstanding environmental and operational hazards, and (3) a utility power interface section, tailoring the power output to the requirements of the electrical grid.

One important economic concept that is used to support capital investment analysis is the capacity factor (CF). The capacity factor measures the operational hours, seasonal constraints, and source and demand fluctuations that limit the full utilization of the installed power plant by dividing the actual energy produced during a specified time period by the amount of the energy that would have been produced under full power.

$$CF = \frac{Actual\ Energy}{Energy\ at\ Full\ Use} \tag{13.3}$$

A wind turbine rated at 100 kW would produce 876,00 kWh of electricity per year if operating 100% of time. However, wind velocity fluctuation, maintenance down time and even demand matching will constrain the annual output power to a lower value found by multiplying CF by the energy at 100% use. For average production, the average input must be computed, and the given average shaft velocity would determine a gearbox that optimizes the power transference. However, the operational generator speed would oscillate about this optimum point, and the electronic system must compensate for such deviation. Therefore, there is a typical cost of energy generated by the system that must be taken into consideration for the amortization of the capital, operation, and maintenance. Since hydropower and wind power systems do not require fuel for operation, the following equation suggested by Ramakumar and Hughes[6] expresses the cost of energy, neglecting taxes, surcharges, and insurance:

$$C = \left[\frac{r(1+r)^n}{(1+r)^n - 1} + m \right] \frac{P}{87.6(CF)} \tag{13.4}$$

where
   C = generation cost in U.S. cents per kilowatt hour
   CF = capacity factor
   m = fraction of the capital costs needed per year for operation and maintenance of the unit

n = amortization period in years
P = capital cost in U.S. dollars per kilowatt
r = annual interest rate per unit, (0.01 times the annual percentage rate)

If the induction generator operates with a hybrid system, for example, diesel, synthetic fuels from biomass, gas, or fuel cell, there is another term to be added to Equation 13.4, shown in Equation 13.5, where F is the fuel input per volume and D is the fuel cost in U.S. cents per volume.

$$C = \left[\frac{r(1+r)^n}{(1+r)^n - 1} + m\right]\frac{P}{87.6(CF)} + FD \tag{13.5}$$

Multiple connection of renewable energy sources will have an average cost given by Equation 13.6.

$$C_{avg} = \frac{\sum_i \left[\frac{r(1+r)^n}{(1+r)^n - 1} + m\right] P_i R_i}{(87.6)\sum_i (R_i CF_i)} \tag{13.6}$$

where
C_{avg} = average generation cost in U.S. cents per kilowatt hour
i = index related to the device
CF_i = capacity factor for the ith device
m_i = operation and maintenance of the ith device
n_i = amortization period in years for the ith device
P_i = capital cost in U.S. dollars per kilowatt for the ith device
r = annual interest rate per unit (0.01 times the annual percentage rate)
R_i = rating in kilowatts of the ith device

If the amortization period and the operation and maintenance charge rates are taken to be the same for all the devices, then Equation 13.7 applies.

$$C_{avg} = \frac{T_{cr}I}{(87.6)R_{eq}} \tag{13.7}$$

where
I = total investment in U.S. dollars = $\sum_i P_i R_i$

R_{eq} = equivalent continuous rating in kilowatts = $\sum_i R_i CF_i$

T_{CR} = total charge rate = $\dfrac{r(1+r)^n}{(1+r)^n - 1} + m$

If the assumption of the same amortization period and operation and maintenance charge rates for all is not valid, a conservative approach is to use the smallest n_i and the largest m_i for both n and m in the equations above.

13.6 Future Directions

With the present worldwide situation as volatile as it is, the need to allocate capital efficiently has become more apparent for companies so as to both preserve good credit ratings and invest in growth opportunities at the same time. There have been some educational efforts related to energy decision making.[7] Risk-adjusted return on capital approaches[8] and chance-constrained programming[9] have been introduced recently where the investor might wish to obtain the highest possible expected return but also protect against too-risky projects. The accepted risk level is based on the decision maker's risk aversion. The extent to which violations are permitted to happen is a matter of policy—for example, how far will the government allow companies to go above the environmental tax limitation? Such recent approaches formulate an optimization problem with probabilistic constraints and seem suitable for economical evaluation of renewable energy systems.

13.7 Problems

1. Create a flowchart to be used by a decision maker to determine his or her investment in a renewable energy project.
2. If a site has an average wind speed of 10 m/s and turbines with a diameter of 12 m are be installed, what is the capacity factor for 10 turbines rated for 14 m/s. Plot a seasonal power generation graph and discuss ways to compensate for fluctuations.
3. Neglecting taxes, surcharges, and insurance, what would be the capital cost of energy over 20 years of a wind power system whose generation cost is $0.01/kWh, at a annual interest rate of 0.075, annual maintenance and operation costs of 2.5% of the capital cost needed and a capacity factor of 0.5? Extend this cost to five identical units under similar overall conditions.
4. Compare the decision to install a small power (up to 20 kW) synchronous generator with an induction generator. In this analysis, you should include initial purchase price, cost of repairs and maintenance, operational costs, overall efficiency, product availability and unit replacement.
5. Implement the cost equation (Equation 13.4) with a spreadsheet program. Make what-if simulations for several economical scenarios.

References

1. Goldemberg, J., *Energy Environment and Development*, Earthscan Publications, London, 1996.
2. Kaya, Y., *Impact of Carbon Dioxide Emission Control on GNP Growth: Interpretation of Proposed Scenarios*, IPCC Energy and Industry Subgroup, 1990.
3. Cline, W.R., *The Economics of Global Warming*, Institute for International Economics, 1992.
4. Nakicenovic, N., Grübler, A. and McDonald, A., *Global Energy Perspectives*, Cambridge University Press, 1998.
5. Harris, J.M. and Codur, A.M.. Microeconomics and the Environment, http://www.ase.tufts.edu/gdae/education_materials/course_materials.html.
6. Ramakumar, R., Hughes, W.L., Renewable energy sources and rural development in developing countries *IEEE Trans. on Education*, E-24(3), August 1981, pp. 242–251.
7. Shaten, R.J., *The Energy Resources and Economics Workbook*, Society for Energy Education, June 2003.
8. Keers, G., Risk Adjusted Capital Allocation For Energy Project Appraisal KWI white-paper, http://www.kwi.com.
9. Gurgur, C. and Luxhoj, J.T., Application of chance-constrained programming to capital rationing problems with asymmetrically distributed cash flows and available budget, *The Engineering Economist*, 48(3), 2003.

APPENDIX A

Introduction to Fuzzy Logic

Fuzzy logic has been making a name for itself as the most successful application of artificial intelligence. Fuzzy logic has advanced significantly in recent years and has found widespread applications. It has also caught the attention of the public in the past few years. Now, people ask what it is, mathematicians study how it works, engineers work on what it does, and marketers wonder what it can do. Engineering applications generally use fuzzy logic for controllers in noisy, nonlinear and time-variant systems, tuning problems, and estimation of multiple input/output relations. However, the range of applications is amazing. Fuzzy systems can be integrated with neural networks to extract hidden rules out of numerical data. Fuzzy reasoning can be used to model events in politics, military planning, and appraisal processes. There are fuzzy computers being constructed. The topic is so vast that this book can only introduce some concepts.

The theoretical foundations of fuzzy logic were laid in 1965 when Prof. Lotfi Zadeh published the paper "Fuzzy Sets," in the journal *Information and Control*;[1] fuzzy set theory was born, but no one paid much attention to it. After all, the term "fuzzy" had (and still has) a sort of negative connotation in English. Before the publication of this paper, Zadeh was already well known in linear systems theory for his contributions to the analysis of discrete systems. However, he believed that as the complexity of a system increases, the ability to make precise and significant statements about its behavior diminishes. In his words, "The closer one looks at a real-world problem, the fuzzier becomes its solution."

Linear systems theory assumes several properties that are not actually found in real applications. Therefore, conventional methods of control are good for simple systems, while fuzzy systems are suitable for complex problems or in applications that involve human descriptive or intuitive thinking. A fuzzy system can estimate input-output functions and can be trainable for pattern recognition applications and dynamic systems control. Unlike statistical tools and conventional (continuous or binary) logic, it can estimate functions without a mathematical model of the system. Therefore, fuzzy systems are model-free estimators able to "learn from experience" with a linguistic description of the system's operation.

The first application of fuzzy logic for control of a dynamic process was reported by Mamdani[2] in a steam engine. The problem was to regulate the engine speed and boiler steam pressure by means of the heat applied to the boiler and the throttle setting of the engine. Since the system was nonlinear, noisy, and strongly coupled, and no mathematical model was available, the fuzzy control was designed from the operator's experience and performed much better than expected. Following this work, Mamdani tried unsuccessfully to secure funding for his research from the British granting agencies. Unable to obtain support, he ultimately abandoned this line of investigation. However, his work did not go unnoticed in Japan.

Ten years later the Sendai Subway Automatic Train Operations Controller captured the attention of control engineers around the world.[3] The Hitachi team designed, developed, and compared fuzzy and conventional PID-type controllers in 300,000 simulations tests, as well as 3,000 riderless subway runs with real hardware. Strategies used by experienced train operators were implemented in fuzzy rules that performed a "predictive fuzzy control." Speed control during cruising, braking control near station zones, and switching of control were determined by fuzzy rules that processed sensor measurements and considered factors such as the comfort and safety of travelers. In operation since 1986, this controller has reduced the stopping distance by 2.5 times, doubled the comfort index, and reduced power consumption by 10%.

While Western scientists and engineers ignored or attacked fuzzy logic during the 1970s and '80s, Japanese companies applied fuzzy technology to air conditioners; auto engines, brakes, transmissions, and cruise control; cameras and camcorders; washing machines; dryers, and elevators. By the time the first U.S. fuzzy conference was held in Austin, Texas, in June 1991 (at Microelectronics and Computer Technology Corp.) the Japanese had already held international meetings and established two fuzzy research centers: LIFE (Laboratory for International Fuzzy Engineering) and FLSI (Fuzzy Logic Systems Institute).

A.1 Natural Vagueness in Experimental Observations

Most everyday situations and observations are imprecise, carrying a certain degree of fuzziness in the description of their nature. This vagueness may be associated with variables like position, color, texture, shape, or even the language used to describe phenomena. Very often the same concept will have different meanings in different contexts or at different times. A "hot" day in winter is not the same as a "hot" day in summer. A car going "fast" on a highway does not have the same speed as a car going "fast" in a parking lot. Such imprecision or fuzziness associated with descriptions and observations of natural events is common in all fields of study addressed by scientists, philosophers, and business analysts. We are able to formulate ideas, make decisions and recognize concepts that have a high level of vagueness. Consider the following statements as examples:

- The room is hot.
- Mike is quite tall and Mary somewhat short.
- My son said that I am old.
- My grandmother said that I am young.
- That computer's price is around twice the competitor's price.
- You should buy a less expensive tire, which will last as long as a brand-name tire.

The propositions above are incompatible with traditional computer-based modeling and information systems, which require precise numerical data. A common-sense statement like, "When the engine temperature is hot increase the fan speed," can be readily understood by a fuzzy system, even if there is no exact threshold point where the engine is considered to be warm or hot. A fuzzy system can evaluate descriptions like the ones above and make decisions for necessary actions. The power of fuzzy logic is to bridge the gap between machine systems and human nature.

Conventional computers and digital systems follow either-or (Boolean) crisp logic where the answer is always yes or no. Take Figure A.1(a) for

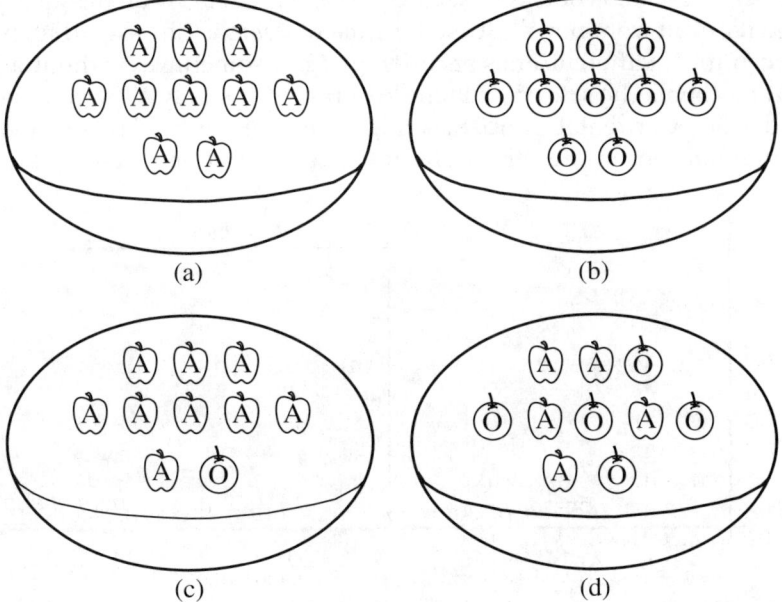

FIGURE A.1
Is each one of these a bowl of apples?

example. Is this a bowl of apples? The answer is yes. How about Figure A.1(b)? Is it a bowl of apples? The answer in this case is no, since it is a bowl of oranges. This is an example of crisp logic in which the bowl contains either apples or oranges.

Now, take the case of the bowl in Figure A.1(c) where someone has swapped an apple for one of the oranges; is it a bowl of apples? The answer is "It is a bowl of mostly apples." If we keep switching the apples and oranges and asking the same question, there will be a situation where the bowl will contain half apples and half oranges as shown in Figure A.1(d). Such a bowl is neither a bowl of apples nor a bowl of oranges. This is the fundamental difference between crisp logic and fuzzy logic: the violation of the Law of Noncontradiction where the set A must not contain the set not-A. Here is a situation where the bowl belongs to both the set of bowls of oranges and the set of bowls of apples.

In conventional logic, the number 1 can indicate that an object belongs to a set, while 0 can indicate that the object is not a member of a set. On the other hand, in fuzzy logic there are degrees of membership that vary from 0 to 1. To see how this definition is used in constructing an actual fuzzy set, let us consider the concept of "adult."

We usually agree that somebody who is 14 years old in most cases is barely if at all an adult. Another person, about 25 years old, might be already established in some profession, be independent, or be married. He or she could be considered an adult. In classical set theory, we are forced to choose an arbitrary cutoff point, say 18 years old. Since the boundaries between what is in a set and what is outside a set are sharp, we can draw an imaginary line that splits adult from nonadult on the scale of years as shown in Figure A.2.

We can draw different lines at different ages, sometimes without a good reason for them. The sale of alcoholic beverages is usually allowed to people 21 and older. Car rental companies might rent cars only to people over 25. The common sense is that the truth of the concept of being an adult grows

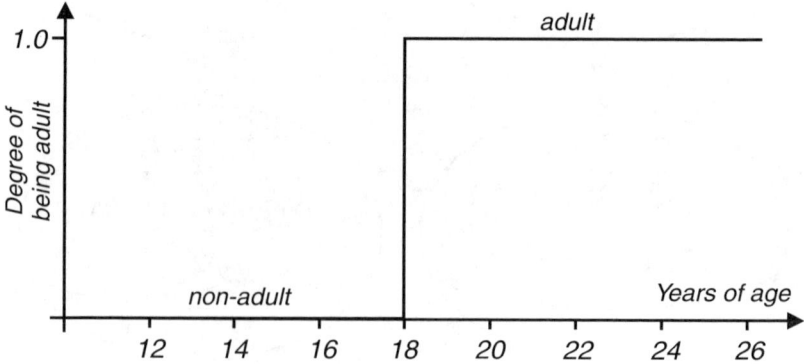

FIGURE A.2
Boundary between adult and nonadult definitions.

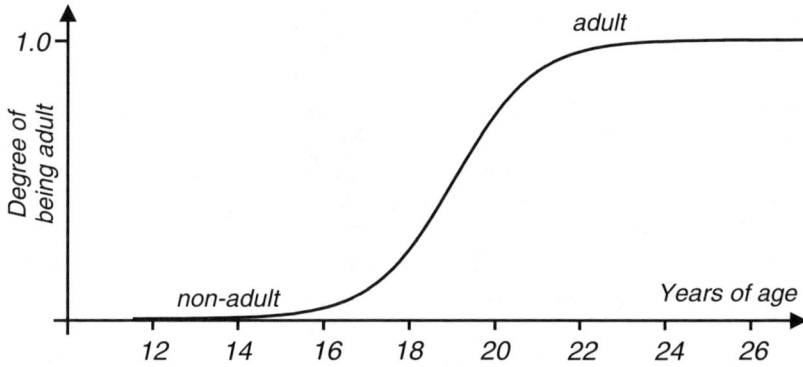

FIGURE A.3
Various degrees of being adult.

with age. So the fuzzy principle views adult as a curve, as depicted in Figure A.3, and not as a sharp line as indicated in Figure A.2. Such a curve that relates how much something belongs to a fuzzy set is called a membership function.

Fuzziness should not be confused with probability. The two can occur together, but they are different concepts. Fuzziness describes the ambiguous or vague case, whereas probability assigns a degree of certainty to a statement. With probabilities, an object either is or is not something; it cannot be both. For example, tomorrow it will either rain or not rain. The uncertainty lies in which of the two possible outcomes will occur. However, the degree of raininess that occurs is described by fuzziness. It may downpour, rain moderately, lightly, or drizzle.

A.2 Fuzzy Logic Operations

People intuitively use logical operations everyday. A typical thought before leaving the office in the afternoon could be: "If I do not get stuck in traffic today and it's not raining, I will play a little tennis." The connective *and* assumes the truth of both statements, *not* stuck and *not* raining, to conclude that a tennis game will be possible. Logic operations like *and, or,* and *not* originated in Aristotle's syllogistic logic and came of age when George Boole published *An Investigation of the Laws of Thought on which are Founded the Mathematical Theories of Logic and Probabilities* in 1854.[4] Boole believed that the processes of human reasoning could be reduced to an algebra or other formal system, modeled after two valued (true/false, 1/0) logic. Although his vision was never verified, he furnished the conceptual basis for a formal system in which logical propositions could be represented and manipulated like algebraic symbols setting the stage for the invention of computers a century later.

Of course, the missing link was that human reasoning is never restricted to yes/no, true/false, and black/white answers. There are degrees of truth and shades of gray that are not handled by crisp logic. Therefore, Boolean logic operations must be extended to fuzzy logic to manage the notion of partial truth, truth values between "completely true" and "completely false."

The fuzzy nature of a statement like "X is low" was presented above. Suppose now that there are two fuzzy sets, low and high, where a typical logical operation could be given as

X is low *and* Y is high

What is the truth value of the *and* operation? The logic operations with fuzzy sets are performed with the membership functions. Although some researchers have explored several interpretations for fuzzy logic operations, the following definitions are the most convenient.

$$truth(X \; and \; Y) = Min(truth(X), truth(Y))$$

$$truth(X \; or \; Y) = Max(truth(X), truth(Y))$$

$$truth(not \; X) = 1 - truth(X)$$

The *and* operation between two fuzzy sets takes the minimum between the membership functions of the two fuzzy operands; *or* takes the maximum between the membership functions of the two fuzzy operands, and *not* is performed by subtracting the membership function from 1. Note that if we plug the values zero and one into the above definitions, we get the same truth table as expected from conventional Boolean logic. This is known as the Extension Principle, which states that the classical results of Boolean logic are recovered from fuzzy logic when all fuzzy membership grades are restricted to the traditional set (0, 1). This effectively establishes fuzzy sets logic as a true generalization of classical set theory logic, so there is no conflict between fuzzy and crisp methods.

To illustrate how the fuzzy logic operations work, let us assume an air-conditioner vent. Suppose the vent has blades that control the openings and can be moved so that the blades tilt downward or upward. This blade angle sends the air flow towards the floor or the ceiling. Figure A.4(a) and Figure A.4(b) show the two fuzzy sets (downward and upward) that describe the position of the vent blades. If the blades are totally rotated to −45° with respect to the horizontal then they are completely downward. On the other hand, if the blades are totally rotated to +45°, they are completely upward.

The range of the input variable (angle) is called universe of discourse, which could be conveniently scaled to per-unit values (values between −1 and +1) by dividing the input angle by the maximum value of the range.

DOWNWARD

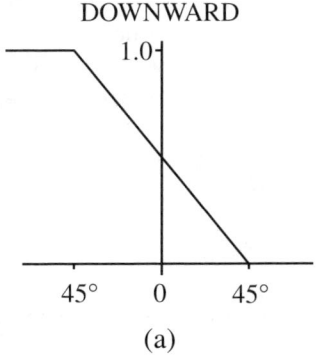

(a)

UPWARD

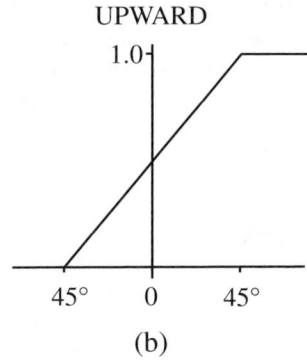

(b)

UPWARD AND DOWNWARD

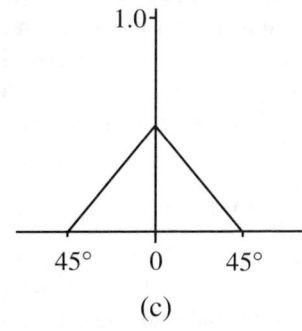

(c)

UPWARD OR DOWNWARD

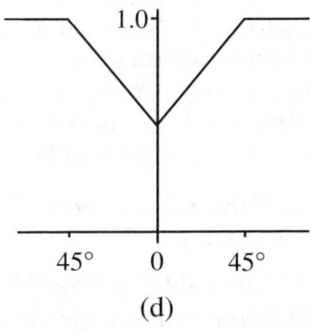

(d)

FIGURE A.4
(a) Fuzzy set downward, (b) Fuzzy set upward, (c) Fuzzy set upward *and* downward, (d) Fuzzy set upward *or* downward.

Any position in between has degrees of being downward or of being upward. The *and* operation of the two fuzzy sets (upward *and* downward) is shown in Figure A.4(c). The *and* operation is the minimum between the two membership functions. Figure A.4(d) shows the *or* operation between the two fuzzy sets (upward *or* downward). It is easily seen that upward = *not* downward, since the membership function of upward is equal to the subtraction of the membership function of downward from the value 1.

A.3 Rules of Inference

Statements assert facts or states of affairs; they give descriptions that can be organized in rules of reasoning. The application of mathematical descriptions and the use of logical rules for formulating hypotheses were developed by several philosophers, mainly Descartes, Newton, Galileo, and Bacon. They were influenced by the earlier syllogistic logic, in which premises were manipulated to produce true conclusions. A typical syllogistic rule of inference that goes by the Latin name *modus ponens* (affirmative mode) can be given as

If A is true
And A implies B
Then B is true

where the connectives *and*, *or*, and *not* are essential to derive the truth of above rules.

In Boolean logic the laws of thought are expressed by rules of operations based on symbols, which are based on an algebra of the numbers 0 and 1. However, there are real-world situations where the statements given by rules are only partially true; they are "fuzzy." A fuzzy rule applies common sense and helps to get the knowledge out of the system's operation. As an example, let us assume two cars traveling at the same speed along a straight road. The distance between the cars becomes one of the factors that leads the second driver to brake his or her car to avoid collision. The following rules might be used by the second driver.

Rule 1. *If* the distance between two cars is medium *and* the speed of the car is medium,

Then brake medium for speed reduction.

Rule 2. *If* the distance between two cars is small *and* the speed of the car is medium,

Then brake hard for speed reduction.

How short the distance is and how fast the speed is judged by the second driver and may incorporate factors like weather conditions, day or night, and the driver's fatigue. Therefore, rules like those above have a degree of truth with respect to other related statements. A set of statements is known as a rule base, and the process that combines all the statements of a rule base to produce a meaningful conclusion is called "inference." Although concepts like height, speed, age, or temperature are fuzzy in nature, there is an actual number that indicates how many feet, miles/hour, years, or degrees are related to the variable that is being measured or controlled. Thus, the input information for a fuzzy system needs to be initially converted into the correspondent fuzzy sets. This is done by the "fuzzification" process that is the computation of the degree of membership of the input variables to the given fuzzy sets.

Figure A.5 shows how the two rules above are processed into a fuzzy system. The top of the figure shows the evaluation of Rule 1, where the distance D between the two cars is medium, the speed S is medium, and the braking action should be medium. The *and* operation between the distance

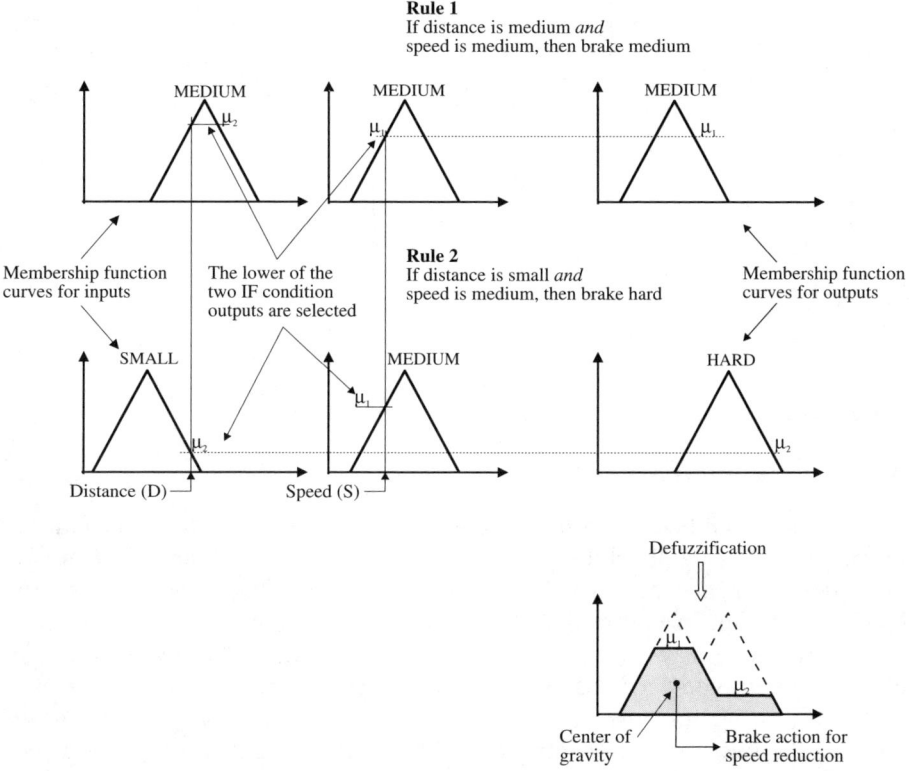

Rule 1
If distance is medium *and*
speed is medium, then brake medium

Rule 2
If distance is small *and*
speed is medium, then brake hard

Membership function curves for inputs

The lower of the two IF condition outputs are selected

Membership function curves for outputs

Defuzzification

Center of gravity

Brake action for speed reduction

FIGURE A.5
Fuzzy inference for two rules that relate speed, distance and brake action for a vehicle in a highway.

and the speed is performed by selecting the minimum truth value ($\mu 1$). The output fuzzy set of Rule 1 is cut off at the strength indicated by the rule firing (height $\mu 1$), turning the triangle into a trapezoid shape. The bottom of the figure shows the evaluation of Rule 2, where the distance D is small, the speed of the car is medium and the braking action should be hard. Again the lower membership function ($\mu 2$) is selected to perform the *and* operation for Rule 2 giving an output fuzzy set for Rule 2 with height $\mu 2$, where the area above the height $\mu 2$ is lopped off.

Each rule in the rule base is fired in accordance with the truth value for the fuzzy logic operation of the rule. Since any rule can be fired, there is an *or* configuration for such input information. This *or* operation is performed at each output fuzzy subset by taking the maximum of the membership functions. Figure A.5 shows the construction of the total output fuzzy set. The final fuzzy set output has to be converted to a crisp value in order to serve an output to an external interface. In the figure it is done by the centroid method, which takes the center of the gravity of the geometric figure as the final crisp value.

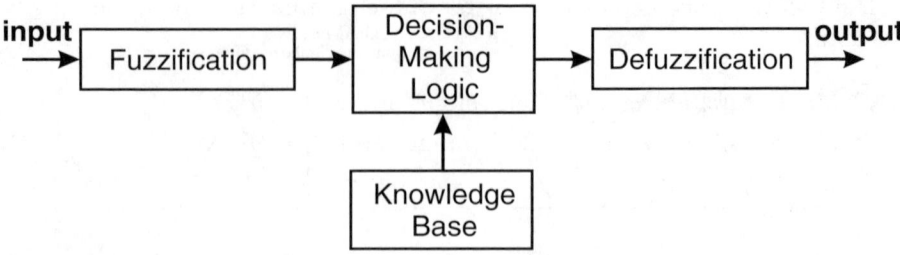

FIGURE A.6
Block diagram of a fuzzy system.

A.4 Fuzzy Systems

Most commercial fuzzy products are rule-based systems that receive current information in a feedback loop from a device as it operates and controls the operation of a mechanical or other device. A fuzzy logic system has four blocks as shown in Figure A.6. Crisp input information from the device is converted into fuzzy values for each input fuzzy set with the fuzzification block. The universe of discourse of the input variables determines the required scaling for correct per-unit operation. Scaling is very important because the fuzzy system can be retrofitted with other devices or ranges of operation by just changing the scaling of the input and output. The decision-making logic determines how the fuzzy logic operations are performed (Sup-Min inference) and, together with the knowledge base, determine the outputs of each fuzzy *if-then* rule. Those are combined and converted to crisp values with the defuzzification block. The output crisp value can be calculated by the center of gravity as:

$$I = \frac{\int U \cdot \mu(U) dU}{\int \mu(U) dU} \tag{A.1}$$

or the weighted average (height method of defuzzification) can be used:

$$I = \frac{\sum_{i=1}^{n} U_i \cdot \mu(U_i)}{\sum_{i=1}^{n} \mu(U_i)} \tag{A.2}$$

The following six steps are involved in the creation of a rule-based fuzzy system.

1. Identify the inputs and their ranges and name them.
2. Identify the outputs and their ranges and name them.
3. Create the degree of fuzzy membership function for each input and output.
4. Construct the rule base that the system will operate under
5. Decide how the action will be executed by assigning strengths to the rules
6. Combine the rules and defuzzify the output

A.5 An Example of Fuzzy Logic Control for Air Conditioning Systems

An ordinary air conditioner has a thermostat that takes a temperature setting like 20°C and turns the cooling system on or off to keep the temperature within a range, say 17 to 23°C. The currents rushing in and the time delay in bringing down the temperature after the system turns on make such a rigid system dramatically inefficient. Mitsubishi Electric Corp. has developed a room air-conditioning system where the controller allows degrees of operation from on to off for optimal operation. The company reports that this improves the room's temperature variation, is three times as stable as a crisp system, and provides a 24% power savings. The temperature sensor can detect whether the room is occupied and can direct the air upward if people are present or downward if the room is empty. Mitsubishi's fuzzy air conditioner can also learn the room's characteristics and fine-tune its own operation.

The following discussion is a simplified implementation of an air-conditioning system with a temperature sensor. The temperature is acquired by a microprocessor that has a fuzzy algorithm to process an output to continuously control the speed of a motor that keeps the room at a "good" temperature. It also directs the vent upward or downward as necessary. Figure A.7 illustrates the process of fuzzification of the air temperature. There are five fuzzy sets for temperature: cold, cool, good, warm, and hot.

The membership function for fuzzy sets cool and warm are trapezoidal; the membership function for good is triangular; and the membership functions for cold and hot are half-triangular with shoulders indicating the physical limits of the process. (Staying in a room with a temperature lower than 45°F or above 85°F would be quite uncomfortable.) The way to design such fuzzy sets is a matter of degree and depends solely on the designer's experience

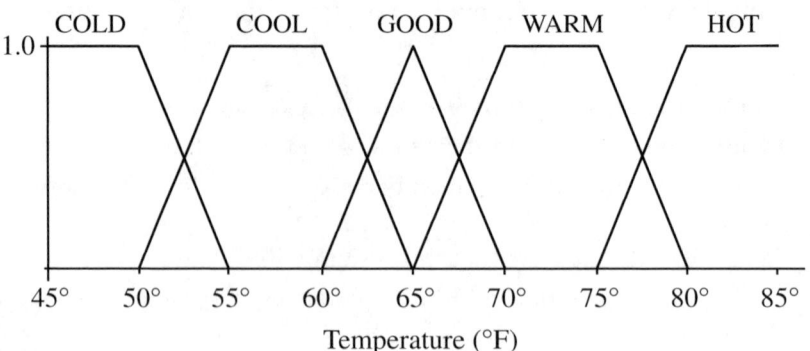

FIGURE A.7
Fuzzy sets for temperature.

and intuition. Most probably an Eskimo and an Equatorian would draw very different membership functions for such fuzzy sets!

To build the rules that will control the motor, we could watch how a human expert would adjust the settings to speed up and slow down the motor on response to the temperature, obtaining the rules empirically. If the room temperature is good, keep the motor speed medium. If it is warm, turn the knob of the speed to fast and blast the speed. On the other hand, if the temperature is cool, slow down the speed. Stop the motor if it is cold. This is the beauty of fuzzy logic: to turn common-sense, linguistic descriptions into a computer controlled system.

Figure A.8 shows how the five fuzzy sets of temperature are related to the five fuzzy sets of motor speed. The input/output pairs are in the regions defined by the intersections, which are defined by the fuzzy rules. The clusters are patches in the geometric plan of temperature–speed. They relate the given information (temperature) with the required action (machine speed set-point). More rules give more clusters, leading to more fuzzy sets and more accuracy.

A fuzzy system can have two or more inputs which are related with *and*, *or*, and *not* operations. It can also have more than one output. This air-conditioning system has an extra output that controls the angle of the blade of a vent. It is known that as the temperature gets hotter, the vent should be directed upward, because the blown cold air would be better distributed in an occupied room. As the temperature gets colder or the room less occupied, the blown air is somewhat warmer. Since the natural convection makes the air go up, the vent should be direct downward. Consequently, a complete rule base for such a system is given by Table A.1.

These five rules will be fired in accordance with the membership functions of the room temperature. Such membership functions give the truth value for each rule and will weight the average of the outputs, controlling continuously

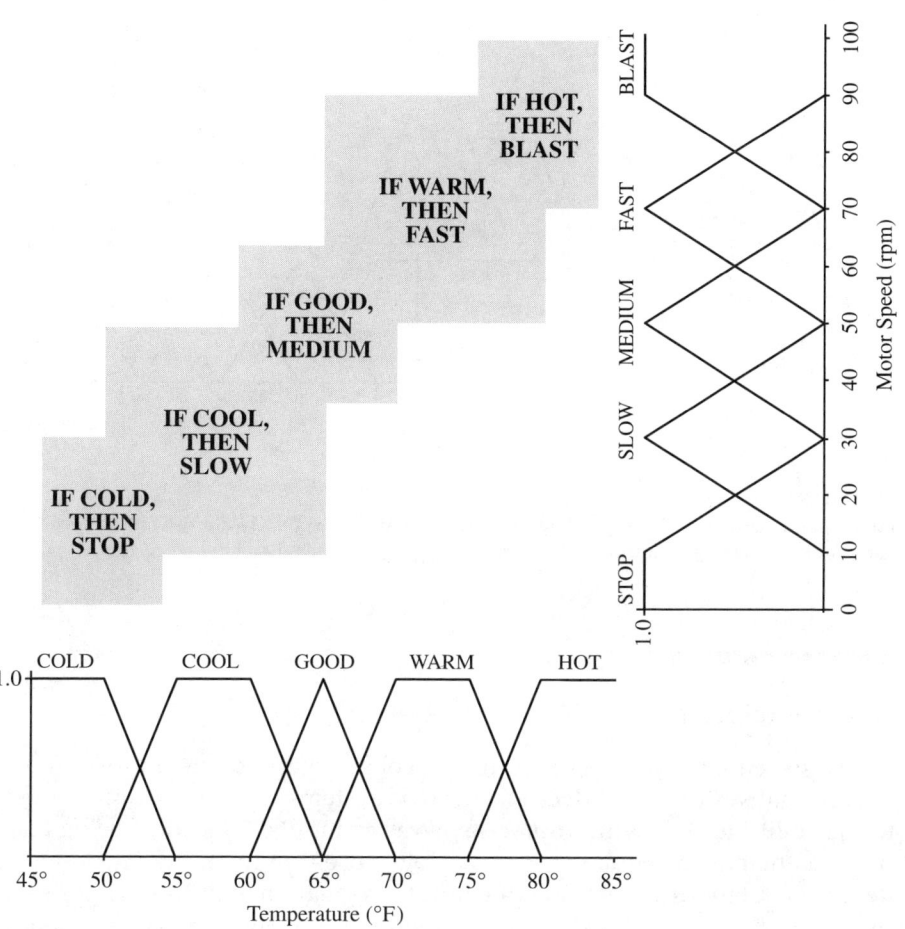

FIGURE A.8
Regions defined by the input–output fuzzy reasoning.

TABLE A.1

Rules for Air-Conditioning Control

Rule 1 If temperature is cold then stop motor, vent is downward
Rule 2 If temperature is cool then motor speed is slow, vent is downward
Rule 3 If temperature is comfortable then motor speed is medium, vent is horizontal
Rule 4 If temperature is warm then motor speed is fast, vent is upward
Rule 5 If temperature is hot then blast motor speed, vent is upward

the motor speed from 0 to 100 rpm and the blade angle from −45° to +45°. Figure A.9 shows an example where the input temperature of 76.5°F fires rules 4 and 5, leading to the two outputs indicated in the figure.

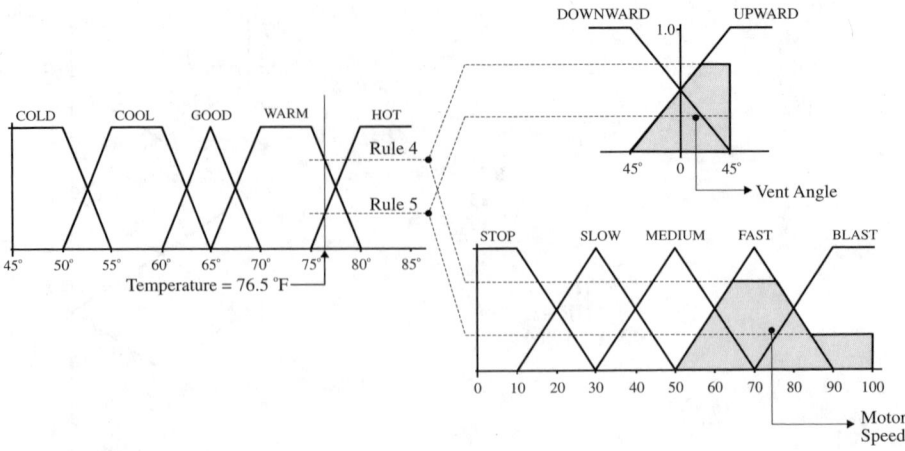

FIGURE A.9

Fuzzy inference where the input temperature fires two rules that relate the motor speed and vent angle for the air-conditioning system.

A.6   Conclusion

Fuzzy systems have a bright future in consumer products, industrial and commercial systems, and decision-support systems. The term "fuzzy" refers to the ability to deal with imprecise or vague inputs. Instead of using complex mathematical equations, fuzzy logic uses linguistic descriptions to define the relationship between the input information and the output action. In engineering systems, fuzzy logic provides a convenient and user-friendly front end to develop control programs, helping designers to concentrate on the functional objectives, not on the mathematics. This appendix introduced a brief history of fuzzy logic, discussed the nature of fuzziness in the real world problems, and showed how fuzzy operations are performed and how fuzzy rules can incorporate the underlying knowledge. The required steps to construct a fuzzy system were demonstrated in a case study of an air-conditioning system. Fuzzy logic is a very powerful tool that is pervading every field with successful implementations.

References

1. Zadeh, L.A., Fuzzy sets, *Inform. Contr.*, 8, 338–353, 1965.
2. Mamdani, E.H., Application of fuzzy algorithms for control of a simple dynamic plant, *Proceedings of the Institution of Electrical Engineers,* United Kingdom, 121, 1585–1588, 1974.

3. Munakata, T. and Jani, Y., Fuzzy systems: an overview, *Communications of the ACM,* 69–76, March 1994.
4. Boole, G., *An Investigation of the Laws of Thought on Which Are Founded the Mathematical Theories of Logic and Probabilities,* (1854), reprinted by Dover Publications, Inc., New York, 1958.
5. Cox, E., *The Fuzzy Systems Handbook,* AP Professional, 1994.
6. McNeill, F. M. and Thro, E., *Fuzzy Logic: A Practical Approach,* AP Professional, 1994.
7. Kosko, B., *Fuzzy Thinking: The New Science of Fuzzy Logic,* Hyperion, New York, 1993.
8. Brubaker, D. I., Fuzzy-logic basics: intuitive rules replace complex math, *EDN,* 111–115, June 18, 1992.
9. Schwartz, D. G., Japanese advances in fuzzy systems and case-based reasoning, *Technical report No. 91–111,* Department of Computer Science, Florida State University, Tallahassee, FL, Nov. 1991.
10. Simões, M. G., Fuzzy Logic and Neural Network Based Advanced Control and Estimation Techniques in Power Electronics and AC Drives, Ph.D. dissertation, The University of Tennessee, Dec. 1995.
11. Simões, M. G., and Bose, B.K., Fuzzy neural network based estimation of power electronic waveforms, *III Congresso Brasileiro de Eletrônica de Potência (COBEP'95),* Dec. 1995, pp. 211–216.
12. Sousa, G.C.D., Bose, B.K., and Simões, M.G., A simulation-implementation methodology of a fuzzy logic based control system, *III Congresso Brasileiro de Eletrônica de Potência (COBEP'95),* Dec. 1995, pp. 223–228.
13. Sousa, G.C.D., Bose, B.K., and Spiegel, R., Fuzzy logic based intelligent control of a variable speed cage machine wind generation system, *IEEE Power Electronics Specialists Conference,* pp. 389–395, Atlanta, GA (USA), June 1995.
14. Simões, M. G., and Bose, B.K., Application of fuzzy logic in the estimation of power electronic waveforms, *IEEE/IAS Annu. Meeting Conf. Rec.,* 1993, pp. 853–861.

Glossary

Degree of membership (μ)   A number between 0 and 1 that expresses the confidence that a given element belongs to a fuzzy set.

Defuzzification   The process of determining the best numerical value to represent a given fuzzy set.

Fuzzification   The process of converting nonfuzzy input variables to fuzzy variables.

Fuzzy composition   A method of deriving fuzzy control output from given fuzzy control inputs.

Fuzzy implication   Same as fuzzy rule.

Fuzzy inference   Same as fuzzy composition.

Fuzzy model   Fuzzy rules and membership functions that describe the fuzzy relationship between the output and input of a controlled system.

Fuzzy rule   If-then rules relating input (conditions) fuzzy variables to output (actions) fuzzy variables.

Fuzzy rule base   A set of rules that define a fuzzy control or fuzzy model.

Fuzzy set (or fuzzy subset)   A set consisting of elements having degrees of membership varying between 0 (nonmember) to 1 (full member). It is usually associated with linguistic values, such as small, medium, large, etc.

Height defuzzification method   A method of calculating a crisp output from a composed fuzzy value, by performing a weighted average of individual fuzzy subsets.

Membership function   A function that defines a fuzzy subset, by associating every element in the set with a number between 0 and 1.

Sup-min or max-min composition method   A composition (or inference) method for constructing the output membership function by using maximum and minimum principle.

Universe of discourse   The range of values associated with a fuzzy variable.

APPENDIX B

C Statements for Simulation of a Self-Excited Induction Generator

```
/* The following program simulates the isolated operation of a self-
excited induction generator (SEIG) */
/* wr1 is the speed of the prime mover. It is assumed to be constant here
but any function of time can */
/* be assigned to it based on load variations */

#include<stdio.h>
#include<math.h>
#include<conio.h>
#include<stdlib.h>

/* Generator1 parameters */

#define c1 180e-6                    /* Self exciting capacitance, F */
#define rs1 0.262                    /* Stator resistance, Ohms */
#define xs1 0.633                    /* Stator reactance, Ohms */
#define rr1 0.447                    /* Rotor resistance, Ohms */
#define xr1 1.47                     /* Rotor reactance, Ohms */

#define tsim 3                       /* Simulation time, secs */
#define deltat 0.0001

double itrmax = 1+tsim/deltat;       /* Maximum number of iterations */
double pi = 22.0/7.0;
double wr1 = 377.1428;                    /* Rotor speed, rad/sec */

void main()
{
    /* Initialization of variables */
    FILE *fp1;                       /* To save the required data */

    int ii;
    double ids1 = 0.0,iqs1 = 0.0,idr1 = 0.0,iqr1 = 0.0,vld1 = 0.0,vlq1 =
0.0,ild1 = 0.0,ilq1 = 0.0;

    /* Initializing residual magnetism */
    double vds1 = 10.0,vqs1 = 10.0,vdr1 = 0.0,vqr1 = 0.0;
```

```
/* h-Runge-Kutta parameter, r1 and l1 are load parameters initialized to
approximate an open circuit */
   double im1,ls1,lr1,k1 = 0.0,m1 = 0.25,xm1,h = 0.0001,r1 = 1000.0,l1 =
100.0,ti;

/* Runge-Kutta parameters */
   double
a1,b1,c11,d1,e1,f1,g1,h1,a2,b2,c2,d2,e2,f2,g2,h2,a3,b3,c3,d3,e3,f3,g3,h3
,a4,b4,c4,d4,e4,f4,g4,h4;

   /* Initializing functions for Runge-Kutta procedure */
   double fids(double,double,double,double,double,double,double,double,
double,double);

   double fiqs(double,double,double,double,double,double,double,double,
double,double);
   double
fidr(double,double,double,double,double,double,double,double,double,
double,double);

   double
fiqr(double,double,double,double,double,double,double,double,double,doub
le,double);

   double fvld(double,double);

   double fvlq(double,double);

   double fild(double,double,double,double);

   double filq(double,double,double,double);

   /* Instantaneous voltage and current values are stored in data.dat */
   fp1 = fopen("data.dat,""w");

fflush(stdin);
printf("Simulating an isolated induction generator\n");

   /* Running the induction generator for tsim seconds */

   for(ii = 0;ii<itrmax;ii = ii+1)
   {
        fflush(stdin);

        ti = ii*deltat;

        /* Making the residual magnetism zero after one iteration */
        if(ii>0)
        {
                vds1 = 0.0;
                vqs1 = 0.0;
```

```
              vdr1 = 0.0;
              vqr1 = 0.0;
      }
      xm1 = wr1*m1;                    /* Mutual reactance, Ohms */
      ls1 = (xs1+xm1)/wr1;    /* Stator inductance, H */
      lr1 = (xr1+xm1)/wr1;    /* Rotor inductance, H0 */
      k1 = 1/((m1*m1)-(lr1*ls1));


      /* Using Runge-Kutta fourth order method to solve for D-Q
parameters */

          a1 = fids(m1,ids1,iqs1,idr1,iqr1,vld1,vds1,vdr1,k1,lr1);
          b1 = fiqs(m1,ids1,iqs1,idr1,iqr1,vlq1,vqs1,vqr1,k1,lr1);
          c11 = fidr(m1,ids1,iqs1,idr1,iqr1,vld1,vds1,vdr1,k1,ls1,lr1);
          d1 = fiqr(m1,ids1,iqs1,idr1,iqr1,vlq1,vqs1,vqr1,k1,ls1,lr1);
          e1 = fvld(ids1,ild1);
          f1 = fvlq(iqs1,ilq1);
          g1 = fild(vld1,ild1,r1,l1);
          h1 = filq(vlq1,ilq1,r1,l1);

      a2 = fids(m1,(ids1+h*0.5*a1),(iqs1+h*0.5*b1),(idr1+h*0.5*c11),
(iqr1+h*0.5*d1),(vld1+h*0.5*e1),vds1,vdr1,k1,lr1);

      b2 = fiqs(m1,(ids1+h*0.5*a1),(iqs1+h*0.5*b1),(idr1+h*0.5*c11),
(iqr1+h*0.5*d1),(vlq1+h*0.5*f1),vqs1,vqr1,k1,lr1);

      c2 = fidr(m1,(ids1+h*0.5*a1),(iqs1+h*0.5*b1),(idr1+h*0.5*c11),
(iqr1+h*0.5*d1),(vld1+h*0.5*e1),vds1,vdr1,k1,ls1,lr1);

      d2 = fiqr(m1,(ids1+h*0.5*a1),(iqs1+h*0.5*b1),(idr1+h*0.5*c11),
(iqr1+h*0.5*d1),(vlq1+h*0.5*f1),vqs1,vqr1,k1,ls1,lr1);
          e2 = fvld((ids1+h*0.5*a1),(ild1+h*0.5*g1));
          f2 = fvlq((iqs1+h*0.5*b1),(ilq1+h*0.5*h1));
      g2 = fild((vld1+h*0.5*e1),(ild1+h*0.5*g1),r1,l1);
      h2 = filq((vlq1+h*0.5*f1),(ilq1+h*0.5*h1),r1,l1);

      a3 = fids(m1,(ids1+h*0.5*a2),(iqs1+h*0.5*b2),(idr1+h*0.5*c2),
(iqr1+h*0.5*d2),(vld1+h*0.5*e2),vds1,vdr1,k1,lr1);

b3 = fiqs(m1,(ids1+h*0.5*a2),(iqs1+h*0.5*b2),(idr1+h*0.5*c2),(iqr1+h*
0.5*d2),(vlq1+h*0.5*f2),vqs1,vqr1,k1,lr1);

      c3 = fidr(m1,(ids1+h*0.5*a2),(iqs1+h*0.5*b2),(idr1+h*0.5*c2),
(iqr1+h*0.5*d2),(vld1+h*0.5*e2),vds1,vdr1,k1,ls1,lr1);

      d3 = fiqr(m1,(ids1+h*0.5*a2),(iqs1+h*0.5*b2),(idr1+h*0.5*c2),
(iqr1+h*0.5*d2),(vlq1+h*0.5*f2),vqs1,vqr1,k1,ls1,lr1);
          e3 = fvld((ids1+h*0.5*a2),(ild1+h*0.5*g2));
          f3 = fvlq((iqs1+h*0.5*b2),(ilq1+h*0.5*h2));
          g3 = fild((vld1+h*0.5*e2),(ild1+h*0.5*g2),r1,l1);
          h3 = filq((vlq1+h*0.5*f2),(ilq1+h*0.5*h2),r1,l1);
```

```
        a4 = fids(m1,(ids1+h*a3),(iqs1+h*b3),(idr1+h*c3),(iqr1+h*d3),
(vld1+h*e3),vds1,vdr1,k1,lr1);

b4 = fiqs(m1,(ids1+h*a3),(iqs1+h*b3),(idr1+h*c3),(iqr1+h*d3),(vlq1+h*
f3),vqs1,vqr1,k1,lr1);

    c4 = fidr(m1,(ids1+h*a3),(iqs1+h*b3),(idr1+h*c3),(iqr1+h*d3),
(vld1+h*e3),vds1,vdr1,k1,ls1,lr1);

    d4 = fiqr(m1,(ids1+h*a3),(iqs1+h*b3),(idr1+h*c3),(iqr1+h*d3),
(vlq1+h*f3),vqs1,vqr1,k1,ls1,lr1);
        e4 = fvld((ids1+h*a3),(ild1+h*g3));
        f4 = fvlq((iqs1+h*b3),(ilq1+h*h3));
        g4 = fild((vld1+h*e3),(ild1+h*g3),r1,l1);
        h4 = filq((vlq1+h*f3),(ilq1+h*h3),r1,l1);

        /* Instantaneous D-Q parameters obtained after solving the
differential equations */
        ids1 = ids1+(a1+2*a2+2*a3+a4)*(h/6.0);
        iqs1 = iqs1+(b1+2*b2+2*b3+b4)*(h/6.0);
        idr1 = idr1+(c11+2*c2+2*c3+c4)*(h/6.0);
        iqr1 = iqr1+(d1+2*d2+2*d3+d4)*(h/6.0);
        vld1 = vld1+(e1+2*e2+2*e3+e4)*(h/6.0);
        vlq1 = vlq1+(f1+2*f2+2*f3+f4)*(h/6.0);
        ild1 = ild1+(g1+2*g2+2*g3+g4)*(h/6.0);
        ilq1 = ilq1+(h1+2*h2+2*h3+h4)*(h/6.0);

        /* Calculating the magnetization current */
        im1 = sqrt((((ids1+idr1)*(ids1+idr1))+((iqs1+iqr1)*(iqs1+
iqr1))));

        /* Calculating the mutual inductance
        /* This relation between the mutual inductance and magnetization
current
        /* is obtained from experimental data */
        m1 = (0.0423*exp(-0.0035*im1*im1))+0.0236;

        /* Switching the load at ti seconds */
        if(ti>2)
    {
            r1 = 20;
            l1 = 20e-3;
        }

        /* Storing terminal voltage and current in a file */
        fprintf(fp1,"\n%5.41f\t%5.41f,"vlq1,iqs1);
    }
    printf("simulation complete");
        fclose(fp1);
```

```
    fflush(stdin);
    getch();


}


/* Functions to evaluate expressions for D-Q parameters */

double fids(double m1,double ids1,double iqs1,double idr1,double
iqr1,double vld1,double vds1,double vdr1,double k1,double lr1)
{
    double dids;
    dids = k1*((rs1*lr1*ids1)-(wr1*m1*m1*iqs1)-(rr1*m1*idr1)-
(wr1*m1*lr1*iqr1)+(lr1*vld1)-(lr1*vds1)+(m1*vdr1));
    return dids;
}


double fiqs(double m1,double ids1,double iqs1,double idr1,double
iqr1,double vlq1,double vqs1,double vqr1,double k1,double lr1)
{
    double diqs;
    diqs = k1*((wr1*m1*m1*ids1)+(rs1*lr1*iqs1)+(wr1*m1*lr1*idr1)-
(m1*rr1*iqr1)+(lr1*vlq1)-(lr1*vqs1)+(m1*vqr1));
    return diqs;
}
double fidr(double m1,double ids1,double iqs1,double idr1,double
iqr1,double vld1,double vds1,double vdr1,double k1,double ls1,double lr1)
{
    double didr;
    didr = k1*(-
(rs1*m1*ids1)+(wr1*m1*ls1*iqs1)+(rr1*ls1*idr1)+(wr1*ls1*lr1*iqr1)-
(m1*vld1)+(m1*vds1)-(ls1*vdr1));
    return didr;
}


double fiqr(double m1,double ids1,double iqs1,double idr1,double
iqr1,double vlq1,double vqs1,double vqr1,double k1,double ls1,double lr1)
{
    double diqr;
    diqr = k1*(-(wr1*m1*ls1*ids1)-(rs1*m1*iqs1)-
(wr1*ls1*lr1*idr1)+(rr1*ls1*iqr1)-(m1*vlq1)+(m1*vqs1)-(ls1*vqr1));
    return diqr;
}


double fvld(double ids1,double ild1)
{
        double dvld;
        dvld = (ids1/c1)-(ild1/c1);
        return dvld;
}
```

```
double fvlq(double iqs1,double ilq1)
{
        double dvlq;
        dvlq = (iqs1/c1)-(ilq1/c1);
        return dvlq;
}

double fild(double vld1,double ild1,double r1,double l1)
{
        double dild;
        dild = (vld1/l1)-r1*(ild1/l1);
        return dild;
}

double filq(double vlq1,double ilq1,double r1,double l1)
{
        double dilq;
        dilq = (vlq1/l1)-r1*(ilq1/l1);
        return dilq;
}
```

APPENDIX C

Pascal Statements for Simulation of a Self-Excited Induction Generator

Pascal program for simulation of the self-excited induction generator;
{Post-Graduation Program in Electrical Engineering - Federal University of Santa Maria
Solution for the state equations of the transient model for the self-excited induction generator using a 4th order Runge Kutta method.}

```
USES CRT, PRINTER;

type
register1 = record
   diD,diQ,diqr,didr,dilD,dilQ :real
end;
register2 = record
   dVcD,dVcQ,va : real;
end;

var
   arq1 : text ;
   arq2 : text ;
   reg1 : register1;
   reg2 : register2;
   X0,Y0,h,x, teste : real ;
   k1,k2,k3,k4 : real ;
   q1,q2,q3,q4 : real ;
   p1,p2,p3,p4 : real ;
   r1,r2,r3,r4 : real ;
   s1,s2,s3,s4 : real ;
   t1,t2,t3,t4 : real ;
   u1,u2,u3,u4 : real ;
   z1,z2,z3,z4 : real ;
   VD,Vdr,VQ,Vqr : real ;
   Pm,Rs,Rr,Ls,Lr,Xs,Xr,M,w,ws,Xm,Im,R,L,C : real ;
   dVcD,iD0,iQ0,iqr0,idr0,VcD0,VcQ0,ilD0,ilQ0 : real ;
   A1,A2,A3,t0,tf,t,difiD,dif,ma : real ;
   difference,y,npoints,TQ,theta,modulev : real ;
```

```
   k,j : integer ;
   line : integer ;
   step : real ;
   a : char ;

FUNCTION f1 (VD,Rs,diD,dVcD,Vdr,w,M,diQ,Lr,diqr,Rr,didr,A1,A2 : real) :
real ;
   begin
   f1: = 0;
   f1 : = A1*(VD-Rs*diD-dVcD)+A2*(Vdr-w*M*diQ-w*Lr*diqr-Rr*didr) ;
   end ;
FUNCTION f2 (VQ,Rs,diQ,dVcQ,Vqr,w,M,diD,Rr,diqr,Lr,didr,A1,A2 : real) :
real ;
   begin
   f2: = 0;
   f2 : = A1*(VQ-Rs*diQ-dVcQ)+A2*(Vqr+w*M*diD-Rr*diqr+w*Lr*didr) ;
   end ;
FUNCTION f3 (VQ,Rs,diQ,dVcQ,Vqr,w,M,diD,Rr,diqr,Lr,didr,A2,A3 : real) :
real ;
   begin
   f3: = 0;
   f3 : = A2*(VQ-Rs*diQ-dVcQ)+A3*(Vqr+w*M*diD-Rr*diqr+w*Lr*didr) ;
   end ;
FUNCTION f4 (VD,Rs,diD,dVcD,Vdr,w,M,diQ,Lr,diqr,Rr,didr,A2,A3 : real) :
real ;
   begin
   f4: = 0;
   f4 : = A2*(VD-Rs*diD-dVcD)+A3*(Vdr-w*M*diQ-w*Lr*diqr-Rr*didr) ;
   end ;
FUNCTION f5 (diD,di1D,C : real) : real ;
   begin
   f5: = 0;
   f5 : = (diD/C)-(di1D/C) ;
   end ;
FUNCTION f6 (diQ,di1Q,C : real) : real ;
   begin
   f6: = 0;
   f6 : = (diQ/C)-(di1Q/C) ;
   end ;
FUNCTION f7 (dVcD,R,di1D,L : real) : real ;
   begin
   f7: = 0;
   f7 : = (dVcD/L)-(R*di1D/L) ;
   end ;
FUNCTION f8 (dVcQ,R,di1Q,L : real) : real ;
   begin
   f8: = 0;
   f8 : = (dVcQ/L)-(R*di1Q/L) ;
   end ;
```

```
begin
    assign(arq1,'irleber.m');
    settextbuf(arq1,reg1.diD);
    settextbuf(arq1,reg1.diQ);
    settextbuf(arq1,reg1.diqr);
    settextbuf(arq1,reg1.didr);
    settextbuf(arq2,reg2.dVcD);
    settextbuf(arq2,reg2.dVcQ);
    settextbuf(arq1,reg1.dilD);
    settextbuf(arq1,reg1.dilQ);
    rewrite(arq1);
    clrscr ;
    writeln('Input the initial values:');
    writeln;
    write('Enter with the calculation step (h):');
    readln(h) ;
    write('Enter with the initial time:');
    readln(t0) ;
    write('Enter with the final time:');
    readln(tf) ;

    R: = 1000;
    L: = 100;
    C: = 160e-6;
    Rs: = 0.575;
    Xs: = 1.3074;
    Rr: = 0.0404;
    Xr: = 1.3074;
    Pm: = 4;
    w: = 2*pi*60;
    id0: = 0;
    iQ0: = 0;
    idr0: = 0;
    iqr0: = 0;
    VcD0: = 0;
    VcQ0: = 0;
    ilQ0: = 0;
    k: = 0;
    t : = t0 ;
    dif: = 1;
    Ma: = 0.0659;
    M: = 0.0659;
    im: = 0;
    reg1.diD : = iD0 ;
    reg1.diQ : = iQ0 ;
    reg1.diqr : = iqr0 ;
    reg1.didr : = idr0 ;
    reg2.dVcD : = VcD0 ;
    reg2.dVcQ : = 0.00001 ;
```

```
    reg1.dilD : = i1D0 ;
    reg1.dilQ : = i1Q0 ;

while t< = tf do
   begin
   k: = k+1;
   if k = 1 then
   begin
   modulev: = sqrt(sqr(reg2.dVcD)+sqr(reg2.dVcQ));
   theta: = arctan(abs(reg2.dVcD/reg2.dVcQ));
   reg2.va: = (modulev/sqrt(2))*sin((w*t+theta*pi/180));
   writeln(arq1,reg1.did/sqrt(2));
   flush(arq1);
   end;
   if k = 20 then
   k: = 0;
   clrscr;
   gotoxy(10,10);
   writeln('tempo ',t);

   k1: = 0 ;
   k2: = 0 ;
   k3: = 0 ;
   k4: = 0 ;
   q1: = 0 ;
   q2: = 0 ;
   q3: = 0 ;
   q4: = 0 ;
   p1: = 0 ;
   p2: = 0 ;
   p3: = 0 ;
   p4: = 0 ;
   r1: = 0 ;
   r2: = 0 ;
   r3: = 0 ;
   r4: = 0 ;
   s1: = 0 ;
   s2: = 0 ;
   s3: = 0 ;
   s4: = 0 ;
   t1: = 0 ;
   t2: = 0 ;
   t3: = 0 ;
   t4: = 0 ;
   u1: = 0 ;
   u2: = 0 ;
   u3: = 0 ;
   u4: = 0 ;
   z1: = 0 ;
   z2: = 0 ;
```

```
z3: = 0 ;
z4: = 0 ;
if t = t0 then
   VD: = 5
else
   VD: = 0;
if t = t0 then
   VQ: = 5
else
   VQ: = 0;
if t = t0 then
   Vdr: = 0
else
   Vdr: = 0 ;
if t = t0 then
   Vqr: = 0
else
   Vqr: = 0 ;

im: = (sqrt(sqr(reg1.diD+reg1.didr)+sqr(reg1.diQ+reg1.diqr))) ;
writeln('dVcD',reg2.dVcD) ;
writeln('dvcQ ',reg2.dVcQ) ;
M: = (0.0423*exp(-0.0035*(sqr (Im)))+0.0236);
TQ: = -1.5*Pm/2*(reg1.diQ*reg1.didr-reg1.diD*reg1.diqr)*M ;
Xm: = M*w;
Ls: = (Xs+Xm)/w;
Lr: = (Xr+Xm)/w;
A1: = -Lr/(sqr(M)-Lr*Ls);
A2: = M/(sqr(M)-Lr*Ls) ;
A3: = -Ls/(sqr(M)-Lr*Ls);
dif : = abs(Ma-M);
Ma: = M;

k1: = h*f1(VD,Rs,reg1.diD,reg2.dVcD,Vdr,w,M,reg1.diQ,Lr,reg1.diqr,Rr,
reg1.didr,A1,A2);
q1: = h*f2(VQ,Rs,reg1.diQ,reg2.dVcQ,Vqr,w,M,reg1.diD,Rr,reg1.diqr,Lr,
reg1.didr,A1,A2);
p1: = h*f3(VQ,Rs,reg1.diQ,reg2.dVcQ,Vqr,w,M,reg1.diD,Rr,reg1.diqr,Lr,
reg1.didr,A2,A3);
r1: = h*f4(VD,Rs,reg1.diD,reg2.dVcD,Vdr,w,M,reg1.diQ,Lr,reg1.diqr,Rr,
reg1.didr,A2,A3);
s1: = h*f5(reg1.diD,reg1.dilD,C);
t1: = h*f6(reg1.diQ,reg1.dilQ,C);
u1: = h*f7(reg2.dVcD,R,reg1.dilD,L);
z1: = h*f8(reg2.dVcQ,R,reg1.dilQ,L);

k2: = h*f1(VD,Rs,reg1.diD+(k1/2),reg2.dVcD+(s1/2),Vdr,w,M,reg1.diQ+(q1/
2),Lr,reg1.diqr+(p1/2),Rr,reg1.didr+(r1/2),A1,A2);
q2: = h*f2(VQ,Rs,reg1.diQ+(q1/2),reg2.dVcQ+(t1/2),Vqr,w,M,reg1.diD+(k1/
2),Rr,reg1.diqr+(p1/2),Lr,reg1.didr+(r1/2),A1,A2);
```

```
p2: = h*f3(VQ,Rs,reg1.diQ+(q1/2),reg2.dVcQ+(t1/2),Vqr,w,M,reg1.diD+(k1/
2),Rr,reg1.diqr+(p1/2),Lr,reg1.didr+(r1/2),A2,A3);
r2: = h*f4(VD,Rs,reg1.diD+(k1/2),reg2.dVCD+(s1/2),Vdr,w,M,reg1.diQ+(q1/
2),Lr,reg1.diqr+(p1/2),Rr,reg1.didr+(r1/2),A2,A3);
s2: = h*f5(reg1.diD+(k1/2),reg1.dilD+(u1/2),C);
t2: = h*f6(reg1.diQ+(q1/2),reg1.dilQ+(z1/2),C);
u2: = h*f7(reg2.dVcD+(s1/2),R,reg1.dilD+(u1/2),L);
z2: = h*f8(reg2.dVcQ+(t1/2),R,reg1.dilQ+(z1/2),L);

k3: = h*f1(VD,Rs,reg1.diD+(k2/2),reg2.dVcD+(s2/2),Vdr,w,M,reg1.diQ+(q2/
2),Lr,reg1.diqr+(p2/2),Rr,reg1.didr+(r2/2),A1,A2);
q3: = h*f2(VQ,Rs,reg1.diQ+(q2/2),reg2.dVcQ+(t2/2),Vqr,w,M,reg1.diD+(k2/
2),Rr,reg1.diqr+(p2/2),Lr,reg1.didr+(r2/2),A1,A2);
p3: = h*f3(VQ,Rs,reg1.diQ+(q2/2),reg2.dVcQ+(t2/2),Vqr,w,M,reg1.diD+(k2/
2),Rr,reg1.diqr+(p2/2),Lr,reg1.didr+(r2/2),A2,A3);
r3: = h*f4(VD,Rs,reg1.diD+(k2/2),reg2.dVcD+(s2/2),Vdr,w,M,reg1.diQ+(q2/
2),Lr,reg1.diqr+(p2/2),Rr,reg1.didr+(r2/2),A2,A3);
s3: = h*f5(reg1.diD+(k2/2),reg1.dilD+(u2/2),C);
t3: = h*f6(reg1.diQ+(q2/2),reg1.dilQ+(z2/2),C);
u3: = h*f7(reg2.dVcD+(s2/2),R,reg1.dilD+(u2/2),L);
z3: = h*f8(reg2.dVcQ+(t2/2),R,reg1.dilQ+(z2/2),L);

k4: = h*f1(VD,Rs,reg1.diD+k3,reg2.dVcD+s3,Vdr,w,M,reg1.diQ+q3,Lr,
reg1.diqr+p3,Rr,reg1.didr+r3,A1,A2);
q4: = h*f2(VQ,Rs,reg1.diQ+q3,reg2.dVcQ+t3,Vqr,w,M,reg1.diD+k3,Rr,
reg1.diqr+p3,Lr,reg1.didr+r3,A1,A2);
p4: = h*f3(VQ,Rs,reg1.diQ+q3,reg2.dVcQ+t3,Vqr,w,M,reg1.diD+k3,Rr,
reg1.diqr+p3,Lr,reg1.didr+r3,A2,A3);
r4: = h*f4(VD,Rs,reg1.diD+k3,reg2.dVcD+s3,Vdr,w,M,reg1.diQ+q3,Lr,
reg1.diqr+p3,Rr,reg1.didr+r3,A2,A3);
s4: = h*f5(reg1.diD+k3,reg1.dilD+u3,C);
t4: = h*f6(reg1.diQ+q3,reg1.dilQ+z3,C);
u4: = h*f7(reg2.dVcD+s3,R,reg1.dilD+u3,L);
z4: = h*f8(reg2.dVcQ+t3,R,reg1.dilQ+z3,L);

  reg1.diD: = reg1.diD+((k1+2*k2+2*k3+k4)/6) ;
  reg1.diQ: = reg1.diQ+((q1+2*q2+2*q3+q4)/6) ;
  reg1.diqr: = reg1.diqr+((p1+2*p2+2*p3+p4)/6) ;
  reg1.didr: = reg1.didr+((r1+2*r2+2*r3+r4)/6) ;
  reg2.dVcD: = reg2.dVcD+((s1+2*s2+2*s3+s4)/6) ;
  reg2.dVcQ: = reg2.dVcQ+((t1+2*t2+2*t3+t4)/6) ;
  reg1.dIlD: = reg1.dIlD+((u1+2*u2+2*u3+u4)/6) ;
  reg1.dIlQ: = reg1.dIlQ+((z1+2*z2+2*z3+z4)/6) ;
  t: = t+h ;
  dif: = 1 ;
  if t>15 then
    begin
    R: = 10;
    L: = 0.009;
    end;
end;
```

```
      close(arq1);
      clrscr ;
      step : = t0 ;
      writeln ;
      gotoxy(25,2) ;
      writeln('            => RESULTS <=') ;
      writeln ;
      writeln('INTERVAL : ',h) ;

      line : = 0 ;
      for k : = 0 to (1000) do
      begin
      gotoxy(5,7+line) ;
      writeln('INTERVAL : ',step) ;
      gotoxy(40,7+line) ;
      writeln('dVcD: ',reg2.dVcD) ;
      step : = step+h ;
      line : = line+1 ;
      if line = 14 then
         begin
         gotoxy(5,23) ;
         write('Press <ENTER> to continue…') ;
         a : = readkey ;
         clrscr ;
         line : = 0 ;
         end ;
      end ;
END.
```

Index

A

Abg induction machine model, 263
  assumptions made in the formulation of,
    264
  saturation effect representation by, 270
  stationary frame, 266
ABM block of PSpice, 269
AC
  -AC conversion, 163
  -AC transformation, 160
  -DC conversion parameters, 143
  -switched inverter, 132
  generators, 4
    conversion of electrical energy in, 5
Active-reactive-based power control, 251
Actuators, 200
Adaptive step approach, 204
Adjustable gains, 221
Aerodynamic torque, 210
Air conditioning, 13
Air resistance
  losses due to, 78
Air windage opposition, 35
Air-gap
  flux, 173
  line, 79
  power, 89
  voltage
    influence of exciting frequency on, 105
    relationship with magnetizing
      current, 95
Alternating current. *See* AC
Anemometers, 200
Angular speed
  variations, 61
Appliances
  electric power consumption due to, 13
Assembly language, 295
Asymmetric magnetic pull
  control of, 118
Asynchronous generators, 2, 5, 11
  during self-excitation, 43
  steady-state representation of, 21

Asynchronous micropower plants
  safety of, 66
Available torque, 28

B

Back-to-back double converters, 244, 247
Batteries
  as storage systems, 8
  chargers, 143
  chopper regulators and, 136
  conversion of chemical energy through, 4
  use of to recover residual magnetism, 75
Battery charge controller. *See* regulators
Benefit-cost ratio, 310
Biomass, 70
  as an energy source, 14
  micropower plants, 10
Bipolar transistors, 132
Blanking period, 159
Blaschke, Felix, 180
  rotor equations, 190
Blocked rotor test, 33
Blondel-Park transformation. *See* Park
    transformation
Boost converters, 140
Brownouts, 11
Buck converter, 139
Buck-boost converter, 141
Butterworth filters, 277
Buyback rates, 307

C

C computer language, 257
  simulation of a self-excited induction
    generator in, 294
C2d function of MATLAB, 293
Cage rotors, 114. *See also* squirrel cage rotors

347

Cage-type induction generator, 209. *See also* squirrel cage rotors
Calculus of variations, 197
Capacitance
    requirements for banks of capacitors, 71
Capacitive current, 75
Capacitive reactance, 75
    reduction of terminal voltage with, 105
Capacitive reactive power, 83
Capacitor banks
    connection of across terminals, 71
    connection of in a series, 83
    induction generators connected to phase banks of, 75
Capacitors, 163
    in converters, 141
Capacity factor, 312
Carbon emissions, 303
Carbonization
    index of, 303
Carnot cycle
    limitations of thermal conversion of energy by, 4
Cascaded connections, 237
Center-tapped single-phase rectifiers, 143
Charcoal field
    storage systems, 8
Charge control, 135
Chemical energy
    conversion of to electrical power, 4
Chopper converters, 139
Chopper regulators, 136
Circuit parameters
    of the abg model, 263
Closed-loop control, 173
Coal, 300
Coefficients
    Demand Factor, 10
    output sizing, 124
    power, 12, 210
Cogging
    reduction of, 115
Common-collector antiparallel IGBT diode pair, 166
Common-emitter antiparallel IGBT diode-pair arrangement, 165
Commutation techniques, 132
Compound DC generators, 11
Computation-intensive embedded systems, 133
Conservation, 1
Consumption
    sufficient energy for, 3
Control

coordination, 225
loops, 173
rules, 219
systems, 261, 286
theory, 197
Controllable converters, 143
Controls
    electronic, 10
Converted torque, 27, 89
Converter applications, 132
    choice of topology, 156
Converters, 138, 144, 172. *See also* specific converters
    current controlled, 188
Copper losses, 82
    in stator components of induction generators, 25
Core losses, 77
Cost function, 198
Costs
    design considerations, 119
    electricity, 308
    overall power supply, 8
    renewable energy technology project analysis, 306
Cuk converters, 141
Current, 180
    commutation, 154
    total harmonic distortion of, 6
Current controlled converters, 188
Current source inverters (CSI), 138
Current-fed link system, 173
Cycloconverters, 132, 163, 173, 244, 247

D

Darlington BJTs, 132
DC
    -AC conversion, 156
    -DC conversion, 139
    -link bus, 227
    -link cascaded converters, 163
    -link voltage control, 225
    -side commutated inverter, 132
    generators, 4, 11
        comparison of with induction generators, 74
        conversion of electrical energy in, 5
    transformation, 8
Dead time, 159
Decision variables, 198
Decoupling signal, 191

Delayed current, 75
Delta connections of phase groups, 118
Demagnetization, 19, 44, 62, 66
Demand Factor, 10
Design
    classes, 122
    stress, 12
Desired rated voltage, 76
Detuning parameters, 190
Diamond, 133
Diesel
    generators, 8
DIFFER block of PSpice, 269
Digital computation-intensive embedded
        systems, 133
Diode-bridge, 165
Diode-bridge single-phase rectifier, 144
Diodes, 140
    in full-controlled converters, 148
Direct current. *See* DC
Direct electrical energy conversion, 3
Direct torque control, 187
Direct vector control, 187, 190, 215
    with rotor flux, 283
    with stator flux, 285
Discounted cash flow, 310
Discrete integration processes, 273
Displacement
    current, 58
    power factor, 6
Disrupting point, 30
Dissipation, 201
Dissipative snubbers, 132
Distributed generation, 1, 299
Distribution
    costs of micropower plants, 3
    network
        induction generators interconnected
            to, 36
        interconnection to or disconnection
            from, 19
    transformers, 11
Divided step approach, 203
    disadvantages of, 208
Double PWM converters, 160
Doubly fed induction generators, 40
    converter technologies for, 247
    sub- and supersynchronous modes, 239,
        244
    with static converters, 238
DSP blocket of MATLAB, 277

E

Economics, 9, 299
Efficiency, 20
    design issues, 125
    of induction generators, 30
Electric
    distribution network
        interconnection to or disconnection
            from, 19
    energy users
        interaction with the environment, 300
    generators
        selection of, 11
    motors, 200
    power transfer
        optimization of efficiency, 198
    showers, 13
Electrical conversion
    principles of, 3
Electrical design, 124
Electrical load, 262
Electrical losses, 125
Electrical power
    definitions, 5
Electrical torque, 28
Electricity
    conversion of from carbon, 304
Electro-gas-dynamic (EGD) systems, 4
Electrochemical energy, 14
Electroelectronic controls, 199
Electromagnetic energy
    conversion of into electricity, 4
Electromechanical controls, 199
Electronic controls, 10, 199, 201
Electronic sensors, 200
Embedded systems
    digital computation-intensive, 133
Energy
    alternative sources of, 1
    availability of, 9
    consumption, 303
    conversion of direct energy sources, 4, 14
    costs, 1
    electrical conversion of, 3
    electromagnetic, 4
    intensity, 303
    production, 308
    quality of, 258
    storage systems, 7
        classification of, 8
Enhanced voltage-link double-PWM
        converter, 173
Environment, 299

interaction of energy users with, 300
policies regarding, 301
ETABLE block of PSpice, 269
Excel, 257
    simulation of steady-state operation of an
        induction generator using, 272
Excitation
    impedance, 77
    loss of in modeled induction generator, 63
Exponential step approach, 204

F

Fan design, 127
Faraday's Induction Law, 28
Feasibility analysis
    renewable energy technology project, 306
Feedback model, 287
Feedforward power signal, 214
Field orientation, 187
Field-oriented control, 179, 186
Filter inductors, 140
Fixed step approach, 202
    disadvantages of, 208
Fixed volts/hertz ratio, 172
Fluid dynamic converters, 4
Fluid speed
    relationship to generated power, 11
Flux, 180
    control, 179
    input, 177
    intensity, 212
        control with fuzzy logic controller
            FLC-2, 219
    loop, 191
Flyback converters, 143
Flywheel
    storage systems, 8
    transient energy storage with, 200
Force-commutated thyristor inverters, 132
Forced commutated cycloconverters. *See*
        matrix converters
Form factor, 6
Forward converters, 143
Fossil fuels, 300
Foucault current, 25
Four-pole induction machine, 269
Frame sizes, 113
Freewheeling diodes, 140
    in full-controlled converters, 148
Freewheeling path, 140
Frequency

controls, 3
    effect on reactance, 71
    factors affecting, 108
Friction, 35
    losses due to, 78
Fuel cells, 137, 304
    conversion of chemical energy through, 4
    converters for, 142
Fuel tank
    storage systems, 8
Full-bridge converters, 143
Full-controlled bridge single-phase
        converter, 145
Fuzzy controllers
    descriptions of, 216
Fuzzy logic, 93
    computation, 219
    control, 197, 233
    vector control, 212
Fuzzy optimization control
    experimental evaluation of, 225
Fuzzy-based control, 193

G

G-rotor equivalent model, 170
G-stator equivalent model, 170
Gases
    as an energy source, 14
Gate drivers, 156
Gate turn-off thyristors (GTOs), 132
    multi-step inverters and, 160
Gearbox
    selection of, 11
General internal combustion engines, 8
Generation
    distributed, 1
Generator torque shaft, 172
Generator-load
    steady-state matching of, 261
Generators. *See also* induction generators
    calculations for determining efficiency of,
        83
    conversion of electricity, 4
    coupling to rotating power turbines, 11
    electric
        selection of, 11
    induction, 1
    interface with primary energy source and
        load, 12
    rotating electrical, 5, 8
    size determination, 260

sizing, 123
speed, 212
Global warming, 302, 305
Governors, 200
Graetz converter, 152
Greenhouse-gas emissions, 305
policies for, 302
Grid
asynchronous injection of power into, 2
connection of generators to, 5
induction generators connected to, 1
injection to, 8

H

Half-bridge converters, 143
Half-converters, 148
Half-wave rectifiers, 143, 149
Half-wave three-phase bridge, 149
Hardware implementation
inverters, 156
Harmonic cancellation, 161
Harmonic components, 5, 148
Hasse, Lutz, 180
Heat
recovery, 1
storage systems, 8
Heuristic techniques, 197
High efficiency induction generators, 39
High-performance inverters, 169
High-pulse bridge converters, 156
Hill-climbing, 93, 233
adaptive step approach, 204
control, 197
algorithm, 202
for induction generators, 199
divided step approach, 203
exponential step approach, 204
fixed step approach, 202
Historic load demand, 299
HVDC converters, 132, 156
Hydro, 70
generators, 8
power plants, 1
induction generators in, 19
performance of, 272
pumped applications, 237
Hydroelectric power plants
sizing, 9
Hydrogen-based economy, 304
Hydropower, 300, 308

Hysteresis
current, 25
losses due to, 78

I

IEC Standards, 121
1000-2-2, 71
IEEE Standards, 121
112, 32
519, 70
Impedance, 72
Impressed rotor voltage, 244
Indirect electrical energy conversion, 3
Indirect vector control, 186, 215, 280
Induced torque, 27
Inductance
detuned leakage, 190
series, 153
Induction generators, 1. *See also* self-excited
induction generators (SEIG)
advantages and disadvantages of, 39, 83
analysis of payback from, 307
cage-type, 209
characteristics of, 123
comparison of with DC generators, 74
components of, 113
computer simulation of, 56
connecting to the distribution network, 38
control of, 193, 257
design
factors to be considered, 119
major issues, 124
doubly fed, 40
during self-excitation, 43
evaluation of performance of, 80
fuzzy control in, 212
generated power, 25
high efficiency, 39
hill climbing control for, 199
magnetic saturation in transient studies
of, 72
measurement of parameters, 32
no-load, 45
operating characteristics from current *vs.*
voltage sources, 240
operation of as a function of the terminal
voltage, 74
output characteristics of, 200
parallel connection of, 64
parameters of, 69
power factor of, 94

prediction of terminal voltage of, 102
reactive power consumption by, 69
representation of losses, 30
robustness of, 21
selection of, 11
state space-based modeling of, 44
steady state model, 19
torque-speed characteristics of, 89
transient simulation of, 61
vector control for, 179
Induction motors
start-up of in a network, 109
Inductive load
influence of on self-excitation capacitance,
79
Inductive reactance parameters, 71
Inductors, 163
Industrial simulations, 257
Inertial flywheel, 8
Inertial modes, 211
Infinite bus bar
generalization of association of self-
excited generators on, 55
Initial costs, 8
Instantaneous load, 71
Instantaneous power theory, 180
Instantaneous torque, 27
Insulated gate-bipolar transistors (IGBTs),
132, 139, 165, 209
Insulation design, 125
Integrated gate-commutated thyristor
(IGCT), 133
Intelligent power modules, 133
Interception point, 75
Interconnected loads, 3
Interconnection guidelines, 20
Interfacing, 133
Intergovernmental Panel on Climate Change,
302
Internal combustion engines (ICE), 8
Internal flux, 180
Internal rate of return, 310
Inverse vector rotation, 182
Inverter-grade thyristors, 131
Inverters, 132, 136, 160, 169. *See also* specific
inverters
topology of, 156, 209
Investments
scheduled, 3
Iron losses, 35
in induction generators, 19, 25
Isolation, 8, 133
IVC rotor models, 280

J

Just-in-time installations, 113

K

KPO, 219
Kraemer system, 237
KWR, 219

L

Lab magnetizing curve test, 96
Laboratory tests
hill-climbing control, 198
minimization of, 98
Large power plants, 13
definition of, 2
Leakage inductance
detuned, 190
Least cost analysis, 311
Lenz's law, 29
Line filters, 163
Line-frequency transformation, 8
Line-power factor, 215
Line-side inverter output, 214
Line-voltage transformation, 8
Load
current, 203
calculations, 83
characteristics of *vs.* mechanical
power, 92
increment, 201
historic demand, 299
inductive, 79
interconnected, 3
interface with primary energy source and
generators, 12
parameters, 70
power
definition, 5
resistance
effect of, 104
sizing, 9
variation, 91
Load-commutated converters, 132
Local inductance, 214
Look-up tables, 173
Lossless rectifier, 5

M

Machine inertia, 203
Machine instantaneous variables, 180
Machine related limits
    rotation effects on, 105
Machine theory, 45
Machine torque, 177
Magnetic
    design, 124
    losses, 125
Magnetizing
    current, 25, 43, 219, 251, 269
        relationship with air-gap voltage, 95
    curves, 74
        determination of, 102
    inductance, 173
        variation of, 95
    reactance, 43
        determination of, 83, 102
        variation of, 57
Magneto-hydro-dynamic (MHD) systems, 4
Maintenance costs, 8
    design considerations, 119
Manufacturing
    optimization of the process, 119
Materials
    design considerations, 119
MATLAB, 61, 257
    c2d function, 293
    evaluation of Simulink program, 288
    simulation of self-excited induction
        generator in, 292
    simulation of vector-controlled schemes
        using Simulink subprogram, 277
Matrix converters, 163, 173, 247
    advantages of, 164
Maximum generated power (MGP), 201
Maximum power
    and torque speeds, 262
McMurray inverter, 132
McMurray-Bedford inverter, 132
MCTs, 133
Mechanical
    controls, 199
    energy
        conversion into electricity of, 4
    losses, 125
    power
        characteristics of *vs.* load current, 92
Mechanical design, 125
Metarule, 199
Micro turbines, 8
Micropower plants

asynchronous
    safety of, 66
biomass, 10
    definition of, 2
Microcontrollers, 1
Microelectronic device evolution, 131
Microsoft Excel. *See* Excel
Mini-power plants
    definition of, 2
Modeling
    of induction generators, 56, 61
Modulators
    implementation of, 189
Monitoring, 136
Monopoly of power distribution, 11
MOSFETs, 132, 139
Motoring operation, 173
    flux in, 193
Multi-level inverters, 161
Multi-step inverters, 160
Mutual inductance, 43, 293
    nature of, 77

N

N-phase-to-m-phase matrix converters, 163
National Electrical Manufacturers
        Association (NEMA), 113, 225
    design classes, 122
    Standard MG-1, 121
Natural gas, 300
Net present value, 310
Neurofuzzy techniques, 198
No-load
    induction generators, 45
    mechanical and rotational speeds, 240
    tests, 35
Noise
    reduction of, 115
Nuclear power, 300

O

Objective, 198
Oil, 300
Open circuit test, 33, 49
Operation costs, 8
Operations research, 197
Optimal control theory, 197
Optimization, 197
    principles, 198

Oscillations, 38
    equation for, 58
    load current, 203
    relationship with torque, 58
Oscillatory torque, 211
Outage time
    allowable, 8
Outer speed control loop, 173
Output
    power, 221
        changes in, 259
    sizing coefficient, 124
    voltage, 43
Oversizing, 10

P

P-q instantaneous power theory, 180
P-n junction of semiconductors, 4
Parallel generators
    DC, 11
    modeling of self-excited induction
        generators, 63, 65
Parameters
    uncorrected, 72
Parasite currents
    losses due to, 78
Park transformation, 44, 62
Pascal, 257
    simulation of self-excited induction
        generator in, 271
Payback period, 311
Peak power tracking, 93
Peltier effect, 4
Petroleum
    prices, 1
Phase belts, 118
Phase changing, 8
Phase groups
    connections of, 118
Phase-control type static VAR compensators,
        132
Phase-controlled cycloconverters, 163
Phir, 280
Photovoltaic, 137
    arrays, 4
    converters for, 142
PI regulators, 173
Pneumatic container
    storage systems, 8
Point of maximum torque, 30
Pole-phase groups, 118

Poles
    definition of number of, 117
Policies, 299
    environmental, 301
Politics, 9
Pollution
    caused by nonrenewable energy sources,
        9
    tax, 301
Polyphase circuits
    half-wave three-phase bridge in, 149
Power
    -speed curves, 91
    asynchronous injection of, 2
    coefficient, 210
    components of systems, 261
    consumption, 13
    control
        active-reactive-based, 251
    converters, 200
    electrical
        definitions, 5
    electronics, 1
        converter circuits and, 135
        history of, 132
        systems for, 193
    factor, 7, 34
    flux reversion, 8
    generated, 25
    generation
        world climate change due to, 302
    interruptibility of sources, 8
    plants, 1
        legal definitions of, 2
        sizing, 9
        stand-alone operation of, 71
    ratings
        inverters, 156
    reactive supplements, 2
    relation of fluid speed to generation of, 11
    theory, 180
Preformed-wound coil, 116
Primary energy, 3
    characteristics of sources, 7
    indicators of potential, 259
    interface with generators and load, 12
    potential of sources, 258
    source types, 70
Private benefits, 300
Problems, 16, 40, 66, 84, 109, 128, 166, 177, 194,
        234, 254, 296, 314
Programming errors in computer programs,
        295
Protection, 133, 136

PSim, 257, 263
  simulation of a self-excited induction
    generator in, 290
PSpice, 257
  ABM block of PSpice, 269
  DIFFER block, 269
  ETABLE block, 269
  simulation of SEIG in, 262
Public network, 262
  interconnection with, 11
Public utilities, 299
Pulse function, 62
Pulse width modulation (PWM), 139
  control, 189
  IGBT, 209
  techniques, 157, 176
Pumped hydro applications, 237

R

Random-wound coil, 116
Rated speed
  selection of, 11
Rated torque, 173
Raw materials, 133
Rayleigh distribution, 260
Reactance
  frequency effect on, 71
Reactive power, 83, 244
  consumption of by interactive generators,
    69
  supplements, 2
Rectifiers, 143, 209
  use of to recover residual magnetism, 75
Regenerative torque, 214
Regulators, 135
Relays, 200
Remnant magnetism, 55
Remote sites
  industrial, 9
  micropower plants, 3
Renewable energy systems, 304. *See also*
    specific renewable energies
  application of WRIG systems to, 237
  classification of, 137
  decision-making criteria for, 299
  induction generator-based, 312
  installation costs of, 197
  policy shift for, 301
Residual magnetism, 49, 62, 66, 74
  recovering, 75

Resistive load, 46
  state equations of self-excited induction
    generators with, 47
Resistive losses, 172
Resonant point, 75
Reverse blocking devices, 132
Reverse-biased square operating area, 133
Ripple factor, 6
RLC load
  modeling, 63
  state equation of self-excited induction
    generators with, 47
Rotating systems, 7, 137
  of storage, 8
Rotation
  characteristics of, 105
  rotor power factor in, 94
Rotatory renewable energy systems, 137
Rotor, 12, 36, 114
  -injected power, 241
  Blaschke equations, 190
  blocked, 22, 33
  current, 237
  cylindrical, 22
  definition of mechanical power in, 92
  determination of torque from current of,
    30
  displacement current, 58, 269
  flux, 186, 188, 219, 280
    direct vector control with, 283
  frequency variation of, 75
  in doubly fed induction generators, 40
  iron losses of, 19, 25, 35
  laminations of, 120
  parameter variation, 43, 58
  position, 183
  resistance, 71
  slip factor and, 24
  squirrel cage, 22
  voltage control, 240
    set points, 251
Runge-Kutta fourth order procedure, 296
Rural energy, 9, 13

S

Safety, 20
Saturation effects, 43
Scalar control, 169
Scalar method, 173
Scheduled investments, 3
Scherbius, Arthur, 238

SCRs, 143
Secondary driving machine
    use of to obtain coherent magnetizing
        curve, 75
Seebeck effect, 4
Self-commutated gate turn-off thyristors
        (GTOs), 132
Self-excitation, 43, 74
    beginning of process, 62
    mathematical description of process, 77
    process of, 70
Self-excited induction generators (SEIG), 45
    frequency effect on reactance in, 71
    generalization of association of, 55
    impedance of, 74
    modeling of parallel grouping of, 63
    simulation of in C, 294
    simulation of in Pascal, 271
    simulation of in PSim, 290
    simulation of in PSpice, 262
    state equations of resistive load in, 47, 49
    voltage regulation of, 104
Semiconductors, 4
    power devices, 131
    raw materials for, 133
Sensitivity analysis, 311
Sepic converter, 142
Series inductance, 176
    effect at input AC side, 153
Series regulators, 136
Servomechanisms, 200
Shaft oscillation
    relationship with torque, 58
Short circuit tests, 33, 49
Shunt regulators, 135
Shutdown modes, 214
Siemens AG, 179
Silicon, 133
Silicon carbide, 133
Simulation of in PSim
    simulation of in MATLAB, 292
Simulink
    evaluation of program, 288
    simulation of vector-controlled schemes
        using, 277
Single-phase full-wave rectifiers, 143
Single-phase H-bridge inverters, 157
Sinusoidal pulse-width modulation (SPWM),
        176
Sinusoidal waveforms, 8
Slip
    factor, 23, 35, 36, 241
    power, 237
    programmed, 173

Small power plants, 13
    design fundamentals of, 258
    induction generator use in, 1
    legal definition of, 2
Snubbers
    large dissipative, 132
Social benefits, 300
Solar energy
    capacity planning, 10
    conversion of into electrical energy, 4
Solenoids, 200
Space vector techniques, 157
    notation for, 184
Speed
    control, 224
        with fuzzy logic controller FLC-1, 216
Speed loop
    control of, 212
    control of with fuzzy logic controller
        FLC-3, 224
Speed reference, 214
Square safe operating area of IGBTs, 133
Squirrel cage rotors, 114, 246
    in induction generators, 21
Stabilization, 8
Stand-alone loads, 3
Star connections of phase groups, 118
State evaluation functions, 198
Static converters, 173
    double fed induction generators with, 238
Static systems, 7, 137
    of storage, 8
Static torque, 211
Static VAR compensators, 215
    phase-control type, 132
Stationary reference axis, 45
Stationary renewable energy systems, 137
Stator, 45, 113
    components, 19
    currents, 269
    efficiency of design, 127
    flux, 191
        direct vector control with, 285
    internal primary voltage of in
        asynchronous induction
        generator, 22
    reference, 237
    resistance, 241
        drop, 173
    three-phase excitation, 115
    two-pole winding of, 116
    winding resistance of, 34
Steam turbines, 8
Step modulation, 201
Stochastic techniques, 197

Storage systems
    classification of, 137
Stray load losses, 125
Subsynchronous modes, 239
Super capacitor
    storage systems, 8
Superconducting magnetic energy storage
        (SMES), 8
Supersynchronous modes, 239
Switching harmonics, 164
Switching mode power supplies (SMPS)
    converters for, 143
    MOSFET use in, 132
Synchronous
    generators, 11, 37
    power plants, 2
    reference frame equivalent circuit, 239
    rotating reference frame, 184
Syntax errors in computer programs, 295
System controllers, 136

T

Terminal voltage, 76, 193
    influence of rotation on, 105
    influence of rotor speed on, 108
    prediction of, 102
    reduction of with capacitive reactance,
        105
    variation, 74
Theorem of the Average Values of Maxima
        and Minima, 94
Thermal
    design, 125
    energy
        conversion of to electrical power, 4
Thermoelectric energy, 4
Thévenin equivalent, 90
Thompson effect, 4
Three-phase controlled full-wave bridge
        rectifier, 152
Three-phase full-wave bridge rectifier, 152
Three-phase half-converters, 156
Three-phase rectifiers, 148
Three-phase utility bus, 214
Three-phase voltage source inverter, 160
Thyristors, 132
    use of to evaluate circuit operation, 143
Tip speed ratio (TSR), 12, 210
Torque, 27, 210
    -slip curve, 241, 246
    -speed characteristic, 91, 113

control, 179, 215
    open-loop, 192
converted, 89
crossover, 172
current, 219
determination of, 30
machine, 177
pulsation, 211
reduction of, 115
reference, 287
relationship with shaft oscillation, 58
speeds
    maximum power and, 262
Total harmonic distortion (THD), 6
Transfer time
    allowable, 8
Transformer utilization factor, 6
Transformers, 6, 22, 23, 140
Transient modeling, 56, 61, 65
Transient response, 170
Transient state representation, 257
Transient switching process, 65
Transmission, 3
Triple-T circuit, 262
Turbine-generator connections
    steady-state matching of, 261
Turbines, 8, 199, 210
    generator group, 8
    power, 216
    reliability of, 308
    selection of rating, 11
    size determination, 260
    wind
        fuzzy optimization control
            experiment, 225
Two-level inverters, 161

U

Uncontrollable converters, 144
Undersizing, 10

V

VAR compensators
    phase-control type static, 132
Vector control, 169, 277
    direct
        with rotor flux, 283
    fuzzy logic-based, 212
    indirect, 280

techniques, 179
theory, 247
Vector rotation, 182
Verhoef inverter, 132
Vertical wind turbines, 210, 216
Vibration
    reduction of, 115
Virtual two-pole machine, 264
Voltage, 43, 75
    buildup, 95
    controls, 3
    generation of, 4
    in groups of induction generators, 57
    modeling of, 63
    quality and frequency of generated by
        small power plants, 2
    regulation, 72, 82, 104
    rotor set-points, 251
    terminal, 193
        prediction of, 102
    total harmonic distortion of, 6
    transformation, 180

energy power coefficient, 12
example of net power from, 14
fuzzy optimization control experiment,
    225
generators, 8
mechanical controls in systems, 200
power plants, 1, 10, 259
    induction generators in, 19
    performance of, 272
shunt regulators, 135
speed distribution, 273
Windings, 22
    damage due to heating of, 80
    impedances, 77
    in Park transformation, 44
    of wound rotors, 114
World climate change, 305
    related to power generation, 302
Wound rotors, 114
Wound-rotor induction-generator (WRIG)
        systems, 237. *See also* doubly fed
        induction generators
    operation of, 242
Wound-rotors
    schemes in induction generators, 1

W

Water dams, 8
Water flow estimation, 273
Weighted torque, 210
Wind, 70
    energy
        income from, 308

Z

Zeta converters, 142
Ziklai BJTs, 132